Of
Minds
and
Molecules

OF MINDS AND MOLECULES

New Philosophical Perspectives on Chemistry

Edited by
Nalini Bhushan
Stuart Rosenfeld

OXFORD
UNIVERSITY PRESS
2000

OXFORD
UNIVERSITY PRESS

Oxford New York
Athens Auckland Bangkok Bogotá Buenos Aires Calcutta
Cape Town Chennai Dar es Salaam Delhi Florence Hong Kong Istanbul
Karachi Kuala Lumpur Madrid Melbourne Mexico City Mumbai
Nairobi Paris São Paulo Shanghai Singapore Taipei Tokyo Toronto Warsaw

and associated companies in
Berlin Ibadan

Published by Oxford University Press, Inc.
198 Madison Avenue, New York, New York 10016

Library of Congress Cataloging-in-Publication Data
Of minds and molecules : new philosophical perspectives on chemistry
edited by Nalini Bhushan and Stuart Rosenfeld.
p. cm.
Includes bibliographical references and index.
ISBN 0-19-512834-6
1. Philosophy—Chemistry. I. Bhushan, Nalini. II. Rosenfeld,
Stuart. (Stuart Michael) 1948–1999.
QD6.034 2000
540'.1—dc21 99-40329

1 3 5 7 9 8 6 4 2

Printed in the United States of America
on acid-free paper

This book is dedicated to

Stuart Michael Rosenfeld
(January 28, 1948–January 21, 1999)

and to our son

Ajay Bhushan Rosenfeld
(born February 11, 1998)

Preface

When my coeditor and husband, Stu Rosenfeld, died suddenly in January of 1999, our editorial work for this volume was essentially complete. Much of it simply would not have been possible without him. And yet, this is not the volume it would have been if he had been here to see it all the way through to completion. It could not be, of course, and, perhaps, it should not. The difference that his absence makes to the volume is a way of acknowledging that he was ineliminable. In another way this difference underscores a conviction that we both have shared from the project's inception—that the philosophy of chemistry involves a special kind of interdisciplinary work, requiring that one stick to the integrity of the two disciplines, chemistry and philosophy, while bringing together specialists with two very different kinds of skills and stock of knowledge in the expectation of a rich and productive dialogue. We hope the essays in the volume, when taken together, showcase the results of such interdisciplinary interaction and offer a justification for and reaffirmation of our initial conviction.

Northampton, Massachusetts
March 2000

N. B.

Acknowledgments

We would like to thank Ray Elugardo, Sam Mitchell, and especially Ernie Alleva, for assisting with reviews of different pieces of the project; John Connolly for (inadvertently) supplying the book's title; Cynthia Read and two anonymous reviewers at Oxford University Press for recognizing the merit of the book proposal to begin with, and Peter Ohlin, Cynthia Garver, and Julia Ballestracci at Oxford University Press for seeing it through to publication. Nalini Bhushan would also like to thank the National Endowment for the Humanities for a summer grant, and, during the difficult final stages of the editorial project, Ruth Haas, Loek Helminck, and James Nelson for logistical support; Sigrun Svavarsdottir and Sidney Gottlieb for unwavering confidence and intellectual support; and both of our families, Doris Libman, Jerry and Jane Rosenfeld, Phyllis Peterson, Bharat Bhushan, Dinesh Bhushan, and Nik Bhushan for the emotional and indispensable practical support of caring for our infant son while editorial work on the anthology was being completed.

Contents

Contributors

Davis Baird received his Ph.D. from Stanford University, and is currently associate professor and chair of the department of philosophy at the University of South Carolina. He comes to an interest in the philosophy of chemistry through his interest in scientific instrumentation, particularly instrumentation developed in analytical chemistry. He has a personal stake in analytical instrumentation—he is the son of Walter Baird, founder in 1936 of Baird Associates, one of the early developers of analytical instrumentation. He currently is working on a book on the history of Baird Associates.

Nalini Bhushan received a B.A. in economics and an M.A. and M.Phil. in philosophy from Madras University, Chennai, India, and her Ph.D. in philosophy from the University of Michigan in Ann Arbor. She is currently associate professor of philosophy at Smith College, Massachusetts. Her research interests lie at the intersection of mind and language, with a special focus on categorization. She has published in the journals *Cognitive Science, Philosophical Investigations*, and *Journal of Chemical Education*, among others. She is currently doing work on the chemical senses and on the philosopher Jiddu Krishnamurti.

Barry K. Carpenter received a B.Sc. in molecular sciences from Warwick University and his Ph.D. in organic chemistry from University College, London. He did postdoctoral research with J. A. Berson, Yale University, and is currently professor of chemistry at Cornell University. He has been a member of the editorial advisory boards of *Accounts of Chemical Research* and the *Journal of Organic Chemistry*, and he is currently associate editor of the *Journal of Organic Chemistry*. He was elected Fellow of the American Association for the Advancement of Science in 1989, and he received the Alexander von Humboldt Senior Scientist Award in 1990 and the ACS James Flack Norris Award in Physical Organic Chemistry in 1999. The unifying theme in Carpenter's research has been the study of reaction mechanisms by joint application of theoretical and experimental techniques. He has applied this approach to a variety of problems. The most recent has involved exploring the extent and consequences of the failure of the statistical approximation in the kinetic modeling of reactive intermediates.

John Christie received a B.Sc. in chemistry and his Ph.D. in theoretical chemistry from the Australian National University. He is currently senior lecturer in the chemistry department at La Trobe University. His research areas are theoretical chemistry—simplistic model approaches, mass spectrometry, and the theory of reactions.

Maureen Christie received a B.A. in philosophy from La Trobe University and her Ph.D. in history and the philosophy of science from the University of Melbourne. She is currently an associate in the history and philosophy of science department at the University of Melbourne. Her research areas are laws of nature, the history and philosophy of chemistry, and the history and philosophy of environmental science.

Clark Glymour is Alumni University Professor at Carnegie Mellon University and Valtz Family Professor at the University of California, San Diego. He received a B.A. in chemistry and philosophy at the University of New Mexico and, after graduate work in inorganic and physical chemistry at New Mexico and Indiana University, received a Ph.D in history and philosophy of science at Indiana. His principal research in recent years has concerned automated learning procedures and topics in mathematical psychology.

Emily Grosholz received a B.A. from the University of Chicago and her Ph.D. from Yale University. She is currently professor of philosophy and African American studies and a Fellow of the Institute for the Arts and Humanistic Studies at the Pennsylvania State University, and a Life Member of Clare Hall, University of Cambridge. She is completing a book entitled *Analysis, History, and Reason: A New Approach to the Philosophy of Mathematics*, and has just finished editing a collection of essays entitled *The Growth of Mathematical Knowledge*. In the near future she plans to write about Nina Fedoroff's rewriting of Barbara McClintock's work in genetics into the language of biochemistry.

Roald Hoffmann was born in Poland and educated at Columbia College and Harvard University; thereafter he has been at Cornell University where he is the Frank H. T. Rhodes Professor of Humane Letters. He won the Nobel Prize for chemistry in 1981. For his colleagues in chemistry, Hoffmann has provided the theoretical frameworks necessary for understanding the geometries and reactivity of all molecules. He also writes poetry and nonfiction at the intersection of chemistry, art, and culture. Roald Hoffmann's latest books are a poetry collection, *Memory Effects* (1999), *Old Wine, New Flasks: Reflections of Science and Jewish Tradition*, with Shira Leibowitz Schmidt (1997), and *The Same and Not the Same* (1995).

Robin Le Poidevin is senior lecturer and head of the School of Philosophy at the University of Leeds, England. He studied psychology, philosophy, and physiology at Oriel College, Oxford. He was subsequently a research student at Emmanuel College, Cambridge, where he received his Ph.D. in philosophy. From 1988 to 1989 he was Gifford Research Fellow in Philosophy and Natural Theology at the University of St Andrews. His main research interests are in the philosophy of space and time. He is currently writing a book on temporal representation.

William G. Lycan received his B.A. from Amherst College, where he was briefly a chemistry major before switching to mathematics. He received his M.A. and Ph.D. in philosophy from the University of Chicago. He is currently the William Rand Kenan Jr. Professor of Philosophy, University of North Carolina. His most recent book is *Consciousness and Experience* (1996). He is currently finishing a book on the semantics of conditionals, called *Real Conditionals*.

Tom Morton was born in Los Angeles, where his forebears (both Hellman and Morton) were employed by the motion picture industry. He received his A.B. from Harvard University in classics (Greek) and fine arts, and his Ph.D. from Caltech in chemistry under the joint direction of R. G. Bergman and J. L. Beauchamp. Since 1972, he has served on the chemistry faculties of Brown and Brandeis Universities and, from 1981, the University of California, Riverside, where he is professor of chemistry. His research centers on the chemical consequences of internal rotation within molecules, including recognition of conformationally mobile structures by chemosensory systems. He is also affiliated with UCR's graduate programs in biochemistry and neuroscience.

Jeffry L. Ramsey holds a B.A. and an M.S. in chemistry and a Ph.D. in the conceptual foundations of science from the University of Chicago. He was assistant professor of philosophy at Oregon State University, Corvallis, and is now at Smith College. His research focuses on how scientists view questions about theory construction, explanation, and reduction when they are faced with problems that are insoluble in practice or in principle (e.g., in the area of semiempirical models). He is also interested in questions of conceptual analysis as they arise in the chemical sciences (e.g., the case of shape). His essays have appeared in *Philosophy of Science, Studies in History and Philosophy of Science*, and *Synthese*, among other journals.

Stuart Rosenfeld received a B.A. in chemistry from Colby College and his Ph.D. in organic chemistry from Brown University. He did postdoctoral work with P. M. Keehn at Brandeis and with M. J. T. Robinson at Oxford University. He was professor of chemistry at Smith College, Massachusetts, at the time of his death in January 1999. He coedited (with P. M. Keehn) a two-volume anthology on cyclophanes, and had published extensively in the *Journal of Organic Chemistry, Journal of the American Chemical Society, Tetrahedron Letters*, and *Journal of Chemical Education*, among others. He was interested in educational issues in chemistry, and had recently published *Basic Skills for Organic Chemistry: A Tool Kit* (1998). He was the recipient of grants from the National Science Foundation, Petroleum Research Fund, William and Flora Hewlett Foundation, and Camille and Henry Dreyfus Foundation, among others. His research interests were in the synthesis and structure of strained compounds and novel tautomeric systems and, most recently, in the metaphorical models used by organic chemists.

Daniel Rothbart has a Ph.D. in philosophy from Washington University in St. Louis, with a specialization in the philosophy of science. He is currently associate professor of philosophy at George Mason University in Virginia. He has published extensively on the philosophical aspects of scientific modeling and on the philosophy of experimentation. In addition to writing *Explaining the Growth of Scientific Knowledge* and editing, *Science, Reason and Reality*, his work appears in numerous anthologies, such as *Analogical Reasoning, Chemistry and Philosophy*. He has also published articles in the journals *Dialectica, Erkenntnis, Foundations of Chemistry*, and *Philosophy of*

Science, among others. He presently serves on the executive committee of the International Society for the Philosophy of Chemistry and on the editorial boards of two journals in the philosophy of chemistry.

Eric Scerri studied chemistry at the Universities of London, Cambridge, and Southampton. He holds a Ph.D. in history and philosophy of science from King's College, London, where he wrote a thesis on the question of the reduction of chemistry to quantum mechanics. He has held several appointments in the United States, including a postdoctoral fellowship at Caltech, and is currently visiting professor in the chemistry department at Purdue University in Indiana. Scerri is the founder of the journal *Foundations of Chemistry* (http://www.wkap.nl/journals/foch), and has published extensively on the philosophy of chemistry in *Synthese*, the *PSA* proceedings, *International Studies in Philosophy of Science*, *British Journal for the Philosophy of Science*, and *Erkenntnis*, as well as in *American Scientist*, *Scientific American*, the *Journal of Chemical Education*, and other chemistry journals. His research interests include philosophical and historical aspects of quantum chemistry and the periodic system, as well as general issues in philosophy of chemistry.

Jaap van Brakel studied chemical engineering and philosophy, and has taught at universities in the Netherlands, Canada, and Belgium. Until 1980, he worked mainly in physical technology; since 1986 he has been a full-time philosopher. He is currently at the University of Leuven, Belgium. Over the years he has published in journals such as *Nature*, *International Journal of Heat and Mass Transfer*, *Logique et Analyse*, *Synthese*, *Erkenntnis*, *Studies in History and Philosophy of Science*, *Minds and Machines*, *Behavioral and Brain Sciences*, and *Ethical Perspectives*.

Stephen J. Weininger received a B.A. from Brooklyn College (CUNY) and his Ph.D. in organic chemistry from the University of Pennsylvania. After a year as senior demonstrator at the University of Durham, England, he moved to Worcester Polytechnic Institute, where he is currently professor of chemistry. Weininger's chemical research centers on the intramolecular transfer of energy and electrons in organic molecules. The complex transfers of meaning between science and the wider cultural and social matrix also began to absorb his interest. He became a founding member and then president of the Society for Literature and Science, and is now chair of the division of history of chemistry of the American Chemical Society. Weininger has written about the role of language and nonlinguistic representations in chemistry, the place of the entropy concept in chemistry, and the development of physical organic chemistry after World War II.

Andrea Woody received a B.A. in chemistry from Princeton and her Ph.D. from the department of history and philosophy of science at the University of Pittsburgh. She is assistant professor of philosophy at the University of Washington. Her current research concerns pragmatic techniques such as model building and alternative forms of representation that scientific communities develop to make abstract theories tractable, investigating how these techniques are relevant to philosophical accounts of explanation, representation, and rational theory change. Quantum chemistry remains a favorite landscape for exploring these issues.

Of
Minds
and
Molecules

1

INTRODUCTION

NALINI BHUSHAN
STUART ROSENFELD

Chemistry is a substantial science by the measures of industry, economics, and politics. As an academic discipline, it underlies the vibrant growth of molecular biology, materials science, and medical technology. Although not the youngest of sciences, its frontiers continue to expand in remarkable ways. And although it shares boundaries with every other field of science, it has an autonomy, both methodologically and conceptually; this autonomy, however, continues to be unappreciated by most philosophers of science. Why is there no philosophy of chemistry?

Although there have been philosophical writings on chemistry, increasingly so during recent years, curiously enough, no coherent discipline analogous to the philosophy of physics, biology, or mathematics has emerged. Indeed, some would argue that there is no subject matter here to begin with because chemistry is in the end reducible to physics and therefore without a distinct methodology or conceptual repertoire of its own worthy of philosophical consideration. One motivation for this anthology is to demonstrate that this view requires serious rethinking, particularly in the context of modern molecular science.

Historical Background

In a 1981 review article entitled "On the Philosophy of Chemistry," J. van Brakel and H. Vermeeren[1] pointed out that although there is a vast amount of literature on the *history* of chemistry, there is precious little in the *philosophy* of chemistry. They observe that "even isolated articles in which the philosophy of science is applied to chemistry are extremely rare: in all cases it is clear that the published work is the outcome of a side interest of the author (most of whom are chemists who developed an interest in the philosophy of science)" (p. 508).

An exception to this lack of interest, cited by van Brakel and Vermeeren, is a strand of scholarly activity in Eastern Europe, particularly Russia, East Germany, and Romania, where books and articles on the philosophy of chemistry have been published since the late 1950s. This is mostly due to the influence of the German philosopher Hegel, who took chemistry as an illustration in nature of how "quantitative changes are turned into qualitative changes" (which was supposed to have implications for the dialectical materialism of Engels and Marx). Apart from this school of thought, however, there has been no other historical trajectory of philosophical discussion in the discipline of chemistry.

More recently, in Germany, since 1993 there have been a series of annual conferences, named after the German chemist Emil Erlenmeyer, devoted to philosophical topics in chemistry.[2] It is significant that this is the first time in recent history that the phrase "philosophy of chemistry" has actually appeared in the title of any major event. In 1994 the International Conference on Philosophy of Chemistry was launched. Since 1996 there have been summer symposia on the philosophy of chemistry and biochemistry sponsored by the International Society for the Philosophy of Chemistry. In 1997 there was a special session held on chemistry and philosophy at the annual conference of the American Chemical Society. In the fall of 1998 there was a special, first-ever session on the philosophy of chemistry held at the Philosophy of Science meetings. Importantly, journals devoted to the history and philosophy of chemistry have been launched: *Hyle*, and, most recently, *Foundations of Chemistry*.

Indeed, during the last 15 years there has been an explosion of writing in both the history and philosophy of science on a broad range of topics pertaining to chemistry.[3] But it has been so dispersed that there is still no recognizable area within the philosophy of science that may be called the philosophy of chemistry.[4] However, now more than ever before there is an urgent need to rectify the situation. We believe that this is developing into a clearly articulated disciplinary framework that warrants recognition within mainstream philosophy of science as a coherent and developing dialogue within the context of an emerging field called the philosophy of chemistry. Our aim is to contribute to this development by this volume. There is clearly a discipline in the making here, and it is time to have an anthology with a broad audience that brings this area of interest to center stage in the philosophy of science.

Conceptualizing a Philosophy of Chemistry

The philosophy of chemistry: What would that be? In what respects would it be similar to a philosophy of physics, philosophy of biology, and/or a philosophy of mathematics? In what respects different? What relationship would it have to each of these? Would it yield up the same sorts of insights, or entirely new, unanticipated ones?

We claim here, as have others before, that the philosophy of chemistry is undeveloped, or at least grossly underdeveloped. This means to us that it requires definition,that its boundaries must be located. It means that its existence must be justified—there must be meaningful questions, the promise of new understandings, powerful motivations to drive the difficult stretching of disciplinary space. We must expect significant new knowledge and even important corrections in the course that has been taken by philosophy of science in general in the absence of a fully realized philosophy of chemistry.

What concerns might the philosophy of chemistry encompass? It would surely include the issues that are general to philosophy of science as a starting point: the question of reduction, the role of explanation, and the relation between theory and data. In places where the view is incomplete or disputed, one hopes philosophy of chemistry will serve as an unmined source of insight. But this would be a very limited philosophy of chemistry, and so one seeks other tools to locate its edges. Here it requires that we look at chemistry's own disciplinary boundaries (apart from the very interesting question of why those boundaries lie where they do), that we probe its practices, attending with special care to any that are unrepresented in the other sciences.

It is fair to also ask whether a philosophy of chemistry should matter to chemists. Might it affect the directions taken (or not taken) by individual chemists or at least the judgments they make? Would it enrich the practice of the science in other ways? Can a philosophical analysis of a current problem/practice in chemistry help us see where we might be misled?

We take one of the unique features of this volume to be that it consists of essays by both chemists and philosophers, some working together as teams. It is our view that chemists are best able to put issues on the table that are distinctly chemical, and only then can philosophers take them up and extract and develop the implications in a thoroughly philosophical way. This inclusion of chemists is particularly important in light of recent developments within chemistry itself that are unlike many earlier developments and hold the promise of new philosophical issues. For example, new instruments like the scanning tunneling microscope are truly molecular-level tools. Our newfound ability to image and manipulate matter at an atomic level makes possible a nanoscale technology. This opens up more and different windows into the physical world, windows that reveal individual molecules and time scales that resolve distinctly molecular events. What we are enabled to "see" and "do" and "know" with the help of instruments is not simply and uncomplicatedly "better" but different as well—conceptually, methodologically, and even linguistically. These developments require a certain *level* of analysis by philosophers of what chemists are doing and how it is different from what has traditionally been done—here the chemist can do much to help the philosopher.

Nevertheless, this volume has its roots in an old philosophical problem that finds its expression here in an antireductionist research focus, with the specific goal of establishing chemistry as an autonomous field—autonomous specifically from physics—with its own methods, laws, and kinds of explanation. This is (and has historically been) an essential strand of research, necessary to take seriously the very possibility of a philosophy of chemistry; and yet, we believe an overly zealous focus on antireductionism could stultify the ability to envision new and exciting philosophical topics in chemistry. It is for this reason that we decided that some of our writers for the volume should be chemists. Also, we had a hunch that some of the most interesting aspects of a philosophy of chemistry would emerge from putting chemists and philosophers together. Indeed, we feel that the seeds of some of the most important ideas in this emerging philosophical discipline will be planted by the chemists. Thus, we view the future development of the philosophy of chemistry as an intrinsically interdisciplinary endeavor where each discipline informs, and requires, the other. This volume constitutes one step in that direction.

Key Topics in a Contemporary Philosophy of Chemistry

The essays collected in this volume address a variety of topics that are central to any articulation of a comprehensive philosophy of chemistry. In what follows we briefly discuss each of the key issues and link them to the specific contributions of our authors. In a short piece such as this introduction, we cannot do justice to the subtleties of each author's position, but we think our categorization and commentary offer useful connections between essays. Our goal is to alert the reader to themes and issues that we believe resonate in interesting and surprising ways between authors, rather than draw any substantive conclusions of our own. Still, we hope that some of our observations, interpretations, and attempts at categorization are contentious and spur further discussion.

Autonomy and Antireductionism

In the opening chapter, entitled "Missing Elements: What Philosophers of Science Might Discover in Chemistry," Andrea Woody and Clark Glymour invite us to engage in a fascinating thought experiment: to consider what would become the central issues in philosophy of science if chemistry rather than physics became the paradigmatic science. For example, they suggest that we would have different models of explanation; that we would have to rethink what it means to be a natural kind (philosophers have traditionally taken chemical kinds to be unambiguously natural kinds); that given the crucial role played by instrumentation and automation in chemistry, it would shift the focus back to a concern with scientific methodology, from the current, almost exclusive, concern with the justification of beliefs (with methodology taken as unproblematic). Serendipitously, several chapters in this volume address these sorts of issues—see, for example, Carpenter (explanation), Rosenfeld and Bhushan (natural kinds), and Rothbart (instrumentation).

To seriously consider the question (as do Woody and Glymour) "What if chemistry was to be the research focus of philosophers" is to imply its autonomy; this relates to the more specific topic of antireductionism. The issue of reductionism—the idea that the entities and/or theories and body of knowledge of chemistry is ultimately reducible to physics—has dominated the small extant literature on the philosophy of chemistry, even as recently as a special *Synthese* issue (1997), and yet, ironically, it is still practically taken as a given among most philosophers of science that there is no real issue here because chemistry, dealing with physical entities as it does, must, in the end, reduce to physics. Thus, although reductionism has been a live area of debate in other fields of science, like biology, with the exception of a small group of writers,[5] it has historically excited little interest with regard to chemistry.

What does it mean to say that chemistry is or is not reducible to physics? Two chapters in the volume pursue this question, choosing to focus on epistemological issues such as whether and in what sense chemical theories and/or practices are reducible to those of physics, rather than on the ontological issue of whether or not the entities with which chemists operate are reducible to the entities of physics. Maureen and John Christie, in the chapter entitled " 'Laws' and 'Theories' in Chemistry Do Not Obey the Rules," make a case for the diverse character of laws and theories in the sciences and, more specifically, for a pluralistic approach to laws and theories in chemistry. At the same time, they stress the similarity in relationship, to physics, of chemistry and related sciences that deal with complex systems such as geophysics, meterology, and materials science. An *in principle* antireductionist argument is much harder to make for these sciences than it would be for, say, psychology or biology. The theme of pluralism that runs through this chapter shows up in several other chapters in this volume (see, for instance, Grosholz and Hoffmann, and Weininger).

In the chapter entitled "Realism, Reduction and the 'Intermediate position,' " Eric Scerri, who has previously written extensively on the issue of reductionism, makes a case for taking seriously the intelligibility and productivity of what he calls the "intermediate" position with respect to chemistry. He uses the case histories of the chemists Mendeleev and Paneth to demonstrate that, in fact, they adopted a distinctly philosophical stance toward their discipline located somewhere between reductionism and (nonreductive) realism, and he argues that their position makes sense when it is applied to contemporary chemistry as well.

Instrumentation

Instruments play a large and growing role in the practice of chemistry. An examination of the changing role of instrumentation in analytical chemistry, for instance, reveals that many of these changes are importantly conceptual; some philosophers even advance the possibility that instruments *are* knowledge.[6] An important consequence of this view is a newfound ability to elucidate the notion of progress in chemistry. Although some of these may be general issues that concern all the sciences,[7] the current significance in chemistry is distinct because instrumentation has just opened the doors to atomic-level images and to time scales that allow resolution of molecular-scale events.

Researchers envision the future prospects of recent developments in the areas of nuclear magnetic resonance spectrometry (NMR), enhancing, for instance, the sensitivity of NMR to observe a single molecular spin; analytical-scale separations, used, for instance, to tackle the vast amounts of samples and mixtures generated by combinatorial chemistry and chemical "libraries"; mass spectrometry (MS), used, for instance, for explosives detection such as measuring buried mines; electrochemical instrumentation, used, for instance, in the electrochemical detection of single molecules; and optical spectroscopy, which benefits from techniques like diamond machining and holographic technology used to improve Raman spectroscopy, which, in turn, has recently had exciting biomedical applications.[8]

One of the interesting questions is to what extent these changes in instrumentation are *philosophically* significant. For example, Daniel Rothbart, who examines the epistemology of chemical instrumentation in the chapter "Substance and Function in Chemical Research," argues that a careful scrutiny of experimental phenomena in relation to the various kinds of instruments used to observe them experimentally serves to undermine the familiar and typically uncontested dichotomy between "real" and "artificial" specimens. In the chapter entitled "Analytical Instrumentation and Instrumental Objectivity," Davis Baird takes us through a particular historical trajectory in analytical chemistry between 1930 and 1960 to demonstrate a crucial transformation "in the concept and practice of . . . objectivity" with the development of certain kinds of instrumental methods. As a result, analytical chemistry now relies upon "instrumental objectivity," an approach whose effectiveness has facilitated and justified its use in measurement problems more generally.

Structure and identity

What are the epistemological criteria by which one identifies two substances as the same or different? Does this simply depend on microstructure? Or, do molar properties play the crucial role? In addition, chemists seek to understand molecular structure, in part, because structure is seen metaphysically as the defining feature of a substance. Even where aspects of structure that might distinguish two molecules are extremely subtle, as is the case in certain kinds of stereoisomerism, chemists hold tightly to the distinction. On the other hand, some features of the structure of individual molecules, such as isotopic composition, are often ignored where the properties of concern do not reflect those distinctions. (Of course, there is good reason for this because a molecule and its enantiomer may exhibit dramatically different properties under certain conditions, while there may be no practical consequence of the presence of two isotopically distinct molecules of a substance in a sample.) This raises the issue of the different roles played by microstructure and experimentally observed or molar properties in establishing the identity of a substance. The relative importance of each may depend to a large degree on context.

Four of the chapters address the issue by focusing respectively on the notion of molecular shape, molecular handedness, the relevance of the dimension of time in ascertaining molecular structure and identity, and finally, on the very notion of chemical substance. Jeffry Ramsey, in the chapter "Realism, Essentialism and Intrinsic Properties: The Case of Molecular Shape," discusses the property of shape against the backdrop of the seventeenth-century philosophical distinction between primary and secondary qualities of objects. Although shape has historically been held up as the paradigm example of a primary property, any molecular level identification of shape must involve time as an integral part of the observation; this has the consequence that the property of (molecular) shape does not fall neatly into the categories of either primary or secondary. The chapter ends by drawing a suggestive analogy between contemporary philosophical debates about the property of color and the kind of debate that molecular-level analysis forces us to have about shape.

Chirality has been an area of enormous interest and significance for chemists. The identification of an enantiomeric excess in some amino acids found in a meteorite (*Science* 1997, pp. 275, 951) has generated a lot of excitement among chemists because chirality even figures in discussions of the origin of life. Robin Le Poidevin, in the chapter "Space and the Chiral Molecule," provides an instance of a reciprocally beneficial relationship between chemistry and philosophy. He examines some implications of the phenomenon of chirality in chemistry—optical isomerism in particular—for standard philosophical positions about space. At the same time, he shows how an understanding of "spatial realism" allows us to better appreciate the often subtle differences between different kinds of isomerism.

Steve Weininger, in the chapter "The Timeless, the Transient, and the Representation of Chemical Structure," makes a case for the indispensability of the dimension of time in representing structural identity. Because the properties of a structure are themselves dynamic rather than static, but identified within the context of a particular (dynamic) time frame, Weininger suggests that a plurality of representation is required. Of course, this leads one to probe the very notion of the chemical entity (of which there can be several, sometimes competing, representations). Jaap van Brakel picks up on this theme in the chapter "The Nature of Chemical Substances," where he focuses on the tension between two quite distinct representations of chemical substance, which he labels, in the language of Wilfred Sellars, "manifest" and "scientific" representations, the latter being in effect a representation of a molecular structure. He argues, contrary to the position taken in many contemporary chemistry textbooks (and, therefore, by many chemists), that the conception of substance at the macroscopic level must be taken as the primary and indispensable starting point of chemistry. Van Brakel's view of what chemistry is depends what chemists work with. Thus, what you *see* and operate with experimentally (rather than what there *is*, metaphysically speaking) is what matters to the practice of chemistry. The theme of antireductionism finds its expression throughout the chapter.

Synthesis

It is with respect to synthesis, the rational (material) construction of molecules, that chemists differ most in their practice from other physical scientists. Synthesis, and especially the multistep synthesis of complex organic compounds, represents a problem with multiple solutions that differ in quality. What are the criteria by which these various solutions are evaluated? Are the criteria for assessing quality constant, or is cost, both monetary and of time commitment, important in one case and sophistication and perhaps even artistry of a solution of overriding significance in another? (Among chemists, Cornforth[9] is particularly skeptical of the motivations that chemists themselves cite for engaging in a synthesis.) Is there an aesthetics of synthesis? In this connection, Hoffmann has suggested that chemists tend to view their work through the metaphor of discovery when (at least in the practice of synthesis) the metaphor of creativity, the lens through which artists typically view their work, is more appropriate.[10]

In the chapter entitled "Chemical Synthesis: Complexity, Similarity, Natural Kinds and the Evolution of a 'Logic,' " Stuart Rosenfeld and Nalini Bhushan open up the field of chemical synthesis to philosophical investigation. We argue that the philosopher's conception of the science of chemistry has, in effect, blocked the development of a philosophy for the science. We expose some of these misconceptions as they play out in the area of chemical synthesis and show that, when rectified, the conceptual framework that sustains this field is ripe for philosophical scrutiny.

Models and Metaphors

A current problem in chemistry is the design and implementation of devices in which molecules or groups of molecules function in analogy with their macroscopic counterparts such as wires, transistors, switches, and so on. What philosophical issues might this program raise? A couple that come to mind are familiar from the philosophy of science more generally—the nature of models and the power of metaphors—but to which interesting new approaches are taken by our chemist and philosopher teams in Chapters 12 and 13 of this volume.

We begin with models. Models are a central feature of the organization and practice of modern chemistry.[11,12] Chemists create and refine models in an effort to deepen the understanding of phenomena and to afford predictions that guide future experiments. But models are not physical reality: they constrain thought and even mislead[13] at times, and this is especially likely where our philosophical understanding of their status is incorrect or incomplete.

One might argue that the area of mechanism and mechanistic models in particular is central to the discipline of chemistry. Mechanism also lacks obvious analogues in other areas of science, and so it is a fruitful area to be mined by philosophers once alerted by chemists to some of the significant issues. In the chapter, "Models and Explanations: Understanding Chemical Reaction Mechanisms," Barry Carpenter links different kinds of explanation to different classes of model, broadly understood to include hypotheses, theories, and laws. He then sets up the conceptual issues pertaining to laws, explanation, and theory as discussed by philosophers Van Fraassen, Cartwright, Boyd, Kitcher, and Hempel in the philosophy of science and shows how these issues and distinctions play out in the context of experimental design and research strategies in chemistry, but specifically in the area of chemical reaction mechanisms. Carpenter's article adds a new dimension to the discussion on the epistemic role of experiments in scientific explanation.

We move next to metaphors. Metaphors pervade the language of chemistry. In recent times, for example, the dialogue of chemistry has expanded to include discussion of "molecular devices" (switches, wires, sensors, motors, etc.) in a way that suggests the ongoing creation of an entirely new technology, a nanoscale technology of sorts. Powerful metaphors like these may even direct the course of research. This may be along both productive and unproductive avenues in their (appropriate as well as inappropriate) service *as* models or through their natural and intimate connection *to* models. Furthermore, it is often these metaphors that convey the principal understanding of developments, for better or worse, to the general public. What is the relationship between models and metaphors? To what degree does the metaphorical extension of concepts provide new knowledge?

In the chapter entitled "How Symbolic and Iconic Languages Bridge the Two Worlds of the Chemist: A Case Study from Contemporary Bioorganic Chemistry," Emily Grosholz and Roald Hoffmann take on a novel task: they focus on an article in chemistry that appeared in a recent prestigious journal (*Angewandte Chemie*) and extract for the reader a detailed analysis of the way in which the authors move seamlessly between very different kinds of languages as they move back and forth between the world of the laboratory, samples, and mixtures and the world of the molecule (and thereby between different levels of reality). One goal of the piece is to demonstrate that even though chemists may speak different languages, all at once even, this practice is explanatorily successful. Another is to make a case for the more basic position: that iconic languages are not simply a heuristic device employed by chemists to quickly sketch what a formalized language would do more completely; rather, they are indispensible to the conceptualization and practice of chemistry.

The chemical senses

Finally, the collection includes two chapters that discuss olfactory experiences, which might be surprising to the reader in that it might be thought that this issue belongs in an area far beyond the reaches of the anthology. Although it is true that many issues having to do with olfactory experience are strictly issues for a philosophy of *perception* rather than for a philosophy of *chemistry*, we believe that their resolution is essential to isolating and understanding the chemistry that anchors the perception.

Philosophers have always been interested in the different ways in which objects are "presented" or "represented" to the senses, but the sense modality of primary interest has been vision. Thus, problems of perception are typically problems of *visual* perception, rather than auditory, tactile, olfactory, or gustatory perception. In cases where the other senses are discussed, it is typically in terms of analogies drawn with the visual sense. Issues that arise in these other domains for which there is no clear analogue in vision are ignored; conversely, the fact that certain perceptual issues in vision do not seem to have a counterpart in the other senses becomes reason to regard those sense modalities as perceptually uninteresting. As a case in point, psychologists (like Trygg Engen at Brown University and others) have demonstrated how the domination in odor research by the kinds of research problems and solutions extracted from vision has resulted in a setback for the field, as the models used for vision have slowly been discovered to be inappropriate in the field of smell.[14]

Of the five senses, smell and taste have generated the least philosophical curiosity. Of course, smell and taste do operate very differently than do sight, sound, or hearing; there is good reason why these are called the chemical senses. Indeed, it can be demonstrated quite powerfully just how odd it is to talk about smell in the same terms with which we talk about vision. The information richness of the visual domain and the kinds of information that are available to us about specific properties like shape, size, or color of visual objects at various degrees of resolution have no easy analogue in the domain of smell. What is the shape and size of a smell, and what are its boundaries? These questions seem oddly out of place; indeed, they seem to involve a category mistake.

Are the cases of smell and taste different in a way that forces us to rethink the epistemology of perception in general? For instance, unlike the case of vision where appearance and reality seem to come apart—one piece of evidence for this being that we talk comfortably about visual illusions—it is very difficult to create such a wedge in the case of smell. Smells are around us, they suffuse consciousness and do not allow for a distance between us and the olfactory object sufficent to where it would make sense for us to ask: This is the smell I perceive, but could I be mistaken? Can my nose be fooled? For, one might argue, smell (or the molecules, strictly speaking) floats right up and into the nose, leaving no room for the possibility of perceptual error.

One area for fruitful exploration is an analysis of the phenomenological differences between sight and smell (or taste) and the discussion of possible candidates for a theory of olfactory and gustatory perception. There is room here for a careful and sustained discussion of the property of smell: Does a smell have a structure? Can we conceive of it as oriented in a certain direction? What is the relationship between the perceptual property of smell and the structure of the molecules that give rise to the perception?

Thomas Morton's chapter "Archiving Odors" vividly illustrates the need for just such a resolution when he argues that to coherently and competently undertake the chemist's task of understanding olfaction and, more specifically that of archiving odors, one must, of necessity, begin with the olfactory experience. Morton examines the possibility of the reproducibility of sensation and concludes that the recipe for such replication remains elusive. Included in his discussion are philosophically rich topics such as the similarity of odors (archiving requires this assessment of similarity, a property whose status and philosophical import for the sciences Rosenfeld and Bhushan discuss in their chapter), the sharability of odor experiences across species, and the possibility of being odor insensitive.

William Lycan focuses on the olfactory experience as he analyses phenomenological differences between vision and smell in his chapter "The Slighting of Smell." He uses these differences to argue that if smell had been taken as the paradigmatic sense modality instead of vision, philosophical theories of perception like direct realism and sense data theories would have seemed far less plausible. Lycan ends the chapter with the interesting and suggestive claim that if chemistry had been taken as the paradigmatic science rather than physics or biology, common-sense scientific realism would have seemed the obvious position to take in the philosophy of science. (Andrea Woody and Clark Glymour give the reader an opportunity to ruminate further on the shift suggested by Lycan by articulating in broader strokes the sorts of questions and research projects that they believe would emerge as one shifts ones focus from physics to chemistry.)

In closing, the topics and many chapters in this anthology reflect the central, crucial concern with reductionism but at the same time situate the reader in a position that allows a look beyond. The anthology addresses a wide audience that includes both philosophy and chemistry; it seeks to move the nascent field of philosophy of chemistry to the next level of development by breaking new ground in that it brings together a group of chemists and philosophers in a structured and field-defining discussion aimed at the assumptions, methods, and practices of modern chemistry. We hope this volume constitutes a significant step in broadening the scope of the humanities in general, and in extending the domain of the objects of study of philosophers in particular.[15]

Notes

1. *Philosophy Research Archives*, 7, 1981.

2. The topic of the first conference (November 1993) was Philosophical Perspectives of Chemistry; the topic of the second conference (November 1994) was the Language of Chemistry. The topic of the third conference (September 1996) was the Autonomy of Chemistry in Relation to the Other Natural Sciences.

3. Recent works by Mary Jo Nye, David Knight, Bernadette Bensaude-Vincent and I. Stengers, and Jan Golinski in the history of chemistry; Roald Hoffman, Pierre Lazzlo, and Jean-Marie

Lehn in chemistry; and Joseph Earley, Eric Scerri, and Joachim Schummer in the philosophy of chemistry are examples of this trend.

4. See J. van Brakel, "On the Neglect of the Philosophy of Chemistry," Paper presented at the First International Conference on Philosophy of Chemistry, London, 1994; J. van Brakel, entry on chemistry in *Handbook of Metaphysics and Ontology*, vol. 1 [A–K]. Philosophia, 1991.

5. In a nice review of an area that bears directly on this question, Weininger ("The Molecular Structure Conundrum: Can Classical Chemistry be Reduced to Quantum Chemistry?" *Journal of Chemical Education*, 61, 939–944 [1984]) has commented on the continuing disagreement between Woolley and Bader regarding the status of the concept of molecular structure, with others like Scerri and Mosini contributing to the debate.

6. See, for instance, Davis Baird, "Analytical Chemistry and the 'Big' Scientific Instrumentation Revolution," *Annals of Science*, 50, 267–290; (1993), Davis Baird and Thomas Faust, "Scientific Instruments, Scientific Progress and the Cyclotron," *British Journal for the Philosophy of Science*, 41, 147–175 (1990).

7. See, for instance, Ian Hacking's discussion of the philosophical implications of the role of instruments in science in *Representing and Intervening* (Cambridge: Cambridge University Press, 1983).

8. For more details, see the report by Stu Borman on instrumentation entitled "Peering into the Analytical Crystal Ball," *Chemical and Engineering News*, March 31, 1997.

9. Sir John W. Cornforth, "The Trouble with Synthesis," *Australian Journal of Chemistry*, 46, 157–170 (1993).

10. Roald Hoffman, *The Same and Not the Same* (New York: Columbia University Press, 1995), ch. 19.

11. Colin Suckling, Keith Suckling, and Charles Suckling, *Chemistry through Models* (Cambridge: Cambridge University Press, 1978).

12. Carl Trindle, "The Hierarchy of Models in Chemistry," *Croatica Chemica Acta*, 57(6), 1231–1245 (1984).

13. An interesting example of a specific problem of this sort is described by Barry Carpenter, "Intramolecular Dynamics for the Organic Chemist," *Accounts of Chemical Research*, 25, 520 (1992).

14. For instance, in the case of color, classification research is based on the assumption that stimuli cause sensations of certain qualities (colors) that subjects can describe and categorize. This assumption has netted good results in the case of color. Odor researchers have extended this to the domain of smell, assuming that the olfactory sense fits such a model, with the result that the classification of odor has been highlighted as the most significant research task. However, analogous success on odor classification has not been achieved, for it turns out that there is, at best, a tenuous connection between words and odor perception, in sharp constrast to color. And yet the case of color is so seductive that odor classification continues to be viewed as the most important research goal, with Henning's "smell prism" presented as analogous to the color solid in most textbooks today, although there is little evidence to support this as a valid model for understanding the smell domain. So here we have an instance of research on the olfactory sense modality suffering a setback because it has taken its research cues from the dominant models of explanation used in vision.

15. Thanks to Ernie Alleva, Sidney Gottlieb, and Sigrun Svavarsdottir for helpful comments on an earlier draft of this chapter.

Part I

Autonomy and Antireductionism

2

Missing Elements

What Philosophers of Science Might Discover in Chemistry

ANDREA WOODY
CLARK GLYMOUR

Ad Hominem

In the late middle ages, chemistry was the science and technology closest to philosophy, the material realization of the method of analysis and synthesis. No longer. Contemporary philosophy is concerned with many sciences—physics, psychology, biology, linguistics, economics—but chemistry is not among them. Why not?

Every discipline has particular problems with some philosophical coloring. Those in quantum theory are famous; those in psychology seem endless; those in biology and economics seem more sparse and esoteric. If, for whatever reason, one's concern is the conceptual or theoretical problems of a particular science, there is no substitute for that science, and chemistry is just one among others. Certain sciences naturally touch on substantive areas of traditional philosophical concern: quantum theory on metaphysics, for example, psychology on the philosophy of mind, and economics and statistics on theories of rationality. In these cases, there is a special interest in particular sciences because they may reform prior philosophical theories or recast philosophical issues or, conversely, because philosophy may inform these subjects in fundamental ways. That is not true, in any obvious way, of chemistry.

So what good, then, what special value, does chemistry offer contemporary philosophy of science? Typically philosophical problems, even problems in philosophy of science, are not confined to a particular science. For general problems—problems about representation, inference, discovery, explanation, realism, intertheoretic and interdisciplinary relations, and so on—what is needed are scientific illustrations that go to the heart of the matter without requiring specialized technical knowledge of the reader. The science needed for most philosophy is familiar, not esoteric, right in the middle of things, mature and diverse enough to illustrate a variety of fundamental issues. Almost uniquely, chemistry fits the description.

In philosophy of science, too often an effort gains in weight and seriousness merely because it requires mastery of an intricate and arcane subject, regardless of the philosophical interest of what it says. Yet, surely, there is something contrived, even phony, in illustrating a philosophical point with a discussion of the top quark if the point could be shown as well with a discussion of the ideal gas law. Familiarity and simplicity make portions of chemistry attractive for illustrating a great range of philosophical points, but compared to quantum field theory, or neuroscience, chemistry has no cachet at all. Arguably, the greater part of contemporary philosophical writing on the neurosciences and on physics are chiefly expository—old theories in new mathematics, or new mathematics dumbed down, or vast arrays of information or recent history summarized—with some philosophical moral tacked on. Much of the philosophy of science has become elevated science journalism, and to journalists and their readers, chemistry offers little excitement: no black holes, no blindsight, no vacuum collapse. Perversely, chemistry may be an unpopular science among philosophers because it offers them little more than occasions for philosophy. Let's consider some of those opportunities.

Intertheoretic Relations

Wedged between the theories of physics and biology, chemistry is one of the best arenas for developing and assessing accounts of intertheoretic relations. Systematic explorations are rare; many people appear to believe the end result is too obvious to be worth the trouble. Such mistakes are the petri dishes of opportunity. Even a cursory consideration of the relations of chemical theory to other theories reveals intricate linkages that strain traditional analyses (see Bogaard, 1978; Liegener & Del Re, 1987; Primas, 1983; Sarkar, 1992; Scerri, 1991a, 1991b, 1994).

Given that most chemists consider quantum mechanics to be the foundational theory of their discipline, our examination begins there: introduction of the Schrödinger and Heisenberg formulations of the new quantum theory in the late 1920s drew immediate attention to the nature of the chemical covalent bond, a phenomenon difficult to comprehend from the vantage point of existing chemical theory. Consequently, the Heitler-London (1927) model of molecular hydrogen was a supreme achievement of the new quantum theory, even given the limited quantitative accuracy of its predictions. During the decade that followed, refined approximation techniques produced increasingly accurate ground-state bond energies for simple diatomic molecules. There were significant consequences of this achievement, both epistemic and sociological (see Nye, 1993). The calculations quickly came to be viewed as strong confirming evidence for the quantum theory and, hence, as poster children for successful theoretic reduction. In 1929 Paul Dirac declared that "the underlying physical laws for the mathematical theory of a large part of physics and the whole of chemistry are completely known." Assuming only mop-up work remained, many contributors to early molecular quantum theory soon departed for the more glamorous landscapes of nuclear physics and elementary particle research.

Suppose we assume that the reduction is in *some* meaningful sense successful. What might we learn from it? What, for instance, is the significance of the achievement for chemistry? Producing an empirically adequate Schrödinger wave function model of the hydrogen molecule could offer little beyond psychological comfort to chemists; they learned nothing about hydrogen they did not already know. It is even unclear whether empirical adequacy could readily have been achieved without experimental bond energies to guide the way through unfamiliar computational terrain. If the computations were valuable in and of themselves, it was not because of what they offered chemistry per se, but because they confirmed aspects of the new physical theory. Indeed, chemistry remains the domain where quantum theory most directly connects with reliably measured classical phenomena.

And there is the philosophical problem. If such intertheoretic connections are important because, in company with experimental data, they test and confirm the "reducing theory," then the inferential steps in such a reduction are critical to the role. In other words, we had better know why our approximations work; it is not adequate to employ any old maneuver that produces agreement with observable data when such agreement serves as the benchmark by which to assess the approximate truth of the reducing theory, in this case, quantum mechanics. Much the same is true if we seek to codify connections to ensure logical consistency between our theories.

The structure of approximate reasoning is not simple. Consider the Born-Oppenheim approximation (separability of electronic and nuclear motions due to extreme mass difference), which in application produces "fixed nuclei" Hamiltonians for individual molecules. In assuming a nuclear skeleton, the idealization neatly corresponds to classical conceptions of a molecule containing localized bonds and definite structure. All early quantum calculations, and the vast majority to date, invoke the approximation. In 1978, following decades of quiet assumption, Cambridge chemist R. G. Woolley asserted:

> Molecular structure makes no appearance in a quantum treatment of molecules starting from first principles. We are thus dealing with a qualitative change in the theory which is expressed in the mathematics by a discontinuous approximation, and one is bound to question whether invoking the structure hypothesis is always 'the right thing to do.' (Woolley, 1978, p. 1076)

Since that time, Woolley's views have been elaborated in a series of articles challenging the validity of well-entrenched modeling procedures in quantum chemistry. As a result, the classical notion of molecular shape, a cornerstone of chemical theory that many assumed to be explained by, or at the very least consistent with, quantum mechanics, has been placed in doubt (see Weininger, 1984; Ramsey, 1997a). Meanwhile, calculations employing the approximation can no longer be taken to provide the tight connection between foundational theory and observables needed for evidential support. Other common approximations have yet to be given such scrutiny. In short, confirmation of the quantum theory via application to molecular systems is problematic in ways we are only now beginning to realize. And the supposed reduction of chemical theory to physical theory is stymied by elementary chemical concepts such as "molecular structure" and even "molecule" itself.

Chemistry's connections to other theories are similarly complex. Sklar's (1993) detailed treatment of statistical mechanics demonstrates the difficulty of translating, without transformation or remainder, phenomenological concepts such as temperature, boiling point, and liquid into the framework of statistical mechanics. Temperature, for instance, is a quantifiable property of an individual system when defined via classical thermodynamic laws. But within a statistical mechanical framework, temperature becomes a defining constraint of a canonical ensemble or a parameter of the micro-canonical distribution (Sklar, 1993, p. 351). In applying the concept to a variety of individual systems, we discover that different features of these systems will be picked out. Although the theory gives us instructions for selecting the appropriate features for a particular system, the concept of temperature functions as a determinable property whose corresponding set of determinates appears heterogeneous from other theoretical vantage points, including that of phenomenological thermodynamics. Thus, for temperature and many other crucial concepts, no simple conceptual identification bridges the divide between statistical mechanics and thermodynamics.

With respect to reduction in toto, Sklar concludes:

> To go from the dynamic world picture to a world theory adequate to ground thermodynamic phenomena seems to require the positing of something *sui generis*, the fundamental probabilistic constraint on initial conditions that is the core of the statistical mechanical approach to non-equilibrium dynamics. (Sklar, 1993, p. 370).

In other words, Sklar's position parallels Woolley's with respect to quantum mechanical reduction. In both cases, the reducing theory can, in principle, provide satisfactory accounts of the phenomena deduced from the reduced theory, but only if certain constraints are invoked that themselves lack rationalization from within the reducing theory. In each case, the theoretical connections between the two frameworks, while clearly fruitful and most likely consistent, are insufficient to allow one theory actually to replace the other. Such a result should not be surprising, although it does stands at odds with much of the philosophical discourse on reduction.

When the problems of strict logical reduction become painfully evident, we must remind ourselves that intertheoretic relations come in all shapes and sizes. Reductions themselves can be logical, ontological, or even methodological in nature. Traditionally, the subject matters of chemistry and biology have had little in common. One encompasses the science of the living, leaving for the other the science of the nonliving. But today the two disciplines are strongly linked, and the middle ground is the terrain of the most rapidly expanding scientific enterprise of the last quarter century (see Levy, 1980; Schaffner, 1993). The connections seem reductive here as well, but the reduction is one primarily of constitution. Organisms are composed of cells that, in turn, are constituted by molecules both large and small. There has been less preoccupation with strict logical relations in this case, in part because most chemical theory was well established prior to the relevant biological theory and because the complexity of biological systems inhibits biological data from playing an evidential role analogous to that of chemistry for quantum mechanics. Instead, portions of biological systems are identified with specific molecules whose chemical properties are used to illuminate biological function and structure. The helical structure of DNA was determined in part by considering the chemical interactions of its components; the operations of the living cell rely upon sodium pumps; and allosteric proteins are explained in terms of hydrogen bonding. On the presupposition that chemical information informs biological understanding, we are investing immense resources to "map" the human genome right down to the last little carbon chain.

The philosophical upshot of this discussion should be obvious. The most commonly cited example of a successful theoretical reduction in the modern sciences, that of chemistry to physics, is problematic, and details of its formal structure are still not well understood. The link between statistical mechanics and phenomenological thermodynamics is no less treacherous, while the role of chemical theory in biology rests, by all appearances, largely outside the bounds of logical derivation. Two paths are available at this juncture: we can question the adequacy of our typically logical conceptions of reduction, or we can ask whether construing intertheoretic relations as paradigmatically reductive is likely to be a fruitful path for current philosophical investigation. The intricacy and richness of relations between quantum mechanics, chemical theory, molecular biology, and observable phenomena suggest we reevaluate how best to frame our discussions of scientific theories, their interrelations, and their functions. Relations of reduction seem natural within a framework that views the epistemological challenges of confirmation (as traditionally conceived) as the central task for philosophy of science and represents theories as deductively closed sets of sentences. It is much less obvious what traditional accounts of reduction have to offer projects concerned more with understanding the process that yields justifiable claims than establishing the justification of the claims themselves. But more on that later.

Explanation

Some of the oldest and most venerable questions in philosophy of science concern explanation: What makes an explanation "scientific"? What makes it adequate or correct? Why should we desire explanatory theories? We can approach these questions from two directions, starting from general philosophical frameworks and corresponding intuitions about the essential natures of both explanation and science, or abstracting from the richness and complexity of actual scientific practice. In this section, we peer down each path, glimpsing what we might learn about explanation from chemistry.

Philosophers remain divided in fundamental ways about what constitutes the essence of the explanatory relation. Accounts of explanation vary in their emphasis on what the explaining theory or hypothesis asserts about causation, constitution, identity, and function; on logical or probabilistic features of relations between theory or hypothesis and data; and on cognitive functions a theory or hypothesis or representational technique serves. These three aspects are not mutually exclusive of course, although most philosophical discussions have turned on details about the first two and have ignored or merely gestured at the third. Aristotle's account of explanation in the *Posterior Analytics* involved both deductive relations and causal and functional content. Hempel's (1965) account was essentially reductive and attempted to replace the causal and functional part of Aristotle's account with almost purely logical requirements, while Salmon's (1984) theories move in the opposite direction, stressing the identification of causal structure. More recent proposals by Van Fraassen (1980) and others emphasize satisfying probability requirements, while an account due to Kitcher (1989) endorses unification via a repeated pattern of logical derivation. Philosophical accounts of explanation that detail cognitive functions of explanations, and relate them to either of the other two aspects, are hard to find, although one would think the growing emphasis on cognitive aspects of science in recent years would have provoked them. For example, Friedman (1974) proposed an almost entirely logical account of explanatory unification under the title "Explanation and Understanding," but the contribution of the logical relation to understanding was scarcely discussed by either Friedman or his critics.

Explanation in chemistry displays a few characteristics worth noting, as follow.

Causation and Control

Although chemistry offers few, if any, functional explanations, it provides paradigmatic causal explanations. Offhand chemical explanations, for example, of why iron rusts, cite both constituent chemicals in the environment and a process (an oxidation reaction). More detailed explanations of reactions unfold the process: they cite a mechanism or give reaction rates, or both, or more. The details are more than just theoretical filling for the particular case of why iron rusts. They permit the explanation of contrary cases (for example, of why galvanized iron does not rust) and of variations in cases (for example, of why rusting is more rapid at higher temperatures or in more moist atmospheres). Unlike typical explanations in dynamics (for example, of the motion of a pendulum, or the motions of the planets), chemical mechanisms specify a causal sequence of discrete changes in the relations of constituents, with explicit dependence on environmental variables. The result is a kind of theory-based control and capacity for planning and predicting the results of alterations of conditions that is enormously rich and diverse. How theoretical explanations facilitate planning and control is a topic of intense interest in artificial intelligence, but little touched on in philosophy.

SANFFBAT

The theory of the chemical bond is one of the clearest and most informative examples of an explanatory phenomenon that probably occurs in some form or other in many sciences (psychology comes to mind): the semiautonomous, nonfundamental, fundamentally based, approximate theory (SANFFBAT for short). Chemical bonding is fundamentally a quantum mechanical phenomenon, yet for all but the simplest chemical systems, a purely quantum mechanical treatment of the molecule is infeasible: especially prior to recent computational developments, one could not write down the correct Hamiltonian and solve the Schrödinger equation, even with numerical methods. Immediately after the introduction of the quantum theory, systems of approximation began to appear. The Born Oppenheimer approximation assumed that nuclei are fixed in position; the LCAO method assumed that the position wave functions for electrons in molecules are linear combinations of electronic wave functions for the component atoms in isolation. Molecular orbital theory assumed a characteristic set of position wave functions for the several electrons in a molecule, systematically related to corresponding atomic wave functions.

The interesting thing is that approximation methods have been retained even after the emergence of direct computational methods more faithful to the quantum theory and more accurate in their predictions. That argues that the explanations given by molecular orbital theory, in particular, serve some function for chemists that numerical methods do not. Everyone who has taken a college chemistry course will at least recall the bulbous pictures of "molecular orbitals." What are they doing still hanging around?

Likely answers turn on the cognitive functions of explanatory schemes. Molecular orbital theory is visualizable, indeed pictorial, but schematic in a manner that facilitates thinking and talking and teaching. It is recursive; molecular orbitals of big things (organic molecules, for example) can be constructed from molecular orbitals of littler things. Therefore, qualitative predictions can be made, more or less in the course of a conversation. Because it is visualizable, and thus cognitively accessible, molecular orbital theory permits chemists to think about molecular structure and its implications for such things as bond energies and bond lengths in a way that numerical methods do not; this is a capacity chemists have rightly been reluctant to surrender. (Contrast that with the fate of Lewis electron diagrams, which have almost disappeared from chemical parlance, perhaps because, although they were visualizable and recursive, they were not informative about the quantities that came to concern chemists.) An apt analogy is roughly this: molecular orbital theory is to quantum mechanics what folk psychology is to neuroscience. Philosophers so captivated by the second relation might benefit from thinking about the first.

Boundaries in Practice

One of the curious aspects of disciplines is the kinds of explanation they do not give. Aside from a few questionable examples, chemistry is almost devoid of functional explanations, even though chemistry figures as an essential adjunct to many prototypical functional explanations. It makes sense to ask, for example, what function hemoglobin serves, and to answer that its function is to transport oxygen to the cells. But the question and the answer are not part of chemistry, as usually understood, even though how the hemoglobin molecule is able to serve that function does receive a chemical explanation. It makes sense to ask what the function of oxygen is in the body, but it makes no sense to ask for the function of oxygen or for the function of oxidation, full stop. They haven't any. Chemistry is particularly interesting because it is invoked just at the boundary between functional and nonfunctional explanations, and is the obvious contrast case for functional explanations in biology. Similarly, historical explanations play almost no part in chemistry, although chemistry has a major role in many historical explanations. For example, chemistry does not explain the distribution of oxygen on the Earth, although chemical regularities would inevitably figure in any such explanation.

We seldom ask questions about the boundaries on explanation, yet arguably, these boundaries mark an essential feature of the modern scientific worldview; the boundaries were quite different in ancient science. There are other boundaries that seem accidental or arbitrary, but perhaps are not. Chemists think it their business to explain why oxygen gas is diatomic, phosphorus gas polyatomic, and helium gas monatomic. The stabilities of these molecules are derivative, not fundamental, facts of chemistry. But the stability of the oxygen, or phosphorus, or helium atoms are not the subjects of chemical explanation at all. What determines our explanatory boundaries by discipline? Is it pure historical accident, or a disparity of experimental techniques, or faith in a kind of logical "screening off," faith that, for example, the behavior of molecules would be the same no matter what the explanation of atomic stability?

These distinctive features of explanation in chemistry might be useful for understanding the essential goals of the discipline. They also provide further evidence that the success of explanations is determined by some complex interaction of the three factors commonly emphasized by philosophers. Investigating how these factors are related is a project still waiting for proper attention.

Representation

Any scientific endeavor makes decisions about what sorts of theories to pursue and develop. Inevitably, some fundamental constraints are placed on what is judged to be legitimate or interesting content, as the preceding discussion of explanatory boundaries illustrates. Moreover, once the subject matter of a discipline has been settled, additional issues arise regarding how theories are, or should be, manifested in practice. All decisions of these sorts are ones of representation, and chemistry offers bountiful opportunity to explore not only the significance of representational choices for scientific inquiry but the very nature of representation itself.

As a discipline, chemistry has given priority to theories that are compositional, structural, and largely static, a choice that has provoked some scrutiny in recent years (Weininger, 1997; Luisi & Thomas, 1990). Although the nineteenth-century chemist Alexander Williamson argued that change was the essence of chemistry, and what distinguished it from physics, chemical reactions were analyzed in terms of mass equivalents and structural manipulation, with little attention to dynamics. Classical thermodynamics similarly stressed equilibrium states, while twentieth-century chemistry has become the landscape of the stationary state Schrödinger equation, complete with transformation of moving electrons into spatially localized electron distributions.

No doubt this orientation is, in part, a byproduct of the discipline's long history of having its empirical practice develop at least as rapidly as did its theoretical practice. Musings about phlogiston may seem strained and unsophisticated from our vantage point, but the nuanced laboratory skills of eighteenth-century chemists who embraced the theory would still garner respect and admiration. And although, in principle, theoretical descriptions are free to dwell on intricate processes of change, hands-on chemistry must fixate on the enduring end points. Thus, chemistry has been a science of "stuff"; identification and classification are at its core.

So it is fitting that in a causal way (remember twin earth), chemistry provides the chief philosophical example of a theory of natural kinds (Schwartz, 1977). Still, most of these discussions exhibit blind spots. Although there are endless debates about what makes for biological species, there are almost none about what makes for chemical species. Chemistry proposes all sorts of constraints and ambiguities concerning natural kinds (Van Brakel, 1997)—for example, which are the natural kinds, elements, or their isotopes, or both, and why? Chemistry suggests that new natural kinds are generated from old by nonlogical combination: Put bars of tin, lead, and copper together and have in the combination no instance of a natural kind; melt them together in the proper proportions, and you do—bronze. Chemistry further suggests that natural kinds have an odd hierarchical structure. Most materials formed from solutions with proportions of metals are not natural kinds, but their type—alloys—is a natural kind.

Besides the classification of the elements and compounds, there are related chemical systems that illustrate the intricacies and not quite hierarchical structure of scientific classification schemes. Molecular systems that crystallize are further classified by crystal structure. Organic molecules have an elaborate system of classification called "nomenclature" based on both their chemical composition and their symmetries. The classifications are correlated with properties that do not define the classification. Thus, the classification of the elements is associated with a numerical property—atomic weight—and molecular symmetries are associated with optical activity. What are the formal structures that make multiple classifications correct, and how does the existence of one classification scheme constrain others? What is the purpose, methodological or otherwise, of corresponding properties? Chemistry provides our most familiar examples of settled scientific classifications, and when we want to know what makes for such certainty, chemistry is where we should look.

The history of chemistry shows a transition from classifications by history to classifications by structure, but historical classifications remain essential in mineralogy. Rocks with identical chemical and crystal structure constitute different minerals according to the history of their creation—for example, whether the rock was formed above or beneath the surface. Why, we might wonder, are historical classifications sometimes abandoned and sometimes retained? Broadly metaphysical imperatives dictating that a domain of inquiry concerns constitution, or history, or laws of behavior surely matter, but so, just as surely, do methodological goals of reliable prediction or intelligible explanation. Chemistry seems a good place to look for answers and illustrations.

Even when scientific communities have decided, in roughest terms, what their theories should be *about*, they face further choices regarding how this content will be manifested and displayed in practice. The diversity of their choices is one of the most fascinating aspects of scientific practice, promising challenging case studies for philosophical conceptions of representation (see, eg. Brodie, 1867).

Consider: Is all content inherently propositional? Pictures insinuate otherwise, although the contrary thesis, that pictures are only a different language, is famous in aesthetics (Goodman, 1976). If one thinks all scientific content is propositional, then issues of natural or effective or pellucid representation are essentially issues of vocabulary, axiom selection, and semantic relations over syntactic structures (synonymy, entailment, presupposition, and so on), the very issues that concerned a lot of philosophy of science before 1970. If one thinks otherwise, then it becomes natural to search for examples in which pictorial representations are robust and to seek in their use a function for which they are arguably essential. Is there any discipline that offers better opportunities than chemistry for addressing such questions? Just imagine the chalkboard scribblings of a university professor teaching introductory chemistry: Lewis dot structures, molecular orbital energy diagrams, "curly arrow" reaction mechanisms, and isomeric structural formulae litter the surface, and, inevitably, somewhere on a wall nearby hangs the periodic table of the elements.

In the periodic table, chemistry offers one of the most successful and enduring nonmathematical, nonlinguistic representations of scientific information. The periodic table is a spatial, two-dimensional array of discrete elements in which the spatial relations of elements model nothing about spatial relations. Instead, spatial nearness and betweenness in each dimension are surrogates for similarity relations of other orderable properties such as atomic number, atomic weight, and periodicity. In turn, these determine higher order identity and similarity relations concerning certain forms of chemical reactivity and spectral behavior. The value of the spatial representation seems principally psychological: it aids pattern detection, both in discovery and in recollection.

Of course, the table was publicly introduced in the late nineteenth century, by Dmitri Mendeleev and Lothar Meyer, to represent the empirical "periodic law" long before an electronic theory existed to explain this periodicity. But the table is only one in a long line of diverse representations of the empirical regularity (Von Spronsen, 1969; Venable, 1896). Most early-nineteenth-century speculators proposed simple numerical relations; Dumas produced the beginnings of a full-blown algebra; and others opted for precise graphs of measured data. Later formulations (continuing to this day) have adopted all manner of spatial arrangement: spirals, helices, two-dimensional arrays, and more complex three-dimensional structures.

Why Mendeleev's table became the norm is a question for serious historical scholarship, but some pieces of the story are clear. The table was a convenient and efficient way to facilitate certain pattern-based forms of chemical reasoning. This power was most forcefully displayed by Mendeleev's bold predictions of new elements to fill in the gaps of his table. But the role of the table in modern chemical practice, while perhaps less spectacular, is no less compelling. How it works should command both philosophical and psychological attention.

Another legacy of the nineteenth century—molecular models—are inverse scale models, where aspects of spatial relations are retained but enlarged. Assuming spatial relations are approximately invariant from reality to superscale molecular models, the easily determined relations among models or features of the models permit inferences about the invisible molecules. We know that for over 30 years Dalton taught using a set of ball-and-stick molecular models manufactured for him by a friend. With the development of structure theory, as attention turned to understanding isomerism and its attendant consequences, such models became critical components of chemical education (see, e.g., Brumlik, 1961; Gordon, 1970). Kekule's use of physical models makes perfect sense, given his efforts to develop a chemical architecture. Yet, unlike the scale models of the engineer, where one macroscopic object stands for another macroscopic object, today's molecular models are physical objects representing an approximation of a theoretical conception of molecules. Our atoms, after all, are quantum mechanical, and a far cry from Dalton's (1808) piles of shot.

In other words, although the models themselves have changed little, our understanding of what they represent has evolved dramatically. So why haven't the models themselves been transformed? Modeling spatial relations within and among molecules is not an end in itself, but a guide to molecular stability and to predicting bonding behavior, reaction pathways, crystal structure, and, hence, optical and X-ray properties. That is part of the value in modeling, rather than describing, spatial relations. Nevertheless, models have interesting conventions, without which they are useless. It will do no good to try to predict the effects of calcining a chemical by burning a model of its molecules. Whether the model is made of wood or styrofoam is irrelevant to determining steric hindrance. And likewise, whether actual atoms are more like Bohr's solar system, Thomson's plum pudding, or a gaggle of strings, has no impact on the utility of balls and sticks for certain types of reasoning—only on which arrays of balls and sticks (or strings) are used. It is curious that, with all the discussion of "models" in philosophy of science, the concrete uses of physical models, and the conditions for their use, are scarcely characterized.

If there is a generalization to be drawn from these examples, it must be something along these lines: chemistry is a discipline whose very subject matter requires extensive organization and classification. Since before chemistry emerged as a professional science, the objects of its interest have often been of such large number, such diverse kinds, and such outright complexity that basic reasoning is a challenge for individual reasoners. Sophisticated representational practices and modeling techniques not only aid reliability and computational power, they make it possible for people to talk with one another, and to make inferences of chemical interest in the course of conversation. Consequently, theory—as manifested in these practices—can function to support reasoning and direct inquiry more broadly.

Instrumentation and Automation

Traditional philosophical theories of knowledge, including those of Plato, Aristotle, Descartes, Spinoza, and Locke, were about the *acquisition* of knowledge and, more specifically, about *methods* for acquiring knowledge. The concern for inquiry and its methods continued in nineteenth-century philosophy of science and even influenced the development of several sciences, but that concern all but vanished in twentieth-century philosophy, where it was replaced by an exclusive focus on the notion of justification of belief. In consequence, except for inadvertence, philosophy has been almost entirely removed from the most striking developments in scientific methods during this century: the instrumentation and automation of inquiry.

Nowhere are those developments more influential than in chemistry. Chemical instrumentation has developed at a wild pace since the late nineteenth century: spectroscopy over many electromagnetic frequencies (X-ray, ultraviolet, visible, near infrared) and under a variety of conditions (Raman and Brillouin scattering), mass spectroscopy, Mossbauer spectroscopy, nuclear magnetic resonance, gas chromatography, and many other techniques. In chemistry, these instruments have functioned primarily as diagnostic tools for recognizing a chemical species or differentiating chemical or biological structures or processes. Often, the utility of the techniques preceded their theoretical explanation; optical spectroscopy is an obvious example, but the details of nuclear magnetic resonance, based on an interaction with nuclear spin that Rabi discovered in the 1940s, are still a subject of research. Sometimes the utility of a form of instrumentation is discovered almost purely empirically. The use of magnetic resonance imaging for scanning bodies and discriminating types of tissue, including cancerous tissue, was an application found and established by a physician originally interested in cellular sodium pumps.

Each kind of instrument produces a community of experts who interpret the output of instruments of that kind. It is often easy to accumulate far more data than the human community can analyze, which creates no difficulty as long as there is no scientific or external pressure to make use of the data. When such pressures arise, however, the only solution is to automate interpretation, or parts of it. Any intellectual and social history of twentieth-century science will find a revolution in methods, beginning in the 1960s and accelerating throughout the rest of the century, prompted by a technology and mathematical theory—the electronic digital computer and the theory of computation—that replaced many sorts of scientific thought about particular problems with thought about *methods* for algorithmic solution of broad classes of problems.

Automation can result from many pressures, including the competition for discovery, as well as economics. The digital computer permitted automating Fourier transforms and so made feasible and ubiquitous a technique for separating periodic signals from noise. Automated planning of organic syntheses is helping to make synthesis more efficient and economical. These processes continue in physics, chemistry, molecular biology, and no doubt elsewhere, and they pose a number of epistemological problems. (Where are the philosophers when you need them?)

Scientists have varying reputations for experimental and analytical care, and varying reputations for generating hypotheses that turn out to be correct. They have varying productivities. If we think of each scientist as a mysterious android, these characteristics can be viewed as aspects of their reliability and of the efficiency of the internal algorithms they execute in doing science. We seldom formalize these judgments in measures about people, but we do try to formalize related judgments about explicit algorithmic procedures. It is not always clear how to do those formalizations, or what the trade-offs are or should be. When large amounts of data must be analyzed, feasible automated procedures typically trade off reliability for productivity: more errors are made, but more data are analyzed. How should those trade-offs be measured and weighed?

Automation of inference sometimes produces fundamental tensions. For example, some automated procedures have no guarantees of reliability—no theorems about their error rates, no confidence intervals—but in practice work much better than procedures that do have such guarantees. Which are to be preferred, and why? In some cases, theoretically well-founded procedures produce results that are inferior to stupid procedures. For example, in determining protein homologue families, a simple matching procedure appears to work as well or better than procedures using sophisticated Hidden Markov models. Which sort of procedure is to be preferred, and why?

Finally, the automation of interpretation offers new twists on issues about the objectivity of science, about a "realistic" interpretation of scientific claims, and about the social construction of scientific belief. Often the reliability of data interpretation procedures can only be assessed empirically, by running them on data, and must be adjusted, selected, or "tuned" from data. This fact creates special problems in judging the reliability of procedures designed to detect rare signals in voluminous data. If examples of the target are rare, the selection or adjustment or tuning of a detection procedure may "overfit" the data, and a procedure that works beautifully on test data may not be projectable and may err when put to work on new data. If there is no independent way to establish whether the signal is actually present, the procedure may create its own facts. For all sorts of reasons—rounding off, cutoffs, resolution, computational complexity, algorithmic error—automated procedures may create "artifacts" from raw data, features that mark no external, physical structure or process. Such artifacts can be lodged within the body of scientific belief, and they may be inextricable, either because the data analysis that would be required to discover and eliminate them is infeasible, or because the inertia of coordination—the losses to researchers' convenience or to their results or to the comparability of results entailed by giving up the algorithms—is too great. In some part, the scientific world may be constructed indeed, if not by us then by the algorithms data must pass through to be turned into scientific fact.

The Nature of Inquiry

Philosophical disinterest in methods, standards, and norms of inquiry has been defended by arguments it is hard to take seriously: Popper (1963) dismissed as subjects for psychology almost everything about the process of acquiring knowledge except the very idea of testing, but the slightest reflection shows that the scheduling of hypotheses and tests, as well as the selection of tests, is critical to reliable discovery, and it interacts with notions of justification. Laudan (1977) claimed that the only normative part of scientific inquiry is hypothesis testing and that the generation of hypotheses is only a matter of manufacturing (like sweaters and sandwiches) and not of philosophical interest. But if scientific inquiry is a kind of manufacture, then hypothesis testing is just an aspect of quality control. Another reason, seldom stated but often evident, is the influence of Kuhn's (1970) distinction between "revolutionary" and "normal" science, along with his view that scientific revolutions have no explicit rational structure. Even were this view correct, and it is not, the rational structure of normal inquiry ought to be interesting to everyone whose curiosity about science is not limited to melodrama (see, eg., Langley et al., 1987).

Kuhn's work has been embraced in part because his depiction of normal science, with its implicit rules for orderly puzzle solving, rings true to so many, and yet it leaves unresolved the issue of how such orderliness may be achieved. Scientific theories that are underspecified, unintelligible, or too complex to apply are of little value. Yet all of the theories heralded as triumphs of modern science originally suffered from such adequacies. Why has such scant philosophical attention been given to how practitioners learn to apply theory, learn to reason with it, and primarily as a result, come to make sense of it? (A notable exception is Ramsey, 1997b.)

Perhaps the best reason for the neglect of methodology by philosophers of science is that methodologists within the sciences are doing just fine, and there is nothing really for philosophy to contribute; in contrast, the foundations of quantum theory remain in dispute, as does the consistency of relativistic and quantum theories, and cognitive psychology persists as a fascinating mess of philosophical ideas. But even that reason is an unconvincing explanation of philosophical neglect: cognitive psychology, to mention only one example, is a methodological disaster, but there are few contributions, positive or negative, to its methodology from philosophical sources. For whatever reasons, the philosophical temper that gave rise to Kant's foundationalism, to Peirce's introduction of randomization, and to Ramsey's rehabilitation of Bayesian reasoning and decision theory has become effete.

Our discussion of chemistry suggests that the philosophy of science pays a high price for this disinterest. Let us point out how greater attention to the nature of inquiry might permit us to get a firmer grip on some of the issues raised in previous sections of this chapter.

Traditional accounts of theoretical reduction, although useful for analyzing confirmation relations for the "reducing" theory, provide few, if any, resources for understanding the value such connections hold for the "reduced" theory. In most accounts the "reduced" effectively disappears. Yet the productiveness achieved by introducing one theoretical domain into another is one of the few robust generalizations in the history of science. A striking example is the emergence of computational chemistry, a direct outgrowth of the introduction of quantum mechanics, as a powerful subdiscipline. Although the cases in which ab initio calculations have actively guided experimental research are not numerous, they do exist (Schaefer, 1984). Semiempirical techniques, on the other hand, have been widely integrated into diverse chemical investigations. More provocative still is the fact that molecular quantum mechanics has seeped into almost every crevice of chemical research and education via numerous formulations of "valence bond" and "molecular orbital" models, connecting the world of the organic lab bench with its presumed foundational theory. Witness the power of SANFFBAT.

On the surface, these theoretical structures have little in common with the neat axiomatizations of Von Neumann, Bohm, or even Hempel. These are not clean computations based on derivations from first principles; they are feasible only through heavy reliance on approximation, idealization, and empirical determination of theoretical quantities. More fascinating still, molecular orbital models are frequently conceived, represented, and manipulated diagrammatically by practitioners who lack the technical training to tackle Hamiltonians and wave equations. They have virtually taken on a life of their own. No doubt these theoretical structures, however precariously linked to fundamental theory, have been highly productive for chemical inquiry. The effects of self-conscious application of chemical concepts to problems of biological structure and function have been no less transformative. To the extent that philosophy aims to understand this productiveness—surely an appropriate goal for methodology—we must develop analytic tools capable of grasping the structure and function of such untidy theoretical connections.

Focus on intertheoretic reduction may also have distracted philosophers into inadvertently ignoring the complexities, and significance, of connections within a given discipline. Every real, active science is diverse, necessitating a means for control. The logical unity beneath this diversity reveals itself in two rather distinct ways. One way is through a set of general principles, which could be formalized. In chemistry, the formal principles would compose a long story involving rules of atomic composition and additivity of mass, lists of elements and their properties, specifications of phases, thermodynamic principles, gas laws, electrochemical principles, optical and spectral principles, and more. Those principles, with illustrations and exercises, are contained in sets of textbooks. Unity shows itself in a more important way in scientific practice, in the computation of variables, parameters, and properties of one sort from those of a superficially very different sort. From spectra of boron nitride, for example, thermodynamic characteristics of the molecule (or of a system of boron nitride molecules) can be calculated. Cannizzaro's great triumph in the middle of the nineteenth century was a method of calculating atomic weights from vapor densities. A system of such calculations for related quantities may transform one regularity into another: under thermodynamic substitutions, for example, one of Joule's laws becomes the ideal gas law. The structure of these interderivations seems ill-understood, but they are a mainstay of real science, and the practical value of its unity.

In other words, much of the structure of chemical theory serves to enhance our inferential capacities. Nor is articulation of this structure the only means by which chemistry supports reliable reasoning. In the eighteenth century, Macquer's tables assisted prediction via an orderly repository of a large number of chemical reaction types. The nineteenth century witnessed the manufacture and commercial sale of several "chemical slide rules" for calculating chemical equivalents, and the technology was frequently combined with a standard slide rule in the first half of this century (Williams, 1992). As we have seen, the periodic table itself supports many qualitative and comparative modes of reasoning.

Today, automated, semiempirical calculations are desktop tools for predicting anything associated with electron densities, including chemical reactivity and stability. Spectroscopy is capable of "fingerprinting" an immense range of chemical compounds, effectively embodying an inference to the conclusion "Substance X has composition Y." Some of the most intriguing scribbles on the chemist's blackboard are molecular orbital pictures in which the content of quantum mechanics has been molded into diagrammatic forms that suppress some quantitative information, thereby reducing complexity, while highlighting basic structural properties, such as symmetry, that are sufficient for rationalizing many chemical phenomena.

Such marvelous technologies can be understood only by grappling with concepts, theories, and instruments as they are manifested in practice. They are truly invisible from within the abstract "rational reconstructions" of much twentieth-century philosophy of science. Issues of representation gain centrality as we attempt to understand the cognitive functions of scribbles on the blackboard and the periodic table's matrix array. Likewise, attention to inquiry underscores the relation between explanation and understanding, suggesting that logical features of theories, or the causal stories theories help us to tell, are explanatory because of the role they play in making our theories intelligible. Some philosophers have eschewed the connection between explanation and understanding because it appeared to reduce explanation to psychological satisfaction. But understanding also allows us to be smart, reliable inquirers—that's what makes it so satisfying.

It is worth reminding ourselves that only through applying theory can we evaluate its evidential support and adequacy more generally. Methods of acquiring knowledge are intimately related to our justifications of it. Philosophers used to know this, but contemporary philosophy of science sometimes offers the appearance of suffering from amnesia.

Conclusion

No science is simple, but some are more accessible than others. For comparatively simple, vivid, and diverse illustrations of representation, classification, reduction, explanatory strategies, approximation, formalization, computation, discovery methods, instrumentation and automation, and, most of all, the relations of all of these to one another, chemistry is our best single source. The technical knowledge presupposed by chemical examples is not trivial, but it is often more accessible than string theory or molecular genetics. And because chemistry stands in the middle, and because it is mature, its structures are rich. Chemistry is full of philosophical questions neither chemists nor philosophers have fathomed; we have given only the briefest and most superficial survey of the issues the subject raises. (See also Theobald, 1976.)

The maturity of chemistry marks its reliability: chemistry works. We cannot escape the conviction that if we understood how chemistry works, we would understand most of what there is to know about how science works.

References

Bogaard, Paul A. 1978. "The Limitations of Physics as a Chemical Reducing Agent." *PSA*, 1978(2):345–356.

Brodie, Benjamin. 1867. "On the Mode of Representation Afforded by the Chemical Calculus, as Contrasted with the Atomic Theory." *Chemical News*, 15:295–305.

Brumlik, George C. 1961. "Molecular Models Featuring Molecular Orbitals." *Journal of Chemical Education*, 38(10):502–505.

Dalton, John. 1808. A *New System of Chemical Philosophy*. London: Peter Owen.

Dirac, P. A. M. 1929. "Quantum Mechanics of Many-Electron Systems." *Proceedings of the Royal Society of London*, A123:714–733.

Friedman, Michael. 1974. "Explanation and Scientific Understanding." *Journal of Philosophy*, 71:5–19.

Gordon, Arnold J. 1970. "A Survey of Atomic and Molecular Models." *Journal of Chemical Education*, 47(1):30–32.

Goodman, Nelson. 1976. *Languages of Art*. Indianapolis: Hackett.

Heitler W., & London, F. 1927. "Wechselwirkung neutraler Atome und homoopolare Bindung nach der Quantenmechanik." *Zeitschrift fur Physik*, 44:455–472.

Hempel, Carl G. 1965. "Aspects of Scientific Explanation." In *Aspects of Scientific Explanation and Other Essays* (pp. 331–496). New York: Free Press.

Kitcher, Philip. 1989. "Explanatory Unification and the Causal Structure of the World." In P. Kitcher & W. Salmon, eds. *Scientific Explanation* (pp. 410–505). Minneapolis: University of Minnesota Press.

Kuhn, Thomas S. 1970. *The Structure of Scientific Revolutions*. Chicago: University of Chicago Press.

Langley, P., Simon, H., Bradshaw, G., & Zytkow. J. 1987. *Scientific Discovery: Computational Explorations of the Creative Processes*. Cambridge, MA: MIT Press.

Laudan, L. 1977. *Progress and Its Problems*. Berkeley: University of California.

Levy, Monique. 1980. "The 'Reduction by Synthesis' of Biology to Physical Chemistry." *PSA*, 1980(1):151–159.

Liegener, Christoph M., & Del Re, Giuseppe 1987. "The Relation of Chemistry to Other Fields of Science: Atomism, Reductionism, and Inversion of Reduction." *Epistemologia*, 10:269–284.

Luisi, Pier-Luigi & Thomas, R. 1990. "The Pictographic Molecular Paradigm: Pictorial Communication in the Chemical and Biological Sciences." *Naturwissenschaften*, 77:67–74.

Nye, Mary Jo. 1993. *From Chemical Philosophy to Theoretical Chemistry*. Berkeley: University of California Press.

Popper, Karl R. 1963. *Conjectures and Refutations*. London: Routledge.

Primas, Hans. 1983. *Chemistry, Quantum Mechanics, and Reductionism*. Berlin: Springer-Verlag.

Ramsey, Jeffrey L. 1997a. "Molecular Shape, Reduction, Explanation and Approximate Concepts." *Synthese*, 111:233–251.

Ramsey, Jeffrey L. 1997b. "Between the Fundamental and the Phenomenological: The Challenge of 'Semi-Empirical' Methods." *Philosophy of Science*, 64:627–653.

Salmon, Wesley. 1984. *Scientific Explanation and the Causal Structure of the World*. Princeton, NJ: Princeton University Press.

Sarkar, Sahotra. 1992. "Models of Reduction and Categories of Reductionism." *Synthese*, 91: 167–194.

Schaefer, H. F. III. 1984. *Quantum Chemistry: The Development of Ab Initio Methods in Molecular Electronic Structure Theory*. Oxford: Clarendon.

Scerri, Eric. 1991a. "Chemistry, Spectroscopy and the Question of Reduction." *Journal of Chemical Education*, 68:122–126.

Scerri, Eric. 1991b. "The Electronic Configuration Model, Quantum Mechanics and Reduction." *British Journal for the Philosophy of Science*, 42:309–325.

Scerri, Eric. 1994. "Has Chemistry Been at Least Approximately Reduced to Quantum Mechanics?" *PSA*, 1994(1):160–170.

Schaffner, Kenneth F. 1993. "Theory Structure, Reduction, and Disciplinary Integration in Biology." *Biology and Philosophy*, 8:319–347.

Schwartz, Stephen P., ed. 1977. *Naming, Necessity, and Natural Kinds*. Ithaca, NY: Cornell University Press.

Sklar, L. 1993. *Physics and Chance: Philosophical Issues in the Foundations of Statistical Mechanics*. Cambridge: Cambridge University Press.

Theobald, D. W. 1976. "Some Considerations on the Philosophy of Chemistry." *Chemical Society Reviews*, 5:203–213.

Van Brakel, J. 1997. "Chemistry as the Science of the Transformation of Substances." *Synthese*, 111:253–282.

Van Fraassen, Bas C. 1980. *The Scientific Image*. Oxford: Clarendon.

Venable, F. P. 1896. *The Development of the Periodic Law*. Easton, PA: Chemical Publishing.

Von Spronsen, J. W. 1969. *The Periodic System of Chemical Elements: A History of the First Hundred Years*. Amsterdam: Elsevier.

Weininger, Stephen J. 1984. "The Molecular Structure Conundrum: Can Classical Chemistry Be Reduced to Quantum Chemistry?" *Journal of Chemical Education*, 61:939–944.

Weininger, Stephen J. 1997. "Butlerov's Edict: The Timeless, the Transient, and the Representation of Molecular Structure." Presentation manuscript, Boston Colloquium for the Philosophy of Science, November 5, 1997.

Williams, William D. 1992. "Some Early Chemical Slide Rules." *Bulletin for the History of Chemistry*, 12:24–29.

Woolley, R. G. 1978. "Must a Molecule Have a Shape?" *Journal of the American Chemical Society*, 100:1073–1078.

3

"Laws" and "Theories" in Chemistry Do Not Obey the Rules

MAUREEN CHRISTIE
JOHN R. CHRISTIE

Most philosophers' discussions of issues relating to "laws of nature" and "scientific theories" have concentrated heavily on examples from classical physics. Newton's laws of motion and of gravitation and the various conservation laws are often discussed. This area of science provides very clear examples of the type of universal generalization that constitutes the widely accepted view of what a law of nature or a scientific theory "ought to be."

But classical physics is just one very small branch of science. Many other areas of science do not seem to throw up generalizations of nearly the same breadth or clarity. The question of whether there are any laws of nature in biology, or of why there are not, has often been raised (e.g., Ghiselin, 1989; Ruse, 1989).

In the grand scheme of science, chemistry stands next to physics in any supposed reductive hierarchy, and chemistry does produce many alleged laws of nature and scientific theories. An examination of the characters of these laws and theories, and a comparison with those that arise in classical physics, might provide a broader and more balanced view of the nature of laws and theories and of their role in science.

From the outset, we should very carefully define the terms of our discourse. The notion of *laws of nature* has medieval origin as the edicts of an all-powerful deity to his angelic servants about how the functioning of the world should be arranged and directed. It may be helpful to distinguish three quite different senses in which laws of nature are considered in modern discussions. On occasion, the discussion has become sidetracked and obscure because of conflation and confusion of two or more of these senses.

In the first, or ontological, sense, laws of nature may be considered as a simply expressed generalization about the way an external world does operate. Laws of nature are often seen as principles of the way the world works. They are an objective part of the external world, waiting to be discovered. The laws that we have and use may be only approximations of the deeper, true laws of nature.

Clearly, there is a second, epistemological, sense of laws of nature that carries very different connotations. In this sense, a law of nature is a simply expressed generalization of our best knowledge and belief about the way the world works. We point out two areas where this difference is important. First, a law in this second sense can change over time with the state of scientific knowledge. Until the beginning of the twentieth century, Newton's Law of Gravitation was the law of gravity in the epistemological sense. It has been replaced by an understanding coming out of the general theory of relativity. But in the ontological sense, the true law of gravity may be a deeper and as yet undiscovered principle, and it certainly does not change with the state of human knowledge. Second, antirealists are able to consider and discuss laws of nature in the epistemological sense, whereas the idea of laws of nature in the ontological sense cannot easily fit in with their worldview.

Finally, there is a historical sense of the notion of a law of nature. A dictum is a law of nature in this sense if, for whatever reason, it has been widely described and considered as such. It is even possible that such a law has never been widely accepted as a generalized statement of the best knowledge and belief in a certain area. In some ways this sense is trivial. The reason it must be separately introduced and considered is that some writers, both philosophers and scientists, have felt free to introduce a particular dictum as an example into a discussion of laws of nature, simply because it is widely known as "X's Law," without explicitly considering whether it qualifies in either of the other senses (Christie, 1994).

The primary concern of this article is with epistemology, and we discuss laws of nature almost exclusively in the epistemological sense.

With the notion of theories, there is a rather different set of difficulties. There does not seem to be complete agreement on the fine distinctions between terms like *theory*, *model*, and *hypothesis*. At this stage, we will not attempt a strict demarcation between theories and models.[1]

Oversby (1998) considers the issue and attempts to sharpen up shades of meaning in the way that chemists use the terms, thus arriving at clearly distinct concepts. We would see this as a laudable program. But the distinctions cannot be made with clarity and certainty unless and until there is some sort of consensus, and we think that Oversby's view on this aspect is rather optimistic. He is correct when he points out that chemists universally refer to Boyle's Law, Raoult's Law, and Henry's Law, but to Arrhenius' Theory. Still, there are clear counterexamples. In one of the classic works of chemistry, Coulson's Valence (McWeeny, 1979), references to "molecular orbital theory," "the molecular orbital model," and "the molecular orbital method" are scattered more or less indiscriminately through the relevant chapters. And this variety of terminology is not McWeeny's quirk. Many authors of articles in textbooks, review articles, and research articles, use these terms interchangably, along with "approach" and "approximation," to describe the molecular orbital "insight" ("insight" being used as a generic term with an epistemological flavor).

So here, the term *theory* will be used in a way that embraces the typical named theories of chemistry: such things as molecular orbital theory, valence shell electron pair repulsion theory, transition state theory of reactions, and Debye Hückel theory of electrolyte solutions. No decisive distinction will be made between theory, model, and other similar terms. But there is one distinction that we do make. The term *theory* is considered in an epistemological sense—as an expression of our best knowledge and belief about the way chemical systems work.

With these understandings of the terms, the main arguments we present are as follows:

1. Laws and theories do play an important part in defining the body of chemical knowledge.
2. The characters of the laws and theories that chemists discuss and use often differ markedly from those of the laws and theories from physics that are usually chosen by philosophers as examples.
3. Scientific laws and theories have a much greater diversity of character than has generally been recognized.
4. The characteristic differences between the laws and theories of chemistry and physics largely arise from differences in complexity between the systems studied in the two disciplines and can be related to specific problems in the reduction of chemistry to physics.
5. The peculiar character of chemical laws and theories is not specific to chemistry. Interesting parallels may be found with laws and theories in other branches of science that deal with complex systems and that stand in similar relation to physics as does chemistry. Materials science, geophysics, and meteorology are examples of such fields.

Laws and Theories in Chemistry

The student of chemistry is not usually confronted with laws to nearly the same extent as is the student of physics. Moreover, many of those that do arise come into the realm of physical chemistry, which is really about the physics of chemical systems. In terms of their relationship to a total body of scientific belief or knowledge, the ideal gas laws, the laws of thermodynamics, and similar generalizations have as much claim to be regarded as part of physics as of chemistry. But in everyday scientific practice they tend to come much more into play as working tools of the chemist rather than of the physicist.

In addition, there are laws that are fundamentally and necessarily chemical. The laws of stoichiometry and of valency, and the so-called Periodic Law on which Mendeleev based his classification, are examples of laws that are clearly a part of the corpus of chemistry as such.

Theory is a term that is very widely used by chemists. To take the area of chemical bonding as an example, chemists widely refer to molecular orbital (hereafter MO) theory, valence bond (VB) theory, hybridization theory, valence shell electron pair repulsion (VSEPR) theory, and ligand field theory. And even those probably do not exhaust the list.

But the list of theories of chemical bonding that has just been presented is rather different from a list of theories from an area of classical physics.

In one sense, the chemical theories are rival theories. Valence bond theory provides a genuinely different explanation of chemical bonding to that provided by molecular orbital theory. According to valence bond theory, two bonded atoms obtain an energy advantage by staying close together because electrons can be exchanged from one atom to the other. According to molecular orbital theory, the advantage comes because the separate orbitals of the two atoms combine into molecular orbitals that spread over a region of space including both atoms, and the electrons can be more economically accommodated in these delocalized orbitals. The explanations are quite qualitatively distinct. They also provide different numerical answers, at least initially.

And yet the chemist adopts a pluralistic attitude toward the various theories and feels quite free to adopt whichever one seems best adapted to a particular problem. VSEPR theory is used to rationalize molecular geometries, ligand field theory to explain the spectral and magnetic properties of metal complexes, and so on. MO theory has almost entirely displaced VB theory over the last few decades. But chemists do not usually claim that the former is "right" in some sense and the latter is "wrong." Instead, the reason for the eclipse of VB theory seems to have been that MO theory is better adapted for incorporation into large and sophisticated computer programs!

The scientifically orthodox resolution of this difficulty is spelled out in many places (e.g., McWeeny, 1979, ch. 5). MO and VB theories are shown to be alternative first steps in a pair of schemes designed to successively approximate to an accurate quantum mechanical solution for the structure of a molecule. One aspect of this process that makes it seem a little less convincing is that the later steps in the approximation schemes differ markedly from the first. For example, with MO theory, a series of MO-like steps of successive approximation leads not to an accurate solution of the equations that arise from quantum theory but to the Hartree Fock limit. It is only by taking a rather different type of step, quite out of tune with the basis of the molecular orbital method—configuration interaction or some alternative—that the quantum theory solution can be approximated.

The essence of a molecular orbital is that it represents a spatial distribution for an individual electron that is not perfectly centered on a single atomic nucleus but is distorted from a spherical shape by the proximity of other atomic nuclei in the cluster of atoms that constitutes the molecule. But to step beyond the Hartree Fock limit, it is necessary to recognize that the notion of an orbital as the spatial distribution of a single electron in a many-electron system is, at best, an approximation and cannot be accommodated in an exact quantum calculation.

Interestingly, VB theory recognizes this fact in its first step of concentrating on electron exchange, but runs into difficulty because it is based on atomic distributions, which are each spherically symmetric about one of the several atomic nuclei in the molecular cluster.

For most theoretical chemists, debates about the theory of chemical bonding center around the relative merits of the several named theories and their variants. But there is another viewpoint, from which the strongest theory of chemical bonding is not a named numerical theory like any of those discussed so far but a robust graphical calculus that involves drawing and manipulating lines between pairs of atoms in the structural formulas of molecules. The initial idea is that such a line means a pair of valence electrons shared between the atoms at the ends of the bond. This basic idea then undergoes a number of extensions.

The first such extension is the introduction of double and triple bonds, where four or six valence electrons are shared. Then there are coordinate bonds, where both electrons are donated by one of the two atoms, instead of a formal donation of one from each atom. There are resonance structures and fractional bonds—to explain why the unusually stable benzene molecule, C_6H_6, has a ring structure that has hexagonal rather than triangular symmetry, or why the acetate ion, CH_3COO^-, has two symmetrically equivalent oxygen atoms. We need three-center bonds to explain the structures of the hydrides of boron—two electrons are supposed to be shared by three atoms.

Curved arrows with full heads are used to describe the breaking of bonds and the formation of new ones in chemical reactions. Half-headed arrows are introduced to cover reaction situations where pairs of electrons forming bonds are supposed to move in different directions and join up in new pairings.

The actual status of a theory of this sort is very difficult to pin down in any philosophical sense. Understanding the properties and reactions of different materials in terms of the behavior of individual atoms rests, above all, on an accurate description of the machinations of electrons. Most chemists accept in principle that modern quantum mechanics has "pretty much got this right." But this entails seeing electrons as quantum mechanical entities whose properties are in most ways more wavelike than particle-like. Insofar as their behavior is particle-like, they move at very great speeds throughout a molecule, and every electron exchanges with, and is indistinguishable from, every other electron. And yet, on another level, chemists make a commitment to the clear distinction between valence electrons and core electrons, and particular electrons are considered to be associated with particular bonds: indeed, some bonding electrons are explicitly described as "localized"!

It is also accepted at a basic level that every electron has a repulsive interaction with every other electron. Yet at another level, electrons are supposed to "hunt in pairs" in forming bonds and in rearranging bonds in the course of a chemical reaction.

It is tempting to suppose that the whole of this elaborate graphical and spatially expressed calculus is a merely formal accounting scheme with no direct ontological content. It is a successful scheme in accounting for most of the features of molecular structures and many of the features of chemical reactions. It has often been used in making predictions and deductions about such of these features as are directly measurable for comparison. But for all of its empirical success, it may be only a formal scheme.

There are many indications that seeing it as a formal accounting scheme fails to capture the whole richness of the content of the theory. Parts of the bonding calculus are in direct correspondence with spatial features of molecular structures. Molecular geometries are expressed in measurable form in X-ray or neutron diffraction patterns and in microwave spectra. They correspond directly to features that arise very simply in the bonding calculus. The attractive forces that hold the atoms of a molecule together are expressed in measurable form as infrared and raman spectra and as bond dissociation energies. The simple bonding calculus is based, for example, on the idea that one carbon–hydrogen linkage in one molecule is pretty much like any other carbon–hydrogen linkage in any other molecule. There is ample experimental evidence that carbon hydrogen linkages in different molecules are remarkably similar in terms of how far apart the atoms are, how much energy is needed to sever the linkage, what types of chemical reagent will achieve the severance, what is the resonant vibrational frequency associated with the linkage, and so on. The observation that a CH bond in one molecule is very like a CH bond in any other has never been evinced as a necessary entailment of the supposedly underlying quantum theory, though it has been verified by approximate quantum calculations for hundreds of such molecules.

Why should a C—H bond in methane be so remarkably similar to a C—H bond in sugar? Quantum mechanical calculations can confirm the similarities but cannot provide a simple explanation neither for this case nor for the general principle. The simple bonding calculus, which had already become quite sophisticated and recognizably modern half a century before the origins of quantum theory, takes this generalization as axiomatic and marches on from there.

Difference between Laws and Theories in Physics and Those of Chemistry

The laws of physics have been one of the major concerns of philosophers of science. The laws of chemistry have been largely neglected. A philosophical analysis of the laws of physics by Molnar (1969) produced a set of criteria for a "law statement." In this analysis, a law statement had to be a proposition that (1) was universally quantified, (2) was true, (3) was contingent, and (4) contained only nonlocal empirical predicates. There are problems with this set of criteria, and nobody, including Molnar, seems to want to accept them at face value.

The first issue, universal quantification, is a requirement that there be no exceptions to a law. It must be "in all cases" or "in no case." Even "in the vast majority of cases" will not do. The problem with this clause is an implicit ceteris paribus requirement for most laws (Cartwright, 1983). Newton's law of gravitation, for example, is a perfectly respectable law of nature in the epistemological and historical senses. It purports to quantify the force of interaction between a pair of massive objects. It has been empirically verified by measurement of the force between a pair of suspended metal spheres in a laboratory. But the law does not quantify the force, even as an approximation, if we take the trouble to provide the spheres with an electrostatic charge.

The issue of truth has largely been debated at the level of exactness. Do laws have to be exact, or do we have laws of nature that are necessarily approximate? Does the inexactitude arise because the laws themselves express inexact relationships—or because they relate to the properties of ideal entities that are not precisely realized (Ellis, 1991)?[2]

Contingency is problematic because the attitude one adopts to contingency largely depends on whether one's focus is primarily ontological or epistemological. It is not possible to decide the issue of whether the truth expressed in a law of nature is a necessary truth or a contingent truth without delving into the muddy realms of counterfactuals and alternative worlds. For Ellis (1991) and others, laws of nature are necessary rather than contingent truths because they express relationships between entities that are part of the essence of those entities. But for the scientist who sets out to measure the charge on an electron for the first time, the result is a contingent rather than a necessary truth because there does not seem to be any a priori reason why the result should come out at one value rather than another—except insofar as that is the way that electrons happen to be. We would submit that the difference is a difference between an ontological and an epistemological attitude. Laws of nature may well express what are necessary truths, ontologically speaking. We have no grounds for supposing that a universe could work if electrons had a different magnitude of charge. And even if it could, we would not have the right to call such particles "electrons." But those same truths are contingent, epistemologically speaking. The scientist's experimentation to arrive at the generalizations expressed in the laws could never have been circumvented, neither by an exercise of logical deduction, nor by a formal scheme, that would produce those same laws as a matter of necessity.

When we turn to the laws that are used by chemists, we find a great broadening of the type of statement that is involved and a different set of issues arising. The main laws of physics—laws such as the conservation laws, Coulomb's law of electrostatic attraction, and the law of gravitation—are challenged by the sorts of considerations discussed here. But they are exact and unexceptioned in a way that many chemical laws do not even pretend to be.

It has been argued elsewhere (Christie, 1994) that there is a difference in kind between any idealization and approximation involved in, for example, the First Law of Thermodynamics (an "exact" law) and the Ideal Gas Law (an "approximate" law).

The chemical Law of Definite Proportions, which expresses the notion that a pure compound always has a fixed and consistent composition in terms of the elements it contains, is an exact law for many, perhaps most compounds. There is nothing approximate in the assertion that pure sodium chloride contains equal numbers of sodium and chlorine atoms and, therefore, 61.72% chlorine, or that pure water contains just twice as many hydrogen atoms as oxygen and, therefore, 11.19% hydrogen. The whole of analytical chemistry is based on the reliability of such ratios. But there are a few classes of material that chemists would prefer to regard as pure compounds that do not show this consistency of composition. The Law of Definite Proportions is an exact law, but one that has clear exceptions.

The Law of Dulong and Petit, which played an important role last century in assigning correct valencies to a number of elements, is a law that is both approximate and exceptioned. More than 90% of solid elements have molar heat capacities at ambient temperature that cluster within 5% of a central value. The four or five clear exceptions have molar heat capacities that are more than 20% away from the value the law suggests. The approximation and the exceptions are quite distinct aspects of this law.

But the most remarkable laws that chemists have come up with are two laws whose statement cannot even be expressed as a proposition. Dalton's Law of Multiple Proportions can be stated:

> If two radicles A and B combine in different ratios to form more than one compound, then the different masses of B that combine with a fixed mass of A in the various compounds are in simple whole number proportions.

This is an imprecise statement, because there is no sharp cutoff between what constitutes a simple ratio and what does not. Nevertheless, it accurately expresses our understanding of an important chemical truth, because in most instances the ratios are very simple indeed—2 to 1, 3 to 1, or 3 to 2—and in instances where the ratios are not as simple, there is ample independent chemical evidence of complexity of structure (Christie, 1994).

The other important law in chemistry that displays a similar vagueness is Mendeleev's Periodic Law:

> 'The chemical properties of the elements are periodic functions of the Atomic Weights.'

It is quite clear that Mendeleev did not mean "periodic functions" in the strict algebraic sense of $f(x) = f(x - T)$ for fixed period T and for all x. Rather, he meant, as he clearly stated, that when variation of chemical properties with changing atomic weight of the elements was examined, patterns kept repeating, with certain minor variations. The law, while vague, and inexpressible as a proposition, was quite definite in its entailments. It clearly pointed to the missing elements whose properties Mendeleev predicted so remarkably. It resolved ambiguities of atomic weight and valence for some of the less familiar elements, and it quite clearly (but erroneously) indicated that the atomic weight of tellurium must be less than that of iodine.

Oversby (1998) has suggested that it is worth distinguishing between "Laws" that "relate data mathematically . . . are usually limited in their applicability . . . are linked with the notion of idealization . . . are commonly theory-laden" and "rules" that "provide approximate quantitative relationships between data, usually expressed in prose or in the form of an inequality." We would not disagree with the spirit of this distinction. We do not share his optimism that such a distinction actually fits well with scientific usage or that a consensus on this usage could readily be obtained. More important, there is a continuum in the realm of epistemological dicta used in science. There is no sharp dividing line along this continuum, nor should the continuum be seen as a single-dimensional scale. The characters of different scientific laws range from something significantly stricter than Oversby's characterization of a law all the way to what he classes as a rule. Along this continuum, chemists tend to work in a broad range that stretches from the softer side of law all the way to Rule, while, on the whole, philosophers of science have been very reluctant to dabble their toes beyond the stricter side of law.

The examples that we have used for illustration have come from nineteenth-century chemistry. This is a possible ground for criticism. Why have we not used more modern examples? There are several reasons.

First, modern chemistry is a huge subject that is largely fragmented into a number of specializations. Its scope is often technological and exploitative rather than directly scientific and exploratory. Laws are seldom used explicitly by the modern working chemist, though "rules of thumb" are legion. It is difficult to find modern laws of chemistry that are recognized beyond a narrow subspecialization within the discipline.

Second, by using historical laws in this way, we are helping to avoid any confusion between the epistemological sense of laws (as statements of our best knowledge and belief at a particular time about the workings of the systems we study scientifically) and the ontological sense of laws (as statements of the way those systems actually work). Perhaps this introduces some risk of a confusion with the historical sense. We stress that each of the laws we have used as examples stood for at least several decades as a statement of best knowledge and played a significant role in the development of the subject as a whole.

Finally, and more controversially, we argue that modern chemistry rests much more strongly on its own foundations of the nineteenth-century and earlier, and much less on the insights of modern quantum physics, than most modern chemists care to believe. To take a very simple example, many would say that there is no place for a "Periodic Law" in the foundations of modern chemistry. Certainly, there is a place for the periodic table, the argument would go, but that is simply based on valence electron configurations. The counterargument is that having similar valence electron configurations is neither necessary nor sufficient to place elements in the same vertical column of the modern periodic table (e.g., Scerri & McIntyre, 1997). Thorium, whose valence electron configuration is d^2s^2, is not placed in IUPAC group 4 (traditional IVB) with the only other three elements that share this configuration (Ti, Zr, and Hf), but rather among a group of actinide elements, where its only vertical relationship is with Ce, configuration fds^2. Meanwhile, IUPAC group 10 (column 3 of group VIII in the traditional numbering) contains just three elements with three different valence electron configurations: Ni d^8s^2, Pd d^{10}, and Pt d^9s^1. The regular but complicated pattern of periodicity and the actual comparisons of chemical properties provide far stronger influences in determining the shape of the modern periodic table than the detail of valence electron configurations.

Mendeleev's Periodic Law was the central law of comparative chemistry of the elements, in the epistemological sense, from its formulation around 1870 at least until Moseley's discovery of atomic number ovĕr 40 years later. With the substitution of "atomic number" for "atomic weight," it continued in this role for at least another two decades. It was displaced as the insights and implications of quantum theory for chemical systems were explored and incorporated into the discipline. But there is a serious argument that the displacement was only partial and that the "periodic law" retains significant validity as a central law of chemistry even to the present.

When we turn from laws to theories, we again find that the sorts of theories that chemists name as such, and use to help describe their systems, are very different in character from the great theories of physics. Already, we have referred to the frequently pluralistic attitude of chemists to what are, ostensibly at least, rival theories. To illustrate, we cited several theories of chemical bonding. A similar picture emerges if we look at theories of reaction rates in the gas phase.

For the reactions of medium-sized molecules we have the following: Lindemann-Hinshelwood theory, RRK theory, Slater's harmonic theory, RRKM theory, phase-space theory, absolute reaction rate theory, quasi-equilibrium theory, and several others. All of those are grouped under the umbrella of "transition state theory" (Robinson & Holbrook, 1972; Forst, 1973). Among these theories, some are regarded as "inaccurate" or "outdated." But several rivals remain as viable alternatives on which to base a theoretical study of a reaction system, at least as far as journal referees are concerned.

One of us (John) was recently involved in a collaboration in which some experimental studies of reactions of a new class of ion were supplemented with some theoretical calculations in an attempt to rationalize observed behavior. Three different theoretical models—RRK, RRKM, and another—were applied and compared. The journal referee suggested that the theoretical section did not really require the RRK calculations because RRK theory had been superseded by RRKM. John's response to the team leader was that (a) the RRK calculations were completely independent of the RRKM calculations and gave a completely different perspective; (b) the RRKM calculations themselves rested on a number of guesses and assumptions and were very questionable in terms of fundamental significance; and (c) in circumstances like this where all of the theoretical treatments were really rather crude models, it might throw rather more light on the issues to examine several of them. The referee's point was reasonable. There should have been no real objection to leaving out the RRK section. But there was an argument, far from compelling, for its retention. The team leader was persuaded by the argument and pressed it with the scientific editor of the journal. Apparently the referee or the journal editor was also persuaded, because the article appeared with the RRK section still included (Chen et al., 1999).

A few of the theories, most notably Slater's harmonic theory, are regarded as "wrong" on the Popperian grounds that they make clear predictions that do not tally with experimental results. But theoretical chemists are not, by and large, good Popperians. RRKM theory is regarded as among the best of the manifestations of transition-state theory. But to attempt to predict the properties of a particular reaction using RRKM theory, one must feed in values for a large number of numerical quantities (between 20 and several hundred, depending on the size of the reacting molecule), which must be inferred by intelligent guesswork. Only a few are experimentally accessible. They are only obtainable to a very rough approximation via a structural calculation based on one of the chemical simplifications of quantum theory. More usually, they are allocated by considerations of analogy with other molecules. If allowance is made for an error range in each of these quantities, any prediction of a RRKM calculation becomes soft indeed. There is a distinct possibility that any experimental result could be rationalized with a judicious choice of "reasonable" values for these well-defined but poorly characterized input parameters. RRKM theory cannot be seen as making bold predictions on which it might stand or fall; it could even be seen as unscientific in Popperian terms, in that it may not be potentially falsifiable. Slater's harmonic theory is regarded as wrong because it stuck its neck out! RRKM theory clearly does not.

But our purpose here is not an evaluation of Popper's standard of falsifiability of a theory, nor of chemists' adherence to that standard. It is rather to contrast the characters of central theories of physics and typical theories of chemistry. We seek to make a comparison between a central theory of physics (e.g., quantum theory as an umbrella notion, with its expression in such formulations as wave mechanics or matrix mechanics) and a body of chemical theory (e.g., transition-state theory, with expression in such formulations as RRKM theory or RRK theory).

The idea of transition-state theory is to consider reactant, transition state, and product as three different entities that transform into one another. Of these three entities, the transition state is somewhat peculiar because it is a "flatland" entity—it is an infinitesimally thin slice that divides reactant geometries from product geometries.

In effect, the three entities are different gestalt classifications of a fixed assembly of atoms according to the relative positions of their nuclei. Each is then considered to have independent and different sets of quantum states of vibration and rotation—in complete contradiction of what quantum theory would really say about such an assembly of particles. The enumeration of these somewhat artificial "quantum states" of the separate geometric classifications forms the basis for a statistical calculation of rates of passing from one region to another, based on a typical statistical mechanical supposition of equal probabilities of various microstates.

It is immediately apparent that a theory like transition-state theory is making no pretensions at stating and describing the underlying principles of the behavior of the system. In any serious analysis in terms of the deeper and more fundamental laws of physics (of quantum mechanics, in particular) the further assumptions in its derivation are arbitrary, artificial, and somewhere between wildly simplistic and quite unsound. Nevertheless, the theory is typically introduced via a complex mathematical argument in which it is 'derived' using a series of 'assumptions' and 'approximations' from the supposedly underlying equations of quantum theory and/or statistical mechanics.

We would suggest that this sort of thing should provide fertile and little explored ground for the philosopher of chemistry. The characters of the laws and theories that chemists use are rather different and rather more diverse than those of the central laws and theories of physics that have served as the usual philosophical exemplars. The chemist typically seems to have a different, more pluralistic, attitude to rival theories and to use them in a rather different way. Are the differences real or apparent? Are there reasons for the diverse and somewhat strange characters of many chemical laws and theories, and for the chemist's distinctive attitude to them?

The Reduction of Chemistry to Physics

The scientific concern of chemists is with classifying, systematizing, and attempting to understand the properties and the transformations of different types of material. Their other concerns are largely technological and utilitarian rather than directly scientific—producing new materials, assaying, purifying, and finding more efficient ways of producing desirable materials.

With the advent of modern atomic theory began a new major enterprise in the science of chemistry: that of understanding the bulk properties of materials in terms of the properties and machinations of their component submicroscopic atoms and molecules. The enterprise moved very slowly into its current central role in the science of chemistry. Atomic theory won acceptance among chemists only slowly and, in many quarters, grudgingly. Antiatomist views were common among chemists into the late nineteenth-century, and agnostic views persisted until well into the twentieth (e.g., Smith, 1910, pp. 224–225), despite the experimental work of Thomson, Rutherford, and other physicists on the structure of the atom.

Meanwhile, beginning about 1910, development of the quantum theory in physics promised the theoretical framework for a complete description of the behaviors of atomic and subatomic particles. Application of quantum theory to chemical systems was again an enterprise that started very slowly and won only grudging acceptance by many chemists. But by about the 1960s there was general acceptance among chemists that quantum theory provided a set of underlying principles that had the potential to provide explanations throughout the science of chemistry.

This acceptance leads to an apparent reduction of chemistry to physics, in which the science of chemistry may continue to have some practical importance as a specialized subdiscipline but ceases to have any fundamental significance as an independent science. It might also be perceived as thereby ceasing to be of interest to philosophers of science.

But when we look at the reduction of chemistry to physics in practical terms, we find a program that cannot succeed. Equations of motion, whether classical or quantum, can be solved exactly for systems of one or two particles at most. Any system of three or more particles that move and interact with one another can be solved, at best, by a numerical procedure that produces a sequence of approximate solutions that gradually converge on the true solution to the equations of motion. For the quantum theory description of molecular structures, the algorithms for obtaining these approximate solutions are relatively inefficient in that the computational effort required increases as the fourth power of the number of electrons—twice as many electrons means 16 times as much computer time to do a calculation to a similar level of approximation. Almost invariably, the calculations are performed assuming that the heavy atomic nuclei are at fixed positions in space, and that the electrons are the only moving particles. At first sight, this is a good approximation—an atomic nucleus is between about 2,000 and 100,000 times heavier than an electron and typically has velocity and kinetic energy that are smaller by a similar factor. But often issues in molecular structure hinge on energy differences of the order of one 10,000th of the total electron energy (e.g., whether the ozone molecule, which contains three oxygen atoms, should have those atoms in a straight line, in a bent line, or in an acute-angled triangle). In these circumstances, it is not an approximation that can be trusted.

The 1998 Nobel prize for chemistry was awarded to two scientists whose principal contribution was to devise methods that brought approximate quantum theory calculations for the medium-sized molecules within the realms of practicality. The suite of programs that the modern chemist has available for calculating molecular structures is extensive and sophisticated. But, in practice, a compromise always has to be made in terms of the computational effort versus the level of approximation, and some issues of approximation cannot be avoided within the framework of the suite. One such example is the assumption of stationary nuclei; another is the problem of relativistic velocity effects, which become significant for the electrons of elements heavier than about iron (that is, the heavier two thirds of the elements). The time-independent Schrödinger equation is based on Newtonian rather than relativistic mechanics.

Many modern chemists would claim that currently available calculation methods enable them to obtain quantum theory structures for the molecules that interest them to a good degree of approximation and that, in that sense, chemistry has been reduced in practice to the laws of physics. Any deficiencies in the currently practical level of approximation will be dealt with by advances in computing technology in the near future. Such a claim would not be entirely uncontroversial, and we would certainly feel that there would be room for reasonable disagreement.

But it is more important to remember that chemistry is about dynamics, as well as structures. The transformations that occur in chemical reactions are central to the science. Chemists often talk about "reactivity" as though the propensity of a certain substance to enter into particular reactions is a static property of the substance that might be related to a structural feature of its molecule. And this has certainly been a fruitful approach. But ultimately any scientific understanding of chemistry must go beyond reactivities to a consideration of reactions themselves as dynamic phenomena.

In this enterprise, it rapidly becomes apparent that the sort of approach that has been so fruitful for quantum theory calculations of molecular structures cannot succeed with reactions. For a molecular structure calculation, all that is really required is to determine the arrangement of electrons that corresponds to the particular fixed geometric arrangement of nuclei that represents the stable geometry of the molecule.

To perform a similar calculation for a dynamic problem, electron arrangements would need to be calculated for a representative set of all of the different possible geometric arrangements of nuclei that are likely to be encountered in the course of a reaction. For the simplest type of chemical reaction, which involves just three atoms, that means a thousand times the computational effort—allowing only 10 sample values each for three independent geometric parameters; on the same basis, it goes to a million times the effort for a reaction system involving four atoms, a billion times for five atoms, and so on. It might be possible on this basis to calculate the dynamics of the reaction $\mathbf{H + D_2 \rightarrow HD + D}$, which involves only three electrons and three independent geometric parameters. But even allowing for future improvements in computing technology, a truly fundamental calculation of reaction dynamics for a larger chemical system, in terms of the basic equations of the quantum theory, is likely to remain forever beyond our reach.

So, looked at as a whole, the reduction of chemistry to physics fails, not because of any problems of principle (though these may remain; Scerri & McIntyre, 1997), but because of impracticality. The application of the equations of physics to the questions raised in chemistry is ultimately intractable.

There are few chemists who do not believe that quantum theory provides a good description of the behaviors of atoms and molecules and that the properties of bulk matter are determined by the properties and interactions of the microscopic particles of which it is composed. But the grounds of this belief are essentially analogic.

Quantum theory provides an excellent and detailed account of the structural features of light atoms and very simple molecules like hydrogen (H_2) and methane (CH_4). It provides a satisfactory but rather cruder account of the structural features of larger molecules, or molecules that contain heavier atoms. With really large molecules—those containing atoms heavier than iron—the calculations available in the chemist's suite of programs could not fairly be said to produce "approximations" to the consequences of the equations of quantum theory. But the chemical bonds in these molecules, and the results of the various observations and measurements that can be made and related to their structures, display absolutely nothing to suggest that they differ in any significant regard from smaller and lighter molecules.

Quantum theory works for small molecules involving light atoms, for which its consequences can be directly tested. Chemists have much experience in quantifying, classifying, and generalizing the observed properties of various materials in terms of molecular structures and chemical bonds in those molecules. There are no indications of important nor significant unaccountable differences between compounds of light elements and those involving heavy elements, nor between those involving simple molecules and those involving large molecular structures. So there are grounds, but only analogic grounds, for supposing that quantum theory would work well for large molecules and those involving heavy atoms, if only we could do the calculations properly.

Similarly with reaction dynamics, there is really only one reaction that can be modeled to a high level of approximation with the equations of quantum theory. But chemists feel confident about the compliance of all chemical reactions with the laws of quantum theory because, investigated experimentally, that one reaction is in no way atypical.

It is our contention, then, that in some sense the program of reduction of chemistry to physics fails; one of the grounds for this failure lies in practical issues of intractability rather than in any conceptual incompleteness or incompatibility.

Basis of the Peculiarity of Chemical Laws and Theories

If our analysis in the preceding section is accepted, then the unusual characters of chemical laws and theories can be seen to follow readily as a corollary. The development of laws and theories for use in chemistry is constrained by at least one unusual requirement. All chemical behavior is believed to arise from the machinations of microscopic entities that are described by theories like quantum theory. But the entailments of quantum theory are incapable of being rigorously deduced for genuine systems of chemical interest. In these circumstances, the type of law or theory that is of most value in systematizing chemistry, and in integrating chemical knowledge into a broader corpus of scientific knowledge, must fulfill a bridging role. On the one hand, it should provide the framework for a broad and useful degree of empirical generalization over a diverse range of chemical systems. On the other hand, it should be expressed in such a way that it can be teased out as an approximate or likely implication of the underlying physical theories.

The pluralistic attitudes of chemists to rival theories becomes explicable. The chemical theorist is become a caricaturist. The typical modern chemist is a naive realist, who believes in the reality of atoms, electrons, and molecules and that their motions and behaviors are governed by ontological laws of physics, to which our present quantum theory is, at worst, a very decent approximation. But there is no way to map out the intricate detail of that governance—it is a scene for the eyes of the creator alone. A photograph is not available; the best that can be done is to produce sketch maps and cartoons to provide partial insights. So it is less surprising that the chemist can cling to rival theories and can mix and match when applying them to different problems. Ligand field theory is good at describing the light-absorbing properties of transition metal complexes, while VSEPR theory is good at describing the shapes of a wide variety of simple molecules. But then one political cartoonist can really capture the eyes of a well-known politician with a few well-placed strokes, while another traces a figure of completely different shape, in which the same politician is recognizable by his distinctive walk.

It would not be strictly correct to say that the chemist deals with systems that are more complex than those handled by the physicist. After all most of the early work on gravitation was tested and verified using the motions of planets, satellites, and comets in the solar system—a system large enough to contain nearly all of the chemical systems that have ever been studied as small subsets of itself. But the focus of the physicist is on a simpler level: the laws of physics ignore the complexities of the planets in treating gravitational problems; for most purposes, each might as well be a featureless steel ball.

There is some difficulty with using an intuitive notion of complexity, as we have been. How can we decide more precisely what we mean by "complexity" of a system? The initial intuition is that any system is more complex than any of its proper subsystems.[3] The solar system example shows that this intuition fails when the focus with which the system is considered changes. An alternative intuition is that the complexity of a system might be measured by the number and diversity of its component atomic entities (in this sense, an entity is atomic in any consideration of a structure if we do not inquire into its substructure). This also fails in that it does not deal well with the fact that a static structure, or an instantaneous snapshot of a structure, is a less complex entity than a dynamic structure that contains the same atomic units. There may be an information theoretic measure of the complexity of a system in terms of the number of bits of information required to specify its observable properties at the required focus.

We will not embark on a lengthy debate about how best to define and measure complexity in the systems that physicists deal with. We do make the point that we see it as an epistemological matter. Complexity depends on the fineness of the detail in which the observer seeks to measure and "know about" the system. It is as much a property of the attitude of the observer as of any external collection of entities. And this view of complexity in a natural system may be very much part of a realist ontology; although it sounds antirealist, it has no such implications or connotations.

Many of the physical laws are expressed in relatively straightforward mathematical equations. For systems of real interest, the chemist cannot change focus enough to view them in a way that produces equations that are soluble without severe approximation at best, and significant distortion in most cases.

Another type of problem arises when approximate entailments of an underlying theory must be obtained using large computer models and calculations. The user of such models inputs the equations that represent the theory, and the detail of the various parameters that represent a system of interest. The computer performs its calculation and produces a series of outputs. These would certainly include numerical estimates of some properties of the system that could be checked against the experimental values obtained for the same quantities. They may also include other outputs, less empirically verifiable—bond orders or molecular orbital compositions, for example. But these outputs often have a peculiar property. Because a molecular orbital is a concept that only makes sense within a particular approximation framework (known as the Hartree-Fock approximation), molecular orbital compositions can only be obtained while the calculation remains at this low level of approximation. As the sophistication of the approximation increases beyond this level, this particular output disappears. Chemists are used to framing explanations in terms of such things as bonds, orbitals, and transition states. But each of these concepts appears only while the modeling is rough and simplistic, and becomes more obscure, and eventually quite indiscernible, as the sophistication of the model is increased toward a higher level of approximation. The trouble here is that as these concepts disappear, so does the chemist's explanatory language, and so does the possibility of tracing a chain of causality from physical laws to chemical behaviors. A chemist can use a crude model calculation of the structure of a molecule to provide information about bonds and orbitals that can be used in explanations to rationalize its chemical properties. A more sophisticated calculation provides better values for the actual properties themselves, but it provides them as a fait accompli, with no guidelines for an explanatory framework.

Chemistry and Other Branches of Science

We have argued that the distinctive characters of the laws and theories of chemistry arise largely from the strange relationship between chemistry as such and the more fundamental physics that seeks to account for chemical systems. They arise naturally because the program of reduction of chemistry to physics is neither a success nor a failure but a partial success. We would further suggest that this strange relationship brings with it new problems that ought to be of interest to philosophers of science. Others, most notably Scerri and McIntyre (1997) and Weinberg (1993), have made a similar point, and perhaps gone rather further than we have with the philosophical analysis. There is new and largely uncharted territory in the philosophy of chemistry that has not been encountered by philosophers of science in their consideration of neither physics nor biology.

Chemistry is a rich science, both in terms of its vast and varied content and its colorful history. Its development ran almost entirely independently of physics, and until the twentieth century there were only limited areas of interaction and reconciliation of the two sciences.

But we are arguing that the differences in character of laws and theories in the two sciences arise neither primarily from their separate historical developments nor even from the type of natural system that each addresses. The relationship of the two sciences on the reduction hierarchy is what is crucial.

If this is right, we may find that there are interesting parallels between the types of law and theory that have developed in chemistry and those in another branch of science that stands in similar relationship to basic physics. The area that we have in mind is fluid dynamics and, most particularly, its expression in meteorology and climatology. The subject matter of these sciences is completely different to that of chemistry. Nevertheless, we find significant parallels at every level. These areas are ancient and developed largely independent of physics. They abound with rules of thumb, as well as approximate, exceptioned, and imprecise laws. They are generally believed to be underpinned by fundamental laws of physics, but the complexity of the focus of the system prevents the entailments of the laws from being accurately calculated. Large computer models are used, and they suffer from many of the same drawbacks and a few specific different ones: at least approximate calculations of molecular structures are not beset by mathematical chaos, as weather predictions are. Many of the features that are used as explanatory terms in weather simulations are model-sensitive concepts that can easily disappear when a calculation is upgraded to a better level of approximation. Rival theoretical approaches to the description of certain phenomena are sometimes regarded pluralistically, and chosen for computational or illustrative convenience, rather than for being right or wrong.

It seems likely that many of the issues that might be introduced and taken up in the future development of the philosophy of chemistry will find parallel application in other branches of science only distantly related to chemistry, but sharing some of its epistemological problems and peculiarities.

Notes

The authors are indebted to John Oversby, Arthur Petersen, and Michael Weisberg for stimulating discussions, mostly by email, which have contributed to this article in various ways.

1. An extensive, confused, and rather inconclusive debate took place recently on the PhilChem email newsgroup—of interest because the contributors include philosophers, educators, and scientists. (PhilChem, 1998).
2. The authors acknowledge email correspondence with Michael Weisberg, in which he drew their attention to this aspect of the problem.
3. "Proper" is used here in a mathematical sense. Technically, any system is a subsystem of itself. A proper subsystem is one that includes less than the whole system.

References

Cartwright, Nancy. 1983. *How the Laws of Physics Lie*. Oxford: Clarendon.

Chen, Guodong, Cooks, R. Graham, Bunk, David M., Welch, Michael J., & Christie, John R. 1999. "Partitioning of Kinetic Energy to Internal Energy in the Low Energy Dissociation of Proton-Bound Dimers of Polypeptides." *International Journal of Mass Spectrometry*, 185/186/187: 75–90.

Christie, Maureen. 1994. "Chemists versus Philosophers Regarding Laws of Nature." *Studies in the History and Philosophy of Science* 25: 613–629.

Ellis, Brian D. 1991. "Idealization in Science." In C. Didworth, ed. *Idealization IV: Intelligibility in Science* (pp. 265–284). Amsterdam: Rodopi.

Forst, Wendell. 1973. *Theory of Unimolecular Reactions*. New York: Academic Press.

Ghiselin, Michael T. 1989. "Individuality, History, and Laws of Nature in Biology." In M. Ruse, ed. *What is the Philosophy of Biology* (pp. 53–66). Dordrecht: Kluwer.

McWeeny, Roy. 1979. *Coulson's Valence* (3rd ed.). Oxford: Oxford Univesity Press.

Molnar, G. 1969. "Kneale's Argument Revisited." *Philosophical Review* 78: 79–89.

Oversby, John. 1998. "Theories, Laws, Models, Hypotheses." Paper presented at Australian Science Education Research Association Annual Conference, Darwin.

PhilChem Newsgroup. 1998. Discussion Thread "Theories, Models, Formulas, Hierarchies," Feb–Apr 1998. (For access to the archive, email to listserv@vm.sc.edu the command "index philchem" for a catalogue. A command "get philchem log9802a" emailed to the same site will retrieve the first batch of messages from February 1998, etc.).

Robinson, P. J., & Holbrook, K. A. 1972. *Unimolecular Reactions*. London: Wiley.

Ruse, Michael. 1989. *Philosophy of Biology Today*. (chap. 2). Albany: State University of New York Press.

Scerri, Eric, & McIntyre, Lee. 1997. "The Case for Philosophy of Chemistry." *Synthèse* 111: 213–232.

Smith, Alexander. 1910. *Introduction to Inorganic Chemistry* (rev. ed., Chaps. 9 and 13). New York: Century Press.

Weinberg, Steven. 1993. *Dreams of a Final Theory* (chap. 3). London: Vintage Editions.

4

Realism, Reduction, and the "Intermediate Position"

ERIC R. SCERRI

The editors of this volume have asked me to write something about reduction because I have been focusing on this issue in my research for some time.[1] Rather than rehashing any previously published ideas, I want to consider a new aspect of this question—or at least one that is new to me.

I will draw liberally on the work of my thesis grandfather, the chemist, Fritz Paneth. I use the term somewhat unusually, because Paneth was not the person who advised my own advisor Heinz Post but was, in fact, his natural father,[2] from whom Heinz presumably developed an interest in the philosophical aspects of science.[3] I will touch on such areas as realism, including naive realism, the nature of the periodic system, metaphysical aspects of chemistry, and, as suggested by the editors, the reduction of chemistry.

Realism, Atomic Orbitals, and Chemical Education

Of course, nobody likes to be referred to as a naive anything, and not surprisingly, some chemists are quick to react if it is suggested that they tend to adopt a naively realistic attitude in their work.[4] Nevertheless, I think it is true that chemists are often realists, naive or otherwise, and this may not be such a bad thing, as I will try to explain. The question hinges on the extent to which such realistic views are maintained and in what context they may or may not be appropriate.

Broadly speaking, chemists are frequently accused of unwarranted realism by physicists for taking chemical models too literally. Whereas no chemist nowadays believes that tiny springs connect the atoms in a molecule, such denials of models are less prevalent when it comes to hybridization, electronic configurations, and atomic orbitals. Chemistry professors may well begin by declaring that these are approximate concepts, but they frequently fail to emphasize that, from the point of view of physics, they are strictly incorrect or, philosophically speaking, are nonreferring terms. But I contend that to describe something as an approximation does not carry quite the same ontological force of stating it to be nonreferential, and, according to current physics,

I repeat, only the latter is correct. Another way of putting this situation is to say that, in the case of many-electron atoms, atomic orbitals cannot be reduced to quantum mechanics.[5] However, such orbitals are regularly discussed and pictured in glorious color diagrams by chemists as though they were real and concrete entities. In fact, orbitals have become the lingua franca of modern chemistry at all levels from introductory to advanced research.

It is interesting to examine why such an apparently mistaken view has developed among chemists. To put the question starkly: How is it that if many-electron orbitals cannot be reduced to quantum mechanics, nevertheless, many chemistry textbooks begin with a thorough treatment of orbitals, quantum mechanics, atomic structure, electronic configurations, and the like? Are chemical educators merely oblivious of the work of physicists and philosophers who maintain that this reduction is not tenable? Is it because the reduction of chemistry is one case in which one can ignore these pronouncements, perhaps because chemistry lies so close to physics, and that there exists an intuitive belief that chemistry is, in principle, reducible to physics? Or is it perhaps due to some as yet unarticulated justification which implies that chemical educators know what is best for teaching chemistry regardless of philosophical debates on reduction and what physicists might say?[6]

A popular response has been to claim that chemical educators are wrong to base the presentation of chemistry so closely on quantum mechanics in view of the failure of the reduction of chemistry, including the failure to reduce many-electron orbitals. This response has been urged by several authors, including myself in the past (Scerri, 1991a), but I have recently suggested a revised view (Scerri, 2000). The view that I am starting to advocate is that chemical educators should continue to use concepts like orbitals and configurations, but only while recognizing and emphasizing that these concepts are not directly connected with orbitals as understood in modern quantum mechanics but are, in fact, a relic of the view of orbits in the so-called old quantum theory. One might even consider calling them "chemists' orbitals." Such a view is consistent with the notion of chemistry as an autonomous science. The concepts of chemistry cannot be reduced to quantum mechanics but at the same time we should not conclude that chemistry and physics are disunified—as some authors have implied—about all the special sciences (Galison & Stump, 1996). The chemist may, and perhaps should continue to be a realist about orbitals, but hopefully in not too naive a fashion.[7]

Better still, the view I am proposing now is an intermediate position between the realism of believing in the real and concrete existence of many-electron orbitals and the reductive view from quantum mechanics that banishes all talk of orbitals. I suggest that chemical education can benefit from a form of realism, which is tempered by an understanding of the viewpoint of the reducing science but which does not adopt every conclusion from that science.

Has the Modern Periodic System Been Reduced to Quantum Mechanics?

Another way to ask whether chemistry has been reduced to physics is to consider one of its central laws, the periodic law, to see whether it has been "explained away" by quantum mechanics. As a matter of fact, the periodic law has not been reduced in this way. What the periodic law embodies is the approximate repetition of the elements after certain regular but varying intervals in the atomic number sequence of the elements have been traversed. More specifically, the lengths of the periods are, respectively, 2, 8, 8, 18, 18, 32, etc. The frequently made claim is that quantum mechanics has reduced the periodic system through the use of the Pauli Exclusion Principle and by assigning four quantum numbers to each electron. This approach serves to explain the above sequence in a semiempirical manner, a fact that textbook presentations do a good job of concealing (Scerri, 1998b).

Such an explanation is semiempirical because the order of electron subshell filling must be assumed by reference to experimental data. As the noted theoretician Löwdin (1969) has pointed out, the order of filling has yet to be derived from quantum mechanics. This feature is related to the well-known failure of quantum mechanics—or of any other form of mechanics for that matter—to provide exact solutions to the many-body problem, thus rendering the quantum mechanical reduction of the periodic system approximate rather than exact. Although the failure of quantum mechanics to solve the many-body problem is readily acknowledged, the sleight of hand that admits experimental information into the alleged explanation of the periodic system is seldom even suspected (Scerri, 1997b). In any case, it should still be possible to explain the order of shell filling without solving the many-body problem. It has just not yet been done.

The way the periodic system is displayed in contemporary chemistry also contributes to the false impression that quantum mechanics explains the system deductively. The almost universally displayed form of the table nowadays is the so-called medium—long form in which the rare earth elements are shown as a footnote to the main body of the table (figure 4.1). The remainder of the table is divided into the s, d, and p blocks, respectively, reading from left to right. These labels, which have a spectroscopic origin (sharp and diffuse, from the type of observed lines), refer to the type of orbital that is supposed to contain the differentiating electron in each atom. A casual glance at this table reveals a rather obvious lack of symmetry.[8] This type of periodic table somewhat masks the fact that the length of the periods cannot be deduced from quantum mechanics and makes the s,p,d, and f classification,[9] referring to the filling of atomic orbitals, appear to be fundamental. Given the key role that symmetry principles have played in modern physics and chemistry, it would seem desirable to display the inherent symmetry of the periodic system rather than hiding this feature. However, the rather asymmetrical medium–long-form table, which I claim gives an exaggerated impression of the role of quantum mechanics in chemistry, is now well entrenched.[10]

An alternative form of the periodic table (figure 4.2) consists in the pyramidal form, which has been proposed by many authors[11] throughout the history of the periodic system and, indeed, was favored by Niels Bohr when he gave one of its first approximate quantum theoretical explanations.[12] As I have suggested before, the as-yet unexplained symmetrical aspect of the lengths of periods would be more likely to be addressed if this feature were displayed in the modern periodic table (Scerri, 1998b). In making this suggestion I regard the periodic law to be first and foremost a *chemical* law concerning the point at which the elements appear to recur approximately—or, in other words, that the essential content of the law lies in the sequence of numbers denoting the closing of the periods (2, 8, 8, 18, 18, 32, etc.) This sequence differs from

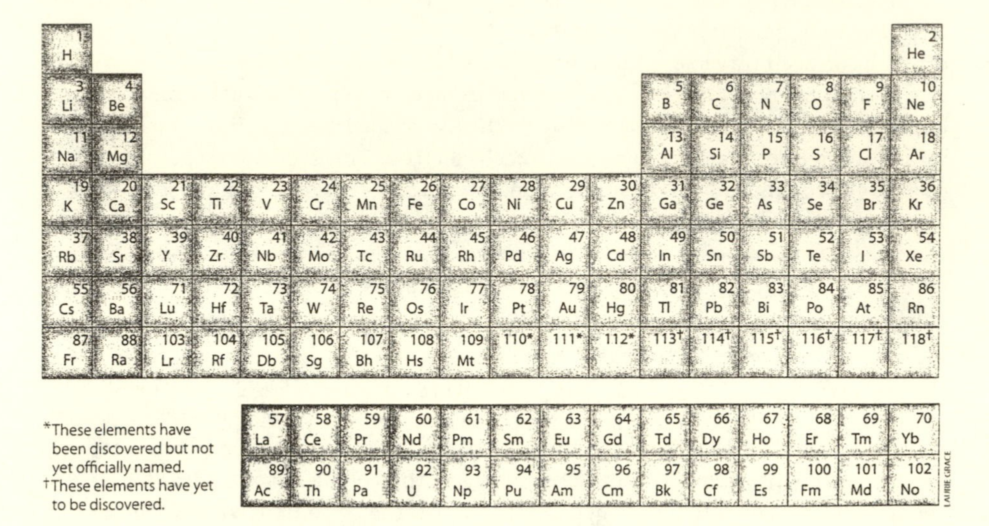

Figure 4.1 Popular Periodic Table—known as the medium–long form—this table can be found in nearly every chemistry classroom and laboratory around the world. This version has the advantage of clearly displaying groups of elements that have similar chemical properties in vertical columns, but it is not particularly symmetrical.

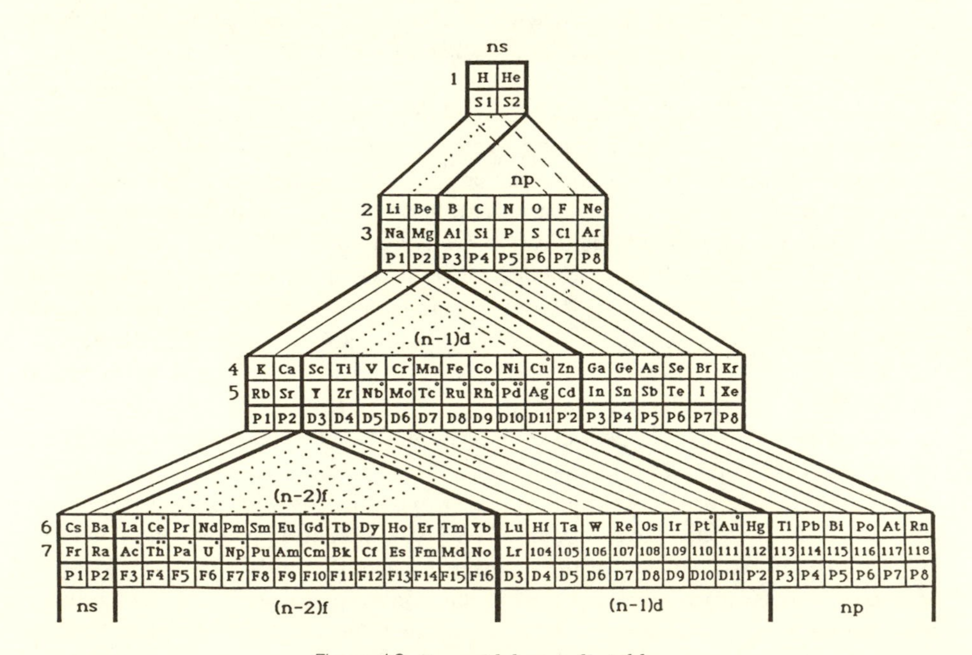

Figure 4.2 Pyramidal periodic table.

the sequence denoting the closing of the shells (2, 8, 18, 32). Only the latter has been deduced strictly by means of the Pauli Exclusion Principle and the relationships that hold between the four quantum numbers (Scerri, 1997b, 1998b).[13]

In insisting on this point I believe that I am adopting a more philosophical approach to the periodic system than is usually encountered in chemistry and physics. I am suggesting that the essence of the periodic law may still be hidden from complete view and that the failure to strictly predict the points at which the periods close is one symptom of our present lack of a full understanding of the law. Perhaps it is even the case that the periodic system which is concerned with the abstract elements will always remain somewhat hidden from full view.[14]

Paneth on Naive Realism and Transcendental Elements

Let me now turn to what Paneth, an inorganic chemist and one of the founders of radiochemistry, had to say about naive realism in chemistry:

> Chemistry, like every natural science, started from the naive-realistic world-view, and gradually found itself compelled to apply corrections to this. It is, however, characteristic of chemistry that it has not advanced as far in the application of these corrections as some other sciences; indeed it is of the essence of its fundamental concepts that they have retained quite an appreciable "naive realistic" residue. . . . Nobody objects, in fact, to speaking of the salty taste of sodium chloride, or the unpleasant smell of hydrogen sulphide. Here we find ourselves, as is surely unnecessary to elaborate, still standing with both feet on the ground of naive realism without, it should be emphasized, being aware that this lack of philosophic clarity entails any disadvantage. Indeed even in the case of that "property of substances" which can be reduced to quantitative determination most easily—color—we usually refrain, for the purpose of chemical characterization, from so reducing it: cinnabar "is" red, gold "is" lustrous. While the introduction of various constants would make a more exact numerical statement possible in such cases, this would be too cumbersome. (Paneth, 1962, p. 5)

And yet Paneth goes to great pains to point out that naive realism is not always the appropriate attitude to adopt in chemistry:

> Taking the philosophically primitive standpoint chemists have usually managed very well in the whole vast range of the subjects of analysis and synthesis. For this reason they have remained unaware that, after all, somewhat deeper epistemological consideration is necessary for the complete understanding of the reactions they have carried out and the theorems they have put forward. (Paneth, 1962, p. 9).

The main case in which Paneth believes we need this deeper outlook concerns the nature of the term *element* and what he calls the law of conservation of elements. Paneth wonders how the elements manage to survive intact in any compound they might find themselves in. To answer this question, he refers to a long-standing notion in chemical philosophy—namely, a view that the elements are the bearers of properties and, being so, are completely unobservable. According to this view, the elements are thought to inhabit what Paneth terms a "transcendental world."[15] Paneth also suggests that this view, which began with the Aristotelian notion of the elements and was upheld by the alchemists, was not altogether banished in the course of Lavoisier's chemical revolution, as most historians of chemistry would have us believe.[16] He analyzes the philosophical writings of Mendeleev to show that even the discoverer of the periodic system held a similar philosophical view on the nature of the elements.[17]

Paneth considers it useful to distinguish two senses in which the expression "chemical element" is used. He refers to these two senses as *basic substance* and *simple substance:*

> I suggested that we should use the term "basic substance" whenever we want to designate that which is indestructible in compounds . . . and that we should speak of a "simple substance" when referring to the form in which such a basic substance, not combined with any other, is presented to our senses. (Paneth, 1965, p. 65).

In the case of element as a basic substance, Paneth claims:

> We cannot ascribe any particular qualities to an element as a basic substance, since it contributes to the production of an infinite variety of qualities which it exhibits both when alone and in combination with other basic substances . . . (Paneth, 1965, p. 65)

> With the concept of simple substance we may remain within the realm of naive realism. When we are concerned with the basic substance, however, we cannot disregard its connection with the transcendental world without getting involved in contradictions. (Paneth, 1965, p. 66)

I believe that Paneth makes an interesting case for this philosophical position that he shares with Mendeleev, and I would like to speculate further along the same line of thought because I think it can answer a specific question in the history of chemistry. As is well known, two men, Mendeleev and Lothar Meyer, independently arrived at the mature form of the periodic system in the late 1860s. Recently, there has been a lot of debate in the history and philosophy of science literature on the question of whether the periodic system was accepted because of its dramatic predictions or because of its successful accommodation of chemical facts (Maher, 1988; Lipton, 1990; Scerri, 1996; Brush, 1996). I happen to believe that accommodation may have counted as much as prediction but will not pursue this question here.[18]

One aspect beyond dispute is that Mendeleev made far more successful predictions than any of the codiscoverers of the periodic system. For example, he successfully predicted new elements, corrected the atomic weights of a number of known elements, and correctly reversed the positions of the elements tellurium and iodine. Why was it Mendeleev and not Lothar Meyer or others who was able to make such notable predictions? Is it simply that the others lacked the courage to do so, as many historians of science state?[19] I suggest that Mendeleev had the advantage of being blessed with a deeply philosophical approach to chemistry, which allowed him to arrive at insights that his less philosophically minded contemporaries could not have reached.

Like Paneth, he believed that the elements are essentially unobservable. He also believed that the periodic system should classify the unobservable basic substances in preference to simple substances. For example, the elements in the halogen group (fluorine, chlorine, bromine, and iodine) are rather different if we focus on the isolable simple substances that are gaseous, liquid, and solid, respectively. The similarity between the members of the group are more noticeable when it comes to the compounds of each one with sodium, for example, which are all crystalline white powders. The point is that in these compounds—fluorine, chlorine, bromine, and iodine—are not present as the simple substance but present in a latent form. This has been described as a metaphysical view of the elements but should perhaps to be termed, as Paneth suggests, a "transcendental view."[20] In other instances, such as his correction of atomic weights, this transcendental view of the elements allowed Mendeleev to maintain the validity of the periodic law even in instances where experimental evidence seemed to point against it. Had he been more of a positivist, he might easily have lost sight of the importance of the law and might have harbored doubts about some of his predictions.

Some support for my speculation comes from the following opening statement by Klutgen writing in the journal *Philosophy of Science* in 1958:

> Dimtri Mendeleev while not creatively a philosopher of science, nor a student of systematic philosophy, was eminently a philosophical scientist. Concern about the nature and foundations of his science is evident throughout the text and footnotes of *Principles of Chemistry*. One has to presume that his conclusions provided him some direction for the study of his great generalizations in chemistry, especially for the greatest fruit of his efforts, the periodic System of the Elements. At least it is apparent that somehow he acquired greater confidence in the feasibility of systematizing extant chemical knowledge than almost any of his contemporaries. (Kultgen, 1958, pp. 177–183)

It would seem necessary at this point to say something about the earlier element schemes, especially the nineteenth-century scheme, and how it was that elements were regarded as transcendental and at the same time were characterized by such properties as their atomic weight. This task will be attempted in the following section.

Early Element Schemes, Especially the Nineteenth-Century Version

For Aristotle, the elements themselves are unobservable and transcendental (in Paneth's terminology), although they give rise to all the variety we see before us. The four elements (fire, earth, water, and air) are regarded as property bearers and are responsible for the properties of substance, although they are themselves unobservable. The elements are immaterial qualities impressed on an otherwise undifferentiated primordial matter and are present in all substances. Furthermore, the properties of substances are governed by the proportion of the four elements present within them.

This view was first seriously challenged by Lavoisier during the course of the chemical revolution. Lavoisier's new chemistry drew on the Aristotelian tradition but also included some new aspects. The new chemistry introduced the concepts of *simple substance* and *material ingredient* of substances. A simple substance is one that cannot be decomposed by any known means. The inclusion of the word *known* is very important because the scheme proposes that simple substances are to be regarded as such only provisionally and that they may lose their status following future refinements in analytical techniques. It is, one might say, a restrained view of scientific progress that recognizes the limitations of chemical analysis. As a result of Lavoisier's work, it thus became a relatively simple experimental question to determine which substances were simple and which were not. A major departure from Aristotle's scheme was that not all substances had to contain every one of these simple substances. There was no longer thought to being one undifferentiated primordial matter but that instead there were a number of elementary constituents or simple substances were now possessed of observable properties such as weight.

A consequence of Lavoisier's scheme is that abstract elements do not necessarily correspond to particular known simple substances. It is possible that what is regarded as a simple substance at a particular stage in history may turn out to be decomposable as a result of subsequent advances in chemical analysis. Stated another way, to be certain of the correspondence between a simple substance and an abstract element one would need to have perfect confidence in one's analytical techniques. To his credit, Lavoisier only provisionally identified simple substances that had been isolated with abstract elements.[21]

And, yet, the transcendental aspect of elements was not completely forgotten and continued to serve an explanatory function in nineteenth-century chemistry but not necessarily a microscopic explanation. A chemist could be skeptical of atomistic explanations, as many were in the nineteenth-century, and yet could readily accept a transcendental explanation, for example, for the persistence of the elements in their various compounds. As was alluded to earlier, one of the benefits of regarding the elements as having a transcendental existence is that it provides a way out of the apparent paradox concerning the nature of elements when combined in compounds. Suppose that sodium and chlorine are combined to form sodium chloride (common salt). In what sense is the poisonous metal sodium present in a sample of white crystalline common salt? Similarly, one may ask how it is that the element chlorine, a green and poisonous gas, continues to exist in common salt. Clearly, the elements themselves, in the modern sense of the word, do not appear to survive, or else they would be detectable and one would have a mixture of sodium and chlorine able to show the properties of both these elements. The response available from the nineteenth-century element scheme is that simple substances do not survive in the compound but abstract elements do.[22]

According to the nineteenth-century scheme, these abstract elements are believed to be permanent and responsible for the observable properties of simple bodies and compounds.[23] However, in a major departure from the Aristotelian view, the abstract elements are also regarded as being material ingredients of simple bodies and compounds. The concept of material ingredient thus serves to link the transcendental world of abstract elements and the observable, material realm of simple substances. For example, the stoichiometric relationships observed in chemical changes are explained in terms of amounts of abstract elements present in the reacting substances through this concept of material ingredient.

Thus, we have three important concepts in the nineteenth-century scheme initiated by Lavoisier considerably earlier. The abstract element is a property bearer and owes its heritage to the Aristotelian element scheme. In addition to being a property bearer, the abstract element is an indestructible material ingredient of substances,[24] and as we saw in the preceding discussion, there is a fundamental distinction between abstract element and simple substance. Abstract elements are unobservable, whereas simple substances such as sodium, chlorine, and oxygen are observed. It should be noted that in contemporary chemistry we only seem to retain the latter notion in that the term *element* now means what to a nineteenth-century chemist would have been called a *simple substance*.[25]

The culmination of the nineteenth-century element scheme occurred with the discovery of the periodic system and with the work of Mendeleev, who more than any other of the discoverers was concerned with the philosophical status of the elements (Mendelejeff, 1871). As stated earlier, Mendeleev was keenly aware of the abstract element/simple substance distinction, and he also realized that abstract elements, or basic substances, were to be regarded as more fundamental than simple substances. The explanation of why "elements" persist in their compounds was to be found in abstract elements and not simple substances, and, as a consequence, if the periodic system was to be of fundamental importance, it would primarily have to classify the abstract elements. Thus, the predictions that Mendeleev made were conceived of with abstract elements in mind. If the available observational data on simple substances pointed in a certain direction, these features could be partly overlooked in the belief that the properties of the more fundamental abstract elements might be other than had been observed up to that point. Of course, contact with observational data would still have to be the test of Mendeleev's predictions, but because the abstract elements were bearers of observable properties or, more specifically, a material ingredient in the form of the atomic weights of the elements,[26] these ideas could be tested empirically. As I have already speculated, it may be that Mendeleev's adherence to a more fundamental realm provided him with greater latitude in making predictions, and was the one of the key features that allowed him to make predictions that were far more successful than those of his competitors.

I will give just one example to illustrate Mendeleev's deep insight.[27] The metal beryllium provided one of the most severe tests for Mendeleev's system. The question was whether to place beryllium in group II above magnesium or in group III above aluminum. Its measured specific heat of 0.4079 indicated an atomic weight of 14, which would place beryllium in the same group as the tri-valent aluminum (Nilson & Petterson, 1878). Furthermore, beryllium oxide is weakly basic, the lattice structure of the metal is unlike that of magnesium, and beryllium chloride is volatile just like aluminum chloride. Taking these facts together, the association of beryllium with aluminum appears to be compelling.

Despite all this evidence, Mendeleev supported the view that beryllium is di-valent using chemical arguments, as well as considerations, based on the periodic system. He pointed out that beryllium sulfate presents a greater similarity to magnesium sulfate, than to aluminum sulfate and that whereas the analogues of aluminum form alums, beryllium fails to do so. He argued that if the atomic weight of beryllium were 14, it would not find a place in the periodic system. Mendeleev noted that such an atomic weight would place beryllium near nitrogen, where it should show distinctly acidic properties, as well as having higher oxides of the type Be_2O_5 and BeO_3, which is not the case. Instead, Mendeleev argued, the atomic weight of beryllium might be approximately 9, which would place it between lithium (7) and boron (11) in the periodic table.

In 1885 the issue was finally conclusively settled in favor of Mendeleev by measurements of the specific heat of beryllium at elevated temperatures. These experiments pointed to an atomic weight of 9.0, in reasonable agreement with Dulong and Petit's law and supported the di-valency of the element (Humpidge, 1885). Above all else, Mendeleev persisted in the belief that beryllium lies in group II because of his faith in the validity of the periodic law, which he believed was essentially a feature of the transcendental elements. All else was rationalized around this central tenet.[28]

By contrast, although some predictions were made by Lothar Meyer, Newlands, Odling, and other pioneers of the periodic system, these pale into insignificance when compared with the predictions by Mendeleev.[29] At the risk of appearing to be proselytizing, I think that this provides a good example of why we need a philosophy of chemistry at present as much as it was needed in Mendeleev's days. (It is to be hoped that this book[30] will help revive the philosophical study of chemistry, or the "central science" as it is often called, and that thinking philosophically about chemistry will reap rewards in real chemistry research and not just serve to build up a new subdiscipline in philosophy.)

Whereas Mendeleev was clearly ahead of his competitors when it came to the prediction of elements, he does not seem to have fared so well with regard to his views on the reduction of chemistry. More specifically, his denial of the reduction of chemistry has generally been held to have been mistaken, especially in view of the subsequent discoveries of radioactivity and the structure of the atom. That such a conclusion has been reached by historians of chemistry is not at all surprising, especially given some of Mendeleev's own pronouncements on the subject:

> The periodic law . . . has been evolved independently of any conception as to the nature of the elements; it does not in the least originate in the idea of a unique matter; it has no historical connection with that relic of the torments of classical thought. (Mendeleev, 1905, vol. 2, p. 497)

> The more I have thought about the nature of the chemical elements, the more decidedly I have turned away from the classical notion of a primary matter, and from the hope of attaining the desired end by a study of electrical and optical phenomena. (Mendeleev, 1905, vol. 1, p. XIV)

> By many methods founded both on experiment and theory, has it been tried to prove the compound nature of the elements. All labour in this direction has as yet been in vain, and the assurance that elementary matter is not so homogeneous (single) as the mind would desire in its first transport of rapid generalization is strengthened from year to year. (Mendeleev, 1905, vol. 1, p. 20)

> The periodic law affords no more indication of the unity of matter or of the compound character of our elements than the law of Avagadro. (Mendeleev, 1905, vol. 2, p. 498)

These quotations stand in marked contrast to Lothar Meyer's statement that the existence of some sixty or even more fundamentally different kinds of primordial matter is intrinsically not very probable" (1884, p. 129).

How could Mendeleev have been so wrong about reduction of the periodic system, or was he indeed wrong? Interestingly, Paneth has argued that, although in some ways Lothar Meyer turned out to be more correct on reduction, Mendeleev would not need to change his view on the basis of current knowledge:

> Yet I believe that something very essential in his fundamental philosophical tenets would have remained untouched by the progress in physics and could be successfully defended even today; and it is just these "philosophical principles of our science" which he regarded as the main substance of his textbook. (Paneth, 1965, pp. 56–57)

What Paneth had in mind is that Mendeleev adopted an intermediate position between realism and reduction to physics:

> The reduction to unity (or quadruplicity) has been successfully achieved by physics; chemistry however will probably preserve, as long as there is still a science of chemistry in the present meaning of the word, a plurality of its basic substances in complete agreement with the doctrine of its old master, Dimitri Ivanovich Mendeleeff. (Paneth, 1965, p. 70)

> Every science must decide for itself the units with which it builds, and the deepest foundation is by no means always the best for the purpose. (Paneth, 1965, p. 67)

Similarly, the French philosopher Bachelard, who began his career as a physical chemist, has written, La pensee du chimiste nous parait osciller entre le pluralisme d'une part et la reduction du pluralisme d'autre part (Bachelard, 1932).

Paneth's Own Intermediate Position: Isotopy and the New Element Scheme

Paneth is clearly not a naive realist because he recognizes two senses of the term *element*. Indeed, he is the last chemist of any note to have stressed the value in drawing this distinction.[31] In fact, Paneth's element scheme is the one still in use to this day, although most contemporary chemists would probably be loath to recognize a metaphysical or transcendental aspect to the nature of the elements, falsely believing that this might smack of alchemy, which they, of course, take to be a gravely mistaken enterprise.[32]

Even though the development of the periodic system was an unparalleled success, especially at the hands of Mendeleev, who predicted new elements and corrected the atomic weights of several existing ones, a number of problems nevertheless began to emerge. First, there was the question of the apparent inversion of the positions of the elements tellurium and iodine. If these two elements were ordered according to atomic weight, they fell into chemically incorrect groups. Mendeleev simply reversed their positions, thus putting more faith in chemical properties and believing that the atomic weights of these elements had been incorrectly measured. But many heroic attempts to show that tellurium has a lower atomic weight than iodine were completely fruitless. This reversal problem, along with a couple of others in different parts of the periodic table, were to remain unsolved for many more years.

Second, the accommodation of the rare earth elements into the periodic system was proving to be quite a challenge for Mendeleev. In broad terms, these elements have almost identical chemical properties such that one might be tempted to put them all into the same place in the periodic table, a result that would contradict the spirit of the entire classification scheme.

The third issue concerned the discovery, in 1894, of the element argon that seemed to be mysteriously devoid of the power to combine with other elements and could not initially be fitted into the periodic system. Several more gases of this type were discovered over the next few years, thus confounding the problem of their accommodation further. It cannot be denied that the discovery of this completely new group of elements, which had not been predicted by anyone, raised some concern regarding the soundness of the periodic system. In 1900 Ramsey, who had isolated these mysterious "noble gases," suggested a new column to be placed between the halogens and the alkali metals in the periodic table, a proposal that Mendeleev readily agreed to.

But even more severe worries were looming in the distance, not just for the periodic system but for the very meaning of the concept of an element. In 1896 Becquerel discovered that certain elements decay radioactively. The intense study of such processes over the following years revealed a host of new "elements;" so many, in fact, that their discoverers scarcely paused to give them new names.[33] They included such things as uranium X and radiothorium, provisionally named after the elements from which they originated. The question naturally arose as to how these substances might be accommodated into the periodic system. An even deeper issue became evident when Rutherford showed that elements could be transmuted into quite different ones, something that only the alchemists had previously imagined possible. Mendeleev, a strong supporter of the nineteenth-century element scheme, could not accept such findings, especially transmutation, because they appeared to contradict the notion of the permanence of the elements. Finally, there was the problem of isotopy, which was eventually to lead to the resolution of some of these problems, but not before a number of scientific, and even philosophical, battles had been fought.

The phenomenon of isotopy refers to the fact that certain elements consist of more than one component (now called isotopes) which seemed, at the time of their discovery, to be impossible to separate. The first attempt to restore order to the situation was made by Soddy, who had discovered the phenomenon in 1911.[34] Soddy announced the theory of isotopy which stated that "isotopes of the same species are inseparable and belong in the same place in the periodic table." However, this new notion implied that one simple substance (something that could not be further purified) would be equivalent to two or more abstract elements in violation of the one-to-one correspondence demanded by the nineteenth-century element scheme. Some argued that isotopes might eventually prove to be separable in which case each isotope would have to be regarded as an abstract element.[35] However, for most chemical purposes, isotopy seemed to make no difference.

Then in 1913 Moseley, working in Manchester, discovered what we now call atomic number (Moseley, 1913). He began by photographing the X-ray spectrum of 12 elements, 10 of which occupied consecutive places in the periodic table. He discovered that the frequencies of features called K-lines in the spectrum of each element were directly proportional to the square of the integer representing the position of each successive element in the table. As Moseley put it, here was proof that "there is in the atom a fundamental quantity, which increases by regular steps as we pass from one element to the next." This fundamental quantity, first referred to as "atomic number" in 1920 by Ernest Rutherford, is now identified as the number of protons in the nucleus of an atom. Moseley's work provided a method that could be used to determine exactly how many empty spaces remained in the periodic table.

At about the same time, Bohr published his trilogy articles in which he introduced the quantum theory of the atom and obtained, by various means, the electronic configurations of many of the elements in the periodic system (Scerri, 1994b). In addition, he solidified the notion that the chemical properties of the elements were due to the orbiting electrons, which corresponds to the atomic charge of each nucleus (Bohr, 1913). Here, then, with the discoveries of Moseley and Bohr, was the solution of the tellurium/iodine inversion question. If the elements were ordered according to atomic charge instead of atomic weight, then everything fell into place. Tellurium, with a lower atomic charge than iodine, could now justifiably be placed before iodine, as its chemical properties demanded it should.

However, up to this point the theory of isotopy had been based on a negative result, namely the apparent inability to separate the isotopes of a simple substance. In 1914 Paneth and von Hevesy set out to examine this notion more directly. Was it just that isotopes could not be separated and so should be regarded as belonging to the same element by default as Soddy's theory demanded, or was it that they really did share the same chemical properties?[36]

Replaceability

Paneth and his collaborator von Hevesy took the view that isotopes might be chemically identical and began to explore this notion experimentally (Paneth & von Hevesy, 1914). They proposed the concept of replaceability of isotopes—that is, they claimed that the replacement of any isotope with another one of the same species would not produce any noticeable chemical effect. They set out to verify the correctness of this view through the law of mass action. For any reaction,

$$\mathbf{aA + bB \leftrightarrow cC + dD}$$

it can be shown that an equilibrium constant, $\mathbf{K_c}$, at any given temperature, for the reaction is given by

$$K_c = [C]^c\,[D]^d/[A]^a[B]^b$$

where the square brackets denote the concentrations of the species A through D.[37] If the replaceability view is correct, it should be possible to take any of these concentrations as the sum of the concentrations of all the isotopes of any particular species. To estimate these concentrations Paneth and von Hevesy used electrochemical experiments and the Nernst equation,[38]

$$E = E^o + (RT/nF)\ln c$$

The main experiment utilized two isotopes of bismuth, $^{209}Bi^{+3}$ and $^{210}Bi^{+3}$, with the result that the observed voltage was found to be constant, regardless of the proportion of the two isotopes present in the sample, within the accuracy of the experiment. Thus, they demonstrated that replaceability was indeed an experimental fact. Isotopes of a single species[39] had been shown to be chemically indistinguishable.[40]

In 1918 Aston discovered isotopes in a number of non-radioactive elements that led to a strengthening of Paneth's case. Paneth pointed out that elements in Fajans's sense of the word seemed to be multiplying each day and were thus endangering the chemists' picture of the world. It would have been far-fetched for chemists to restructure the foundations of their science on the grounds that "elements" had turned out to be composite structures. At about this time Paneth stated clearly what had previously remained as an implicit definition of a chemical element: "A chemical element is a substance of which all atoms have the same nuclear charge" (Paneth, 1925 p. 842).

Even at this time, however, isotopes were suspected to consist of smaller particles. Indeed, the radiochemist Fajans disagreed with Paneth and suggested that each isotope should be regarded as an element, a view, which if taken seriously, would have committed chemistry to research into elementary particles (Fajans, 1914). Chemists were not prepared for this change, nor were they motivated to abandon their existing successful research programs. In 1919 a decisive experiment was conducted by Rutherford, who succeeded in artificially decomposing the nucleus by using α particles, but the method he used could scarcely be described as chemical. Fajans was forced to demarcate between decomposition carried out by physical means (Rutherford's experiment) as opposed to chemical means, and Paneth, who had advocated such a demarcation from the start, was thus vindicated.

Paneth's New Element Scheme

Paneth then took on the more philosophical task of revising the nineteenth-century element scheme to accommodate the new findings, especially the results of his own and von Hevesy's replaceability experiments. One of his aims was to explain to chemists at large the implications of Soddy's theory, which was that chemical analysis does not proceed down to the most fundamental level, a feature that most chemists were finding rather disconcerting at this time.

Paneth was to use the positive results that emerged from the replaceability work to offer a solution. In his new element scheme a mixture of chemically inseparable isotopes was identified with a single element.[41] It will be recalled that in the nineteenth-century scheme an abstract element is a property bearer and also a material ingredient characterized by atomic weight. If Paneth was to reconcile his own element scheme with the more established nineteenth-century scheme he would have to consider the following two points: (1) whether isotopic simple substances were to be regarded as different simple substances—Paneth would answer no, because every *chemical* attempt to separate a mixture of isotopes had failed; and (2) if abstract elements combined the aspect of property bearer and material ingredient (atomic weight), what should be made of isotopes, which had different atomic weights? Would they also have to be regarded as different property bearers and so as different abstract elements? Again Paneth would answer no, but this response required a severing of the connection between the aspect of being a property bearer and also a material ingredient characterized by atomic weight. According to the replaceability experiments, Paneth reasoned that isotopes were the same property bearer but not the same material ingredient because isotopes of the same element differ in atomic weight.[42]

Now this severing of material ingredient, in the form of atomic weight, from property bearer seemed to have other consequences that Paneth also succeeded in explaining. For example, there are several laws based on atomic weight in chemistry to account for the amounts in which elements combine.[43] What would happen if atomic weight, the classical material ingredient, were to be banished from its function of characterizing an abstract element in the observable world? Paneth's response was that the chemical weight laws had already been falsified with the discovery of radioactivity. If a chemical reaction involves a radioactive substance—say A—then it is not the case that atomic weight is conserved as in the kinds of stoichiometric problems that beginning chemistry students are made to perform:

$$A + B \rightarrow C$$

This is because A has lost some of its weight in the time it takes for this reaction to proceed and for weighing to be made. The loss of significance of atomic weight as the material ingredient of an abstract element raises the question of how chemical properties are carried so that different isotopes may show the same chemical reaction. Of course, the answer lay, in making atomic number the new material ingredient.

To summarize, Paneth's scheme retains the analytic limit characteristic of chemistry. A simple substance is one that cannot be further analyzed by *chemical* means, and a mixture of isotopes is, therefore, regarded as a single simple substance. The notion of a one-to-one correspondence between simple substance and abstract element is retained with the result that Paneth is forced to change the material ingredient from atomic weight to atomic number. The nineteenth-century connection between an abstract element as a property bearer (unobservable) and its material ingredient as atomic weight is broken once and for all.[44] In Paneth's scheme, replaceable isotopes correspond to, or are representatives of, the same abstract element. The periodic table that summarizes the properties of abstract elements is preserved instead of needing to devise a complicated new table for the individual isotopes of all the elements.

Paneth's contribution was thus to uphold the autonomy of chemistry and to resist following the reductive path of the physicist, which would have destroyed the periodic system and would eventually have turned chemistry into elementary particle physics. In fact, Paneth was recommending a form of naive realism whereby isotopes of the same element could be regarded as being identical in chemistry, even though we know that they are not strictly identical from the more reductive perspective of physics.

Throughout these developments all the major suggestions made by Paneth were vigorously opposed by Kasimir Fajans, as already mentioned here. Fajans disputed the Paneth–von Hevesy replaceability results and later criticized Paneth's elements scheme, believing that individual isotopes of any particular element should indeed be regarded as distinct simple substances and that they should also be regarded as distinct abstract elements.

Interestingly, Fajans seems to have preferred to use the terms *theoretical element* and *practical element*. Theoretical elements were truly fundamental for Fajans, whereas practical elements were those that had been shown to have resisted decomposition thus far. From the point of view of this essay, this point is significant because it emphasizes that at the beginning of the century the philosophical notion of abstract element (Paneth) or theoretical element (Fajans) was widely held. Indeed, here are two chemists who disagreed on many aspects of the new radiochemistry but took it for granted that the nature of elements consisted in a dual observable and nonobservable nature. It is doubtful whether a contemporary chemist would have arrived at Paneth's element scheme or indeed Fajans's alternative in the absence of such a philosophical appreciation of the subject matter.

The final seal of approval for Paneth's element scheme came in 1923, when a IUPAC committee met in Paneth's absence and, apart from minor differences in terminology, adopted the proposal that he had made as early as 1916 (Aston et al., 1923). Moreover, in an influential article to *Nature* in 1919, Aston used terminology for the elements that he attributed to Paneth.

One further interesting development took place in 1932, when Urey discovered isotopes of hydrogen. In these cases, the masses of the isotopes differ from each other by as much as 100% (Urey & Grieff, 1935). The properties of these isotopes are definitely not the same. For example water (mostly protium oxide) has a boiling point of 100°C, whereas the value for deuterium oxide is 104°C. Strictly speaking, Fajans had been correct to doubt replaceability, but the problem is only significant in isotopes with very low masses, and Urey's discovery did not change the chemical definition of an element in any way.

The conclusion to be drawn from this episode is that chemistry is not obliged to accept every single step toward reduction, which may be suggested by research in physics. Whereas the concept of atomic number represents a beneficial reduction for chemists to embrace, a restructuring of the subject based on the physical difference isotopy would have led to the demise of the periodic system. As Greenaway says in an article to commemorate the 100th anniversary of Mendeleev's periodic table:

> The chief influence of the periodic table on chemistry was to turn it from endless diversification in a search for unknown compounds and elements to a concentration on order and its underlying cause. Chemistry was able to assimilate the new physics—the electron, radioactive transformation etc., without fundamental change. (Greenaway, 1969, p. 99)

The Paneth episode would seem to be another example of the chemist's need to adopt an intermediate position, which, while acknowledging the findings of the reducing science of physics, upholds the autonomy of chemistry and recognizes where to stop:

> Nowadays we know that they [the elements] are composite in two senses: they consist of a mixture of isotopes, and the atoms of every isotope are built up from simpler constituent particles. Since, however, in chemical reactions there is neither an unmixing of the isotopes nor any change in the essential parts of the atom, the law of the conservation of the elements is valid in *chemistry*.
>
> Conviction on this point permits us, in complete agreement with Mendeleeff, to regard the chemical elements as our ultimate building blocks. If we investigate the foundations of chemistry as an independent science, then indeed we do not come up against those primary qualities which for centuries were regarded as the ultimate principles in physics viz. size, shape and motion . . . neither do we encounter the four qualities of physics mentioned above . . . (the neutron, the two types of electron, and the neutrino), but only these eighty-nine chemical basic substances. (Paneth, 1965, p. 68)

Conclusion

The central message of this article, which I believe remains relevant in today's chemistry, has been that the best strategy for the chemist to adopt lies in what I call the intermediate position between reduction and varying degrees of realism. This recommendation would seem to be especially relevant to modern-day chemistry, which has looked increasingly as if it were being overrun by physics. I will now conclude by giving a summary of the main ideas in this article, and in so doing I will attempt to restore some form of historical order to the events discussed.

Mendeleev, the creator of the periodic system of the elements, drew the philosophical distinction between basic substances (abstract elements) and simple substances. Therefore, he cannot be accused of having acted as a naive realist. However, having arrived at the periodic classification by giving emphasis to abstract elements, he resisted the prevalent reductionist tendency of supposing the existence of a primary matter. He considered the elements as distinct individuals and adopted an intermediate position between realism and reduction.

Paneth later followed Mendeleev in insisting on the distinction between basic substances and simple substances. While praising the chemist's use of naively realistic notions in most instances, Paneth pointed out that, to understand how the elements persist in the compounds, chemists must assume a more philosophical position. When faced with the question of isotopy, Paneth maintained the distinction between basic substances and simple substances but did not follow the reductive path of the physicist who would have been inclined to regard isotopes as different elements. Thus Paneth adopted an analogous intermediate position to that of Mendeleev.

In contemporary chemistry, concepts such as orbitals, configurations, and hybridization are frequently used as though they were "real," contrary to the pronouncements of modern physics and philosophers of science who deny the reduction of such typically chemical concepts to quantum mechanics. In this case, the realism I am identifying takes the form of an unwarranted belief in microscopic entities that the theory tells us do not, in fact, exist. My claim is that modern chemistry correctly continues to adopt an intermediate position between realism and reduction in the tradition of Mendeleev and Paneth.

Notes

1. Scerri (1991a, 1991b, 1994a, 1995, 1997a, 1997c, 1998a).
2. Heinz Paneth changed his last name to Post sometime around 1953.
3. Post founded the department of history and philosophy of science at what was then Chelsea College, London University, and which subsequently merged with King's College, London. He has also been one of the most influential postwar philosophers of science in Great Britain. His father, Fritz Paneth, one of the founders of radiochemistry, had a deep interest in philosophical aspects of chemistry, as can be seen in his collected essays (Paneth, 1965).
4. In philosophy, the term *naive realism* is generally taken to mean a belief in macroscopic objects for what they appear to be and independently of any views on what lies below the surface. I will be using the term in this sense but will also use it to mean the adoption of superficial views about microscopic entities when discussing atomic orbitals and configurations.
5. In the case of the hydrogen atom an atomic orbital is well defined in the sense that four quantum numbers can be genuinely attributed to the one electron.
6. If reduction of chemistry as a whole is considered from a naturalistic viewpoint as I have advocated previously (Scerri, 1998a), the question of whether or not chemistry has been reduced to quantum mechanics is more subtle and depends on the present state of computational quantum chemistry and in particular ab initio calculations of chemical properties. In this sense, one might want to concede that chemistry has been approximately reduced to

quantum mechanics, although even the notion of approximate reduction is problematical (Scerri, 1994a).

7. However, the main focus of this chapter is not on the issue of realism, and so I will not be concerned with such issues as warrants for belief.

8. An even more asymmetrical aspect concerns the precise placement of the rare earth element block, which in many tables is still shown to begin one element to the right of the start of the d block. This is despite the fact that several articles have pointed out that the f block truly belongs between the s and d blocks (Jensen, 1986).

9. The f block refers to the rare earth elements. The label f originally stood for the fundamental lines in the spectrum.

10. I am not suggesting that the medium–long form originated with quantum mechanics. Indeed, early versions of this kind were suggested by Werner among others (Werner, 1905).

11. See van Spronsen (1969) and Mazurs (1974) for other pyramidal forms of the table.

12. Of course, there is another pragmatic reason why the medium–long form rather than the pyramidal form has gained widespread acceptance and that is the fact that chemical groups are very easily seen as vertical columns in the medium–long form whereas they are not quite so easily noticed in the pyramid form, although the connecting lines serve to reinforce the group connections.

13. I do not exclude the possibility that future advances may lead to a strict deduction of the points at which the periods close. Similarly, a future generalization of quantum mechanics might enable one to explain the closing of the periods from first principles.

14. The notion of abstract elements is a recurring theme in the remainder of this chapter.

15. The reader will not be surprised to learn that Paneth was familiar with Kant's philosophical views.

16. For a discussion of this point, see Paneth's 1962 article. Also, the mere fact that Mendeleev, and considerably later Paneth, continue to maintain a dual nature for the elements attests to the fact that the chemical revolution did not eliminate the metaphysical view.

17. However, Paneth argues that Mendeleev was somewhat confused in the terminology he chose to discuss this issue. Mendeleev makes the distinction between simple bodies and elements in several passages: "A simple body is something material, for example a metal or metalloid, endowed with physical properties and the ability to show chemical reactions. To the concept of the simple body corresponds the molecule consisting of several atoms . . . On the other hand, element is a term for those material constituent parts of simple and composite bodies which determine their physical and chemical behavior. The concept corresponding to element is the atom". (Paneth, 1871, p. 141). Paneth claims that this is an error on Mendeleev's part. While agreeing that the distinction is a very important one, he believes that the reason it has received so little attention is that the terms used by Mendeleev are not very appropriate. One cannot introduce a distinction between element and simple body since, according to Lavoisier, the definition of element IS a simple body. Second, Paneth thinks that by the association of the terms *element* with *atom* and *simple body* with *molecule*, respectively, Mendeleev seems to have missed the essential point (Paneth, 1965, p. 57). This is because atoms and molecules belong to the same group of scientific concepts, or the same category of concepts, while the fundamental difference which Mendeleev intends to draw between element and simple body is due to their belonging to entirely different epistemological categories.

18. E.R. Scerri & J.W. Worral "Prediction and the Periodic Table", (to appear).

19. For example, see Leicester 1948).

20. If one insists on calling it a metaphysical position, it must be made clear that it is intended in the literal sense of the word, meaning beyond the physical and not the modern philosophical meaning or indeed the Aristotelian sense of the term, meaning what actually exists fundamentally. As is well known, Aristotle's famous book on the subject merely came

after his physics, and this is where modern metaphysics gets its name. Aristotle's book was not called metaphysics because it was intended as an investigation into what lies beyond physics or beyond observation.

21. Such caution began to fade toward the end of the nineteenth-century to the extent that simple substances that had been isolated began to be regarded as the only form of an element and the abstract counterpart to each simple substance was largely forgotten.

22. The more prosaic explanation given in contemporary chemistry is that what survives of each of the elements is the number of protons—in other words, the nuclear charge of the atoms of sodium and chorine. This would also be the case in rather extreme examples such as the Na^{+11} and Cl^{+17} ions. Although this response is surely correct, it also seems a little unsatisfactory
for the identity of chemical elements to depend on the nucleus of their atoms, given that all the chemical properties are supposed to be determined by the configurations and exchanges in the electrons around the nucleus. This paradox has also recently been noted by Noretta Koertge (Koertge, 1998).

23. This is perhaps why Mendeleev, a great defender of the nineteenth-century element scheme, was so reluctant to accept the notion of the transmutation of the elements discovered by Rutherford at the turn of the twentieth-century.

24. Indestructible in the sense that weights are conserved in the course of chemical reactions. This was, of course, the basis of the chemical revolution of Lavoisier.

25. This is despite the fact that our present element scheme, which we owe to Paneth, was arrived at partly by his insistence on the distinction between abstract element and simple substance, as will be discussed later.

26. Of course, I do not mean to imply that the weights of individual atoms can be directly observed. I intend this remark in the chemist's sense that stoichiometric reactions can be rationalized by appeal to atomic weights of participating elements.

27. Mendeleev correctly revised the atomic weight of uranium from 120 to 240. He also predicted intuitively that tellurium should be placed before iodine, thus ignoring the atomic weight ordering for these elements.

28. Whether or not Mendeleev was influenced by Kant's philosophical writings is an issue I will take up in a forthcoming study.

29. A detailed discussion of predictions and the periodic system is given in van Spronsen, (1969). See especially chapter 7.

30. Mention might also be made of the new journal *Foundations of Chemistry* published by Kluwer.

31. The 1962 article is a translation of a lecture given by Paneth in 1931 to the Gelehrte Gesellschaft of Königsberg.

32. For a healthy antidote to this generally held negative view of alchemy, see Pierre Laszlo's "Circulation of Concepts" (1999).

33. As Badash has written, by 1913 over 30 radioelements were known, but there were only 12 places in the periodic table in which to place them" (Badash, 1979).

34. Isotopy was only found in radioactive elements, but in 1913 J.J. Thomson also discovered the effect in the nonradioactive element neon. This was soon shown to be a perfectly general result. Many elements occur as two or more isotopes, although a number of elements only possess a single isotopic form.

35. In fact, isotopes were separated by Aston in 1919, but this fact did not affect Paneth's new element scheme, which became generally adopted, and formed the basis for the definition of the term element by IUPAC in 1923.

36. It would appear that Paneth and von Hevesy were following a rather Popperian strategy.

37. Strictly speaking, the relationship is valid for the activities of substances A to D. In the case of dilute solutions, activities can be equated with concentrations.

38. Where E is the potential of a half-cell, E^o is the standard electrode potential, R the ideal gas constant, T the absolute temperature, n the number of electrons transferred in the course of the half reaction, F the Faraday constant, and c the concentration of the dissolved species.

39. The word *element* has not been used here because this is precisely the issue being addressed. Are we, in fact, dealing with one "element" or more?

40. Another leading radiochemist, Kasimir Fajans, objected to the interpretation of these results on the basis of a thermodynamic analysis and put forward the view that isotopes are different, not just physically but also chemically (Fajans, 1914). Various other experiments were then devised and carried out in an attempt to settle the issue, but the debate continued and, indeed, was widened to the question of what constitutes a chemical element.

41. This was in keeping with Soddy's theory, but, it was now being expressed in more positive terms based on experimental findings.

42. He accommodated the fact that different isotopes have different radioactive properties (typically different half-lives if they are radioactive) by stating that these would not count as chemical properties.

43. This issue is discussed in Kultgen's 1958 article, as well as in most chemistry textbooks.

44. Paneth concluded that if atomic weight was causing problems in this context it was because it was not a chemical property. Since the time of Newton, substances or bodies had become characterized by their weight. This view gained greater acceptance when first Lavoisier and then Dalton emphasized the need to characterize different chemical substances according to weight. refers to S. Toulmin in *Encyclopedia of Philosophy* (Toulmin, 1967).

References

Aston, Frederick W. 1919. "The Constitution of the Elements." *Nature* 104: 393.

Aston Frederick W., Baxter, G. P., Brauner, B., Debierne, A., Leduc, A., Richards, Soddy, T. W. F., & Urbain, G. 1923. "Report of the International Committee on the Elements." *Journal of the American Chemical Society* 45: 867–874.

Bachelard, Gaston. 1932. *Le pluralisme coherent de la chimie moderne*. Paris: Vrin.

Badash, Lawrence. 1979. "The Suicidal Success of Radiochemistry." *British Journal for the History of Science* 12: 245–256.

Bohr, Niels. 1913. "On the Constitution of Atoms and Molecules." *Philosophical Magazine* 6(26): 1–25, 476–502, 857–875.

Brush, S. 1996. "The Reception of Mendeleev's Periodic Law in America and Britain." *Isis* 87: 595–628.

Fajans, K. 1914. "Zur Frage der isotopen Elemente." *Physikalische Zeitschrift* 15: 935–940.

Galison, Peter, & Stump, David J. 1996. *Disunity of Science, Boundaries, Contexts and Power*. Stanford, CA: Stanford University Press.

Greenaway, Frank. 1969. "On the Atomic Weight of Glucinium (Beryllium). Second Paper." *Chemistry in Britain* 5: 97–99.

Humpidge, T. S. 1885. *Proceedings of the Royal Society* 38: 188–191.

Humpidge, T. S. 1986. *Proceedings of the Royal Society* 39: 1.

Jensen, William, B. 1986. "Classification, Symmetry and the Periodic Table." *Computation and Mathematics with Applications* 12B: 487–509.

Koertge, Noretta. 1998. "Philosophical Aspects of Chemistry." In *Routledge Encyclopedia of Philosophy*. London: Routledge.

Kultgen, J. H. 1958. "Philosophical Conceptions in Mendeleev's Principles of Chemistry." *Philosophy of Science* 25: 177–184.

Laszlo, P. 1999. "Circulation of Concepts" *Foundations of Chemistry* 1(3): 225–239.

Leicester, Henry. 1948. "Factors Which Led Mendeleev to the Periodic Law." *Chymia* 1: 67–74.

Lipton, Peter. 1990. "Prediction and Prejudice." *International Studies in Philosophy of Science* 4: 51–60.

Lothar Meyer, Julius. 1884. *Die Modernen Theorien der Chemie*. Breslau: Auflage.

Löwdin, Per-Olav. 1969. "Some Comments on the Periodic System of Elements." *International Journal of Quantum Chemistry* S3: 331–334.

Maher, Patrick. 1988. "Prediction, Accommodation and the Logic of Discovery." In A. Fine, & J. Leplin, eds. *PSA 1988* (vol. 1, pp. 273–285). East Lansing, MI: Philosophy of Science Association.

Mazurs, Edward. 1974. *Graphic Representations of the Periodic System during One Hundred Years*. Tuscaloosa, Alabama: Alabama University Press.

Mendelejeff, D. 1871. "Die periodische Gesetzmäβsigkeit der Chemischen Elemente." *Annalen der Chemie und Pharmacie* VIII Supplementbandes zweites Heft: 133–229.

Mendeleev, Dimitri. 1905. *Principles of Chemistry I*. London: Longmans.

Moseley, Henry G. J. 1913. *Philosophical Magazine* 26: 1024–1034; and 1914, 27: 703–713.

Nilson, L. F. & Petterson, O. 1878. "Über die specifische Wärme des Berylliums." *Berichte* 11: 381–386.

Paneth, F. A. 1962. "The Epistemological Status of the Concept of Element." *British Journal for the Philosophy of Science* 13: 1–14,144–160.

Paneth, F. A, & von Hevesey, G. 1914. "Zur Frage der isotopen Elemente." *Physikalische Zeitschrift* 15: 797–805.

Paneth, F. A. 1920. "Die neueste Entwicklung der Lehre von den chemischen Elementen. *Naturwissenschaften* 8: 839–842; 45: 867.

Paneth, F. A. 1965. "Chemical Elements and Primordial Matter: Mendeleeff's View and the Present Position." In H. *Dingle*, & G. R. *Martin*, eds. *Chemistry and Beyond* (pp. 53–72). New York: Wiley.

Scerri, Eric R. 1991a. "Electronic Configurations, Quantum Mechanics and Reduction." *British Journal for the Philosophy of Science* 42: 309–325.

Scerri, Eric R. 1991b. "Chemistry, Spectroscopy and the Question of Reduction." *Journal of Chemical Education* 68: 122–126.

Scerri, Eric R. 1994a. "Has Chemistry Been at Least Approximately Reduced to Quantum Mechanics?" In D. Hull, M. Forbes, & R. Burian, eds. *PSA* (vol. 1, pp. 160–170.)

Scerri, Eric R. 1994b. "Prediction of the Nature of Hafnium from Chemistry, Bohr's Theory and Quantum Theory." *Annals of Science* 51: 137–150.

Scerri, Eric R. 1995. "The Exclusion Principle, Chemistry and Hidden Variables." *Synthese* 102: 165–169.

Scerri, Eric R. 1996. "Stephen Brush, the Periodic Table and the Nature of Chemistry." In P. Jannich & N. Psarros, eds. *Die Sprache der Chemie*, Proceedings of the Second Erlenmeyer Colloquium on Philosophy of Chemistry (pp. 169–176). Würtzburg: Köningshausen & Neumann.

Scerri, Eric R. 1997a. "Has the Periodic Table Been Successfully Axiomatized?" *Erkentnnis* 47: 229–243.

Scerri, Eric R. 1997b. "The Periodic Table and the Electron." *American Scientist* 85: 546–553.

Scerri, Eric R. 1997c. "The Case for Philosophy of Chemistry." *Synthese* 111: 213–232.

Scerri, Eric R. 1998a. "Popper's Naturalized Approach to the Reduction of Chemistry." *International Studies in Philosophy of Science* 12: 33–44.

Scerri, Eric R. 1998b. "How Good Is the Quantum Mechanical Explanation of the Periodic System?" *Journal of Chemical Education* 75: 1384–1385.

Scerri, Eric R. 2000. "The Failure of Reduction and How to Resist the Disunity of the Sciences." *Science and Education* 9 (No 5): 00–00.

Scerri, Eric R., & Worrall, John W. Forthcoming. "Prediction and the Periodic Table." *Studies in History & Philosophy of Science*.

Toulmin, Stephen. 1967. In Paul Edward's, ed. *Encyclopedia of Philosophy*, (vols. 5 & 6). New York: Macmillan.

Urey, Harold, C., & Grieff, L. J. 1935. "Isotopic Exchange Reactions." *Journal of the American Chemical Society* 57: 321–327.

Werner, Alfred. 1905. "Beitrag zum Aufbau des periodschen Systems." *Berichte Berlin deutsche chemie Gesellschaft* 38: 914–921 and 2022–2027 (Table p. 916).

van Spronsen, Johannes W. 1969. *The Periodic System of Chemical Elements: A History of the First Hundred Years*. Amsterdam: Elsevier.

Part II

Instrumentation

5

Substance and Function in Chemical Research

DANIEL ROTHBART

When chemical instruments are used in the laboratory, a specimen undergoes changes at the microscopic level. Depending on the instrument, the specimen absorbs or emits radiation. Alternatively, radiation is scattered, refracted, or diffracted. We often read that microscopic events produced from chemical instrumentation are real, as opposed to mere artifacts of the experiment. But exactly what does this mean? This philosophical question underlies a continual dilemma for the experimental chemist, whether to declare triumphantly that his/her findings reveal some insight about a chemical substance or to refrain from such a judgment for fear of having produced a mere artificial effect. Of course, a commonplace position is that the artificiality of laboratory techniques can be separated, in principle, from the real effects, because these techniques enable scientists to break the influence of laboratory constructions on experimental "facts." But some commentators have resurrected the fairly skeptical view that such declarations of success are grossly overstated because the interference from various instrumental techniques, laboratory equipment, and theoretical ideas precludes the possibility of exposing properties of independently existing substance.

If we address this philosophical question by exploring techniques of chemical instrumentation, we find that the categories of a laboratory artifact and real effect are not mutually exclusive. As I argue here, the experimental phenomena of chemical research are *both* real *and* artificially produced from laboratory apparatus, manufactured conditions, and sophisticated techniques of researchers. The plan of this chapter is as follows: examine the character of analytical instruments in chemistry (section 1); explain the difference between an artifact and a real effect (section 2); examine the process of virtual witnessing in chemistry (section 3); explore how instruments are designed to mimic known chemical or physical processes (section 4); introduce the philosophical importance of noise-blocking techniques (section 5); and conclude with brief remarks about experimental reduction (section 6). The similarities and differences between absorption spectroscopy and Raman spectroscopy are discussed.

In this chapter I adopt a functional orientation to our understanding chemical substance, according to which a specimen is known by those capacities that technicians try to exploit during laboratory research. Consequently, a specimen functions as both a recipient of and an agent for change at the microscopic level, based on the specimen's dynamic properties. This characterization of the specimen is inseparable from the power of instruments to convert microscopic states to information. In the end, I argue that the metaphysics of chemical substance is conditioned by the systematic unity of specimen and chemical instrumentation.

The Analytical Instrument in Chemistry

A modern analytical instrument is often depicted as an information processor in which certain material states are converted to information about the specimen. For this to occur, a signal must be produced, transformed, and converted into data. For most analytical instruments, coding (that is, production of a signal) takes place immediately after the specimen is bombarded with photons. The resulting effects are translated into signals, which then undergo various transformations and culminate with the inscription of data at the readout.

From another perspective, we can understand an analytical instrument as a machine that confers power to human agents. Because these powers are designed to produce knowledge, the modern instrument can be described as an epistemic technology.[1] What does it mean for a technology to confer a power? Consider the case of an automobile. When we say that the automobile gives power to the driver, we mean that this apparatus provides the *means* in the form of equipment for generating certain *states* of motion toward the realization of a *goal* of transportation.[2] The three components of such powers (means, states, and goal) can apply to analytical instruments. The experimenter acquires a *means*, in the form of equipment, techniques, and environment, for generating certain *states*, which are the experimental phenomena at the microscopic level, toward the realization of the particular *goal*. The goal is information about the specimen.

To develop this idea of analytical instrumentation as sources of power, we should expand our notion of an instrument beyond the material components of copper, glass, aluminum, plastic, and wires. I propose a systematic conception of analytical instrumentation as a multilayer synthesis of three domains: material, phenomena, and abstractions.

1. The material realm of every analytical instrument must include the following manufactured devices: a signal generator, a signal detector, and a readout device. The prototype for every *signal generator* is a reactive system, producing detectable effects from the union of specimen and energy. When using such instruments, the experimenter engages in cunning interventions designed to trick the specimen into revealing its secrets. This process is often analogized to the blind man's probing actions of his cane. But unlike this use of a cane, the specimen is subjected to probing manipulations of its internal state, as Ian Hacking (1983) reminds us. The specimen is poked, dissected, and disturbed from the bombardment of photons. Instrumentally induced actions are performed, various effects appear, and then quickly disappear, and signals are produced.[3] In this way the experimenter uses the apparatus to unconceal properties through a quasi-technological mode of making (Lelas, 1993).

The *signal detector* transforms the energy of the signal into a form that is readable through the use of common devices. For example, the thermocouple converts a radiant heat signal into an electric voltage to prepare the signal for the readout. Amplification is particularly important immediately after signal detection. It is not unusual for electric signals to be amplified by a factor of 12 or more.

The *readout device* converts a processed signal to a form that is understandable by humans. The actual marks on the photographic plates, tables, video displays, or recorded images are data inscriptions at the readout.[4] Such marks can take various forms: a position on the needle, numerals on a digital display, a pointer on a meter scale, or black spots on a photographic plate. Each inscription is the end point of an information retrieval system in which experimental phenomena are converted to messages. Of course, the validity of the inscriptions rests on the integrity of the signal, which, in turn, requires that experimenters block, minimize, or at least compensate for various potential contaminations, as I discuss below.

2. The phenomenal realm comprises a world of instrumentally detectable, and fleeting, events at the microscopic level. (Throughout this chapter, these events are called "experimental phenomena.") These events are usually changes in energy states, produced when the specimen's internal structure is sufficiently agitated by energy from the instruments. This implies the following negative result: the experimenter gets no closer to the world of molecules and atoms than the experimental effects, crafted from the union of manufactured conditions, laboratory skills, and the specimen's internal structure. These microscopic effects constitute the most immediate "subject matter" of the detection system. The old dream of disclosing Nature's "brute existence," unobstructed by instrumental interference, is hopelessly anachronistic. Any claim that the instrument provides a transparent medium for revealing Nature's material essence violates fundamental aspects of instrumental techniques.

3. More than material devices and experimental phenomena, analytical instrumentation includes conceptual abstractions, in the form of an idealized plan, intended to anticipate what will happen when apparatus, specimen, and scientist are united in the laboratory. Usually occurring early in the design process, the plan provides engineers with a conceptual vision of the precise ways in which the material devices unite with the specimen, and the expected response at the phenomenal level. This plan functions as a kind of thought experiment for engineers, one in which a conceptual vision is realized materially in the performance of the experiment.

We often read that thought experiments are confined to the realm of physics.[5] But engineers of modern instruments are readily engaged in thought experiments, based on background theories spanning a wide range of disciplines and conveyed through diagrams, maps, charts, or descriptive language. In a sense, the background theories are put to a test by the instrument's performance during an experiment. What is the source for designing these thought experiments? The answer can be found in the ability of designers to mimic familiar physical/chemical systems. The cognitive plan is constructed by harnessing powers from some separate realm. The designer "lifts" certain properties from a distant setting to construct an information retrieval system. Consider two versatile techniques: infrared absorption spectroscopy and Raman spectroscopy.

Absorption spectroscopy is commonly used for identification, structure elucidation, and quantification of organic compounds. The material realm includes a radiation source, sample, monochromator, detector, and readout. A beam of electromagnetic radiation is emitted from a source and then passes through a monochromator. The monochromator isolates the radiation from a broad band of wavelengths to a continuous selection of narrow band wavelengths. Radiation then impinges on the sample. Depending on the molecular structure of the sample, various wavelengths of radiation are absorbed, reflected, or transmitted. That part of the radiation that passes through the sample is detected and converted to an electrical signal, comprising an event of the phenomenal realm. The idealized plan provides a conceptual replication of the various actions and reactions, based on background principles of the physical sciences.

The success of this technique rests on the following principle: if a specimen absorbs a certain wavelength of light (the wavelength corresponding to a particular energy), then that absorbed energy must be exactly the same as the energy required for some specific internal change in the molecule or atom. Any of the remaining energies in the light spectrum that do not match that change are "ignored" by the substance, and these energies are then reflected or transmitted. The absorbed light energy causes such changes as atomic and molecular vibration, rotation, and electron excitation.

The absorption of ultraviolet or visible radiation by a molecular species *M* requires two steps. First, the species *M* is excited electronically as a result of its reaction to the photon. The lifetime of the excited species is brief, from 10^{-8} to 10^{-9} seconds. Absorption of *UV* or visible radiation generally results from excitation of the bonding electrons of the species. This is possible because organic compounds contain valence electrons that can be excited to higher energy levels. Second, this excited state is terminated by certain relaxation processes, such as the conversion of the excitation energy to heat. Because the wavelengths of absorption peaks can be correlated with the types of bonds that exist in the species under study, this two-step process is often used to identify functional groups in a molecule. Thus, the idealized plan underlying absorption spectroscopy includes, or at least assumes, an understanding of a causal mechanism for generating experimental phenomena (Skoog & Leary, 1992, chap. 12).

For another experimental technique, with a different idealized plan for information retrieval, consider Raman spectroscopy. After a sample is irradiated from a high-intensity monochromatic source, the radiation that is produced undergoes elastic scatter, which is known as Rayleigh scattering. But a small fraction of radiation results from inelastic scattering, producing phenomena known as Raman effects. These phenomena are extremely weak, with the detected intensities being several orders of magnitude less than the intensity of energy from its source.

Raman effects occur because a molecule undergoes vibrational transitions, usually from the ground state to the first vibrational energy level. This vibrational transition involves a net gain in energy, similar to the absorption of photons experienced during absorption spectroscopy. For Raman scattering, the electrons distributed around a bond in a molecule are momentarily distorted, so that the molecule is temporarily polarized. For the molecular vibration to be Raman active, as it were, there must be a net change in the bond polarizability during the vibration. As the bond returns to its ground state, radiation is emitted. Thus, the Raman spectrum is an indirect measure of the vibrational transition based on inelastic scattering of radiation (Coates, 1997, pp. 398–404).

Thus, from this discussion of analytic instruments, we find that the conception of instrumentation advanced by empiricist-oriented philosophers is flawed. For empiricists, instruments are designed to expand the limited sensory capacities, so that the evidence is validated by the same empiricist criteria appropriate for knowledge of middle-sized objects. On this view, the success of every experiment rests on the ability of scientists to remove, or at least minimize, the influence of human constructions in the production of data. The evidence should be cleansed of the biasing effects of equipment, practices, and theories. According to Robert Ackermann, the goal of a scientific experiment is to break the line of influence from (subjective) theory to (intersubjective) facts, so that, once the instrumental data are authenticated, the experimenter can remove from consideration the material conditions and laboratory practices (Ackermann, 1985, p. 128).

In contrast to this conception, instrumental data are rarely reducible to the content of one's perceptual experiences (Bogen & Woodward, 1988). The past dependency on visual data, such as the yellow color of a flame, gives way in modern devices to discursive readouts. Typically, the experimenters read graphic displays, digital messages, or coded charts directly from the inscriptions at the readout. The computer-controlled video display and the common printer/plotter employ language that is accessible to the trained technician. The cognitive skills underlying such acts cannot reduce to a collection of sensory observations. The warrant for such data inscription is conditionalized by the instrumental settings, the specifications of the experiment, and the actions of the experimenter.

Are Experimental Phenomena Real or Manufactured?

Again, when analytical instruments are used, the experimenter gets no closer to a specimen's microscopic properties than the experimental phenomena crafted in the laboratory. But how can such research techniques provide knowledge of independently existing atoms, molecules, and complexes? For Latour and Woolgar, such phenomena are pure artifacts that "*are thoroughly constituted by* the material setting of the laboratory" (Latour & Woolgar, 1979, p. 64; italics theirs). The participants' pronouncements that such artifacts are objectively real is completely misguided. Elsewhere, Latour writes that the attempt to win a scientific dispute by appealing to independently existing entities reflects a tactic of rhetoric rather than a rational use of evidence. The settlement of a controversy over experimental research is not the result of nature's brute existence, but is the cause of nature's representation (Latour, 1987, p. 258). The winner of the dispute has the "authority" to define nature in ways that sustain the victory, based on a complex web of experimental techniques, theoretical ideas, laboratory apparatus, and readable marks (Latour, 1987, p. 88).

But this reading of experimental phenomena as laboratory artifacts, rather than independently existing events, reflects a false dichotomy. As I argue in the immediately following discussion, these events are *both* artifacts of the material conditions of the laboratory *and* real-world occurrences that provide information about a specimen.

Why are experimental phenomena artifacts? The answer centers on identifying a mode of production for bringing some event into existence. An artifact is a product that is generated from skilled manipulation of certain material. This production is often achieved by transforming certain substances from one chemical/physical state to another. For example, the artisan, the pharmacist, and the metallurgist generate new products through various transforming techniques such as decomposing and then synthesizing material. Similarly, when modern spectrometers are used to generate microscopic phenomena, the specimen undergoes transformations through laboratory manipulations (Hacking, 1983). The resulting sequence of actions and reactions is manufactured artificially for the purpose of generating signals. Science discovers because it invents (Lelas, 1993, p. 440).

Why are experimental phenomena real? The answer does not rest with the kind of production process for bringing events into existence. The criterion for a real event in laboratory research centers on the ability of experimenters to reproduce the event in distinct environments. An experimenter is constantly moving back and forth from the actions in the laboratory to some other realm, deciding whether a particular result is a contrived effect of the local environment or something reproducible. The real event can reoccur beyond the material conditions of a particular laboratory, as if this raises the event's status from its immanence in the laboratory environment to a potentially recurring state. The talk of measuring dials, the sources of radiation, and operations of the devices is replaced by talk of reproducible phenomena.[6] Of course, events themselves are never repeated perfectly, because no two events are identical in all respects. A reproduction must always be defined with respect to certain attributes or properties. To claim that the absorption of radiation of particular wavelengths by a chemical compound can be reproduced implies that select attributes of such an absorption process can be realized in another laboratory setting. The experimenter has the responsibility to identify the experimental conditions that must be repeated for such attributes to appear in another experiment. Thus, the reproducibility of an event is definable at the level of properties.

The reproducibility of an effect requires the ability to exploit a specimen's causal properties. Again, for both an absorption spectrometer and a Raman spectrometer, a signal is produced from a complex system in which a specimen reacts to instrumental agitations. Such a reaction is caused by an internal change of the compound's dynamic state. As a specimen in laboratory research, a chemical compound is characterized by its capacities for change.[7] A specimen is known by its tendencies to permit interference from radiation and to react to certain manipulations, leading to the causal production of instrumentally detectable events. In Raman spectroscopy, for example, a specimen is a system of causal properties, which, under appropriate laboratory conditions, produce an inelastic scattering of radiation, known as the Raman effect.

The notion of causal process is pivotal for our understanding of an instrument's signal generator.[8] In this context, a Humean notion of causation, as the constant conjunction of one type of event with another type of event, is inadequate. Raman scattering is not readily reducible to a succession of discrete events, but requires a continuous propagation of causal influence. Wesley Salmon's (1994) notion of causation is promising in this context. For Salmon, the connection between cause and effect is conceived as a continuous process, analogous to a filament or thread, but not a chain. Causation is understood as a network of interactions and processes. Causal interactions are events that occurred prior to the event-to-be explained. So, the events preceding the collision of two billiard balls, or the absorption of a photon by a chemical compound are causal interactions. A connection between the (prior) events and the explained event is provided by causal processes. A causal process is an agent of transmission in which structures are extended to various spatio-temporal regions. Such a transmission implies a kind of "consistency" of characteristics in the form of a transfer of structure. For example, the transmission of light from one place to another is a causal process.

Whether or not this conception of causal process can withstand objections by critics (Dowe, 1992), I believe that Salmon's conception of causal process underlies the kind of production of signals in chemical instrumentation. A causal process associated with a signal generator can be understood to include a transmission of energy from the source to the detected signal.

So, whenever a modern spectrometer is used in chemical research, the probative powers of the instrument are united with the specimen's causal capacities. Both instrument and specimen are characterized by their tendencies to generate change. Again, an analytical instrument has powers to produce certain states that can be converted to information. The specimen functions in laboratory research as a system of dynamic properties associated with the exchange of energy. The unity of instrument and specimen is localized in the signal generator, where the specimen's causal capacities to produce phenomenon are activated by an instrument's manipulative powers.

Virtual Witnessing

One might object to the reproducibility criterion of reality just presented on the grounds that, in practice, experimental findings are often accepted without repeating the experiment. Financial, institutional, and technological pressures frequently impose insurmountable obstacles to the ability of scientists to reproduce an experiment. Occasionally, experiments are repeated when the community of experimenters are faced with stunning results. In other cases, experiments are reproduced so as to improve apparatus, augment data, or refine instrumental techniques. But typically, no one actually repeats an experiment (Hacking, 1983, p. 231). And yet after the results of a study are published in journals or disseminated through conferences, acceptance is quite common. This raises serious questions about research practices: Can the experimental findings of a particular study be validated without the actual reproduction of results by the scientific community? If so, at what point in the process of validation does the community of scientists participate in the public hearing of these results? Is public participation removable from the process of validation?

The answers to these questions can be found in a process called virtual witnessing. According to Shapin and Schaffer (1985), virtual witnessing captures the kind of communal evaluation process used to assess Boyle's experiments. For Boyle, the multiplication of witnesses to an experiment was needed as testimony for a true state of affairs. This could be achieved in three ways. First, the experiment could be performed in a social space, a practice that was common in seventeenth- and eighteenth-century England. But this became rather cumbersome because of the use of large equipment and, of course, could not reach a wide audience. Replication of the experiments was a second way to multiply witnesses. By repeating the experiment, and disseminating the experimental findings through reports and letter, Boyle ensured distant but direct witnesses. But actual replication was rarely achieved, and so replication was impractical. A third and common means of testimony was made possible through virtual witnessing. From this method the experimenter may invite an engraver to create visual images of the experimental scene, such as the schematized line drawings used to imitate Boyle's experiments with the air pump. The viewer of such observable impressions was encouraged to generate mental images as a kind of conceptual simulation of the experiment and to critically evaluate the experiment. Through virtual witnessing, readers may endorse the methodology and accept the findings without actually reproducing the experiment (Shapin & Schaffer, 1985, pp. 60–62).

Returning to the contemporary scene, some readers of an experimental study typically become virtual witnesses by engaging in the critical assessment of the laboratory report. As a tool of persuasion, the report provides the reader with a cognitive vision of the laboratory events in ways that recommend endorsement. The reader is invited to critically evaluate the results, the methodology, and the original plan of action.

As the virtual witness reads about the experiments, he or she formulates a thought experiment designed to replicate the experimental environment. The thought experiment provides a visualization, through various mental manipulations, of some theoretically possible states of affairs. The effectiveness of virtual (vicarious) witnessing involves persuading readers that they could reproduce in practice the same processes and would get the same correspondence of concepts to perceptions (Gooding, 1990). Persuasion rests on more than simply acceptance of data, for it necessitates a comprehensive endorsement of the experiment. The reader is expected to follow a narrative that selects and idealizes certain steps of a procedure (Gooding, 1990). This narrative transports the reader from the actual to the possible by reenacting the significant features of the experiment and by focusing on the instrument's design, material apparatus, and microscopic phenomena. Experimental results are evaluated as plausible or implausible products of the entire experimental design, methodology, and technical "know how."

One might object that "witnessing" is the wrong metaphor because the experiment is not being observed, either actually or virtually. On this objection, the metaphor "reader as jury member" provides a more compelling description; the reader of the report is invited to evaluate the described events by appeal to direct testimony of witnesses. The reader is judging the experiment, as a jury member assesses evidence.

But the "jury member" analogy fails to capture a vital component of evaluation: the report is directed toward highly trained participants in laboratory research. Even if the reader never performed the experiment described in the report, the reader typically appeals to direct participation in experiments that are similar to those described. At least privately, the reader rehearses the kind of malfunctions of apparatus, mistakes of implementation, and interfering effects that plagued past experiments. Are such dangers relevant to the present experiment? If so, are they avoidable? The answers to these questions require a command of the subject matter that is usually reserved for expert witnesses. The metaphor of "reader as jury member" fails to capture such a requirement of the reader. In the context of the legal system in the United States, jury members are rarely trained in the topics examined by expert witnesses.[9] So, because a virtual witness is usually well versed in the techniques of laboratory research, the notion of reader as virtual (expert) witness captures the evaluation process better than the "reader as jury member" metaphor.

The virtual witness evaluates instrumental techniques based in part on whether the experiment "mimics" processes found outside of the laboratory. When the instrument is used, the experimental phenomenon simulates various events found outside of the laboratory, as if the entities "in captivity" are modeled on "nature in the wild." Although such phenomena are realized only under material conditions, the laboratory environment constitutes an approximation, rather than an exact duplication, of "wordly" processes. The (real) phenomenon is presumably similar, but never identical, to one detected under the same experimental conditions. But, as I argue in the preceding discussion, an experimental phenomenon cannot be described in isolation from the various conditions of its generation, including elements of the material and abstract realms. Each experimental phenomenon should be understood at the level of an experimental system. Known similarities between microscopic events are grounded on the analogical connections drawn between two experimental systems. In this respect, one system functions as an analogical prototype for characterizing another.

The Mimetic Function of Instruments

I argue here that a virtual witness evaluates the current laboratory results by comparing the experiment to those in his/her repertoire of past research. This comparison rests on exploring analogical connections between past and present experiments. Moreover, the ability to establish and evaluate analogical connections between disparate experimental phenomena is commonplace in another context—that is, during the design of new instruments. The discovery of new experimental techniques often rests on articulating and exploring similarities to known phenomena. In this way, the power of an instrument to retrieve information about the specimen is established by mimicking certain physical or chemical processes that are familiar to scientists at the time of the instrument's development.

For nineteenth-century spectroscopy, one technique for generating spectral lines was revealed through analogies to acoustics. The spectral lines of an element were conceived as the harmonics of molecular or atomic vibrations (McGucken, 1969). For twentieth-century research using absorption spectrometers, the molecular structure is characterized as a system of point masses joined by vibrating bonds, analogous to simple harmonic motion of two masses connected by a string. The sample's reaction to the probing action of radiation takes the form of discrete photons, similar to impulses of light. In this context, models of electromagnetic radiation were extended by analogy to the development of absorption spectrometers (Rothbart & Slayden, 1994).

Returning to the case of Raman spectroscopy, Raman developed his instrumental techniques by exploring analogical associations between certain unexplained phenomena of scattered light and the Compton effect. By 1923 experimenters discovered that light could be scattered in ways that were not explainable by the classical electromagnetic theory of light because the scattered light does not have the same wavelength as the incident radiation. Raman proposed that the puzzling phenomenon could be understood as the optical analogue to the Compton effect (Raman, 1965). The wavelength of radiation could be degraded so that the scattered radiation appears with diminished frequency. Raman writes:

> In interpreting the observed phenomena [of molecular scattering of light], the analogy with the Compton effect was adopted as the guiding principle. The work of Compton had gained general acceptance for the idea that the scattering of radiation is a unitary process in which the conservation principles hold good. [So], . . . if the scattering particle gains any energy during the encounter with the quantum, the latter is deprived of energy to the same extent, and accordingly appears after scattering as a radiation of diminished frequency. (Raman, 1965, pp. 272–273)

Such methods opened up a new field of experimental research concerning the structure of matter.

In another case, the ability to exploit the mimetic function of instruments had a profound effect on the development of the Wilson cloud chamber, as documented in an illuminating study of Galison and Assmus (hereafter "G&A") (1989). Actually, Wilson (1965) designed his cloud chamber from the confluence of two conflicting research traditions, one from late-nineteenth-century meteorologists, which G&A label Mimeticists, and the other from the Cavendish physicists during the same era. The Mimeticists devised techniques to duplicate phenomena of nature, such as cyclones, glaciers, and cloud formation. For example, to study the dirty air of the nineteenth-century industrial cities of England, John Aitken designed a dust chamber by reproducing the reaction of the cloud of condensed vapor to dust. Thus, he imitated the "natural" processes of condensation by removing all artificial influences of the laboratory setting (Galison & Assmus, 1989, p. 231).

When Wilson directed his energies toward the reproduction of thunderstorms, coronae, and atmospheric electricity, he abandoned the Mimetic tradition in favor of the Cavendish style of analytic physics, reflecting his Cambridge education. This transition constitutes a "profound shift in material culture and conceptual structure," as G&A write (Galison & Assmus, 1989, p. 245). Through techniques developed by Thomson, nature was not duplicated in the laboratory, but artificially divided into its basic elements (Galison & Assmus, 1989, p. 265). Although Wilson was not engaged in ion physics per se, he moved freely back and forth between questions of ionic charge and the nature of atmospheric events (Galison & Assmus, 1989, p. 257). Wilson created artificial laboratory conditions in his study of atmospheric electricity by using air that was specially treated through various filtering devices, avoiding all tracing of the "dusty" air.

To simulate the formation of clouds, Wilson constructed a device in which samples of moist air were expanded suddenly and then supersaturated. Cloudlike features could be reproduced in the chamber, and the trajectories of the water droplets could be photographed. This can be achieved through the following steps: first, the production of the necessary supersaturation of air by sudden expansion of the gas; second, the passage of the ionizing particles through the supersaturated gas, and third, the illumination of the cloud condensed on the ions along the track. The resulting photographs agreed remarkably with Bragg's ideal picture of paths of alpha rays, as he claimed (Wilson, 1965).

However, this contrast between the Mimetic and Cavendish traditions is somewhat exaggerated in the interpretation by G&A. What is lost in their rendition of the mimetic function of instruments is the importance of creating a manufactured environment that *approximates*, but does not *duplicate*, the atmosphere. No exact imitation can be attained. For example, Aitken constructed his dust chamber to simulate dustlike particles in the form of water droplets in supersaturated air. Similarly, Wilson did not exactly produce clouds, rain, or fog, but he constructed an experimental technique to approximate certain atmospheric phenomena. Cloudlike properties were simulated in the production of moist air from the expansion and supersaturation of dust-free air. Wilson's work was eventually integrated into Cavendish research, prompting Thomson to suggest ionic explanations of cloud-chamber phenomena. The rainlike properties are produced when the supersaturated air causes negative ions to grow into visible drops.

The claim by G&A that only one of these traditions developed techniques to imitate real-world conditions is quite misleading. Both traditions used the cloud chamber to manufacture an artificial environment that approximates known phenomena. For the Cavendish physicists, the cloud chamber became one of the defining instruments of particle physics, precisely because the laboratory phenomena were modeled on the movement of the charged particles. The knotty clouds blended into the tracks of alpha particles and the "threadlike" clouds simulate beta-particle trajectories (Galison & Assmus, 1989, p. 268). Of course, G&A are correct that these physicists aspired to dissect nature into its fundamental components, reflecting the long tradition of the corpuscular conception of matter.

So, scientists from both traditions devised techniques to mimic, but not imitate perfectly, phenomena found outside of the laboratory. The difference between these two traditions centers on the prototypes used as the source of instrumental techniques. The nineteenth-century meteorologists used observable atmospheric conditions as the prototype for the dust chamber, while Thomson and his colleagues used the unobservable qualities of fundamental particles to define the electron.

On the Causes of Noise

One could object that the preceding criterion for objective reality of a phenomenon, based on the reproducibility of an effect in other experiments, is utopian, because from the experimenter's perspective all the detectable effects are indistinguishable from the manufactured conditions of the experiment. How can we be reasonably sure that these effects are not the result of extraneous interference from external sources? This objection can be overcome by exploring the problems associated with the occurrence of noise. To my knowledge, in the vast literature surrounding the realism/antirealism debate, there is no extensive study of the topic of noise. The importance of such a study is suggested next.

The category of noise can be divided by whether its source is internal or external to the desired signal. External sources of noise may originate from almost any electrical device in close proximity, emitting electromagnetic waves as a byproduct of their operation. Even the AC power lines may corrupt the desired signal. Of course, atmospheric conditions of thunderstorms may emit radiation of up to 30 MHz. The sun is a source of low-level noise that begins at approximately 8 MHz. Notice that once an external source of contamination has been detected during a particular experiment, that source must be included in the experimental foreground and the boundaries of the experimental space have shifted. The distinction between foreground conditions of the experiment and background events that remain relatively stable is subject to change during the experiment. The internal/external division is a moving limit during the life of the experiment.

Internal sources of noise arise from any power-dissipating device. Of course, every laboratory apparatus is a potential source of contamination. For example, Johnson noise arises because the electrons that carry an electric current always have a thermal motion, causing small fluctuating voltages across any resistor in the electrical circuit. One source of environmental noise is the "pick-up" of a 50-Hz signal due to the main supply in the laboratory.

Noise can affect the system's performance in three areas. First, the reader of the message might misunderstand the original signal, or, more seriously, the signal may be unintelligible. Second, noise can cause the receiving system to malfunction, regardless of the meaning of the data bits. Third, the efficiency of the communication system is threatened, requiring, for example, extra transmitting power to the original signal (Schweber, 1991, pp. 64–65).

Of course, such contaminations are often difficult to detect. From the perspective of electronic circuitry per se, there is no qualitative difference between the desired signal and the noise signal; they are each defined as a variation of voltage with time. Not immediately evident in the electrostatic properties of the signal, the technicians must explore the causal agents that underlie the signal's production. Although information from the desired signal is causally grounded on the union of apparatus and specimen, the noise signal is generated from a source that is extraneous to this system. Consequently, the signal generator is not a completely closed system. Because both types of signals are detectable effects of an underlying process for generating signals, the technician engages in a kind of causal hypothesis testing in the identification of noise.

The very identification of some contaminating agent leads experimenters to explore analogical connections to processes found outside of the laboratory setting. If the experimental environment is known to behave in ways that are sufficiently similar to a known physical system, then experimenters construct certain material to block the interference. The system of forces, entities, and processes is attributed by analogy as the possible source of extraneous signal.

One common noise-blocking technique involves electrostatic shielding, in which some of the wires in the instrument are surrounded by a conducting material. Shielding is designed to minimize the effects of uninvited intrusion that arise from the instrument circuitry. Electromagnetic radiation is absorbed by the shield rather than by the enclosed conductors. The conductors are wrapped with aluminum or copper foil, with a drain wire in electrical contact with the foil, providing almost perfect covering.

The topic of noise has a bearing on arguments for scientific realism, necessitating a more nuanced interpretation of our commitments to reality than one often finds in the philosophy literature. For example, based on his manipulability criterion, Hacking proposes that our commitments to real-world processes are revealed in the use of certain entities as manipulating tools for the study of other, more hypothetical, processes. We know that electrons, for example, are real, because experimenters exploit the causal powers of such entities in the agitation of the specimen's internal structure (Hacking, 1983, chap. 16). But Hacking's manipulability criterion is too narrow. Our beliefs about existence can be revealed in our awareness of contaminants to the desired signal. Although the sources of noise generate unintended obstacles to information retrieval, these sources have a causal influence on the character of the experimental phenomena. Such an influence is neither fabricated nor entirely manufactured. In this way real-world processes are revealed by the damage inflicted on the signal's creation and evolution during experimental inquiry.

Substance Portrayed and Reduced

Again, during an experiment using analytical instruments, chemical substance, and experimental instrument are united in a complex system of capacities and powers. Such research constitutes a positive expression of our beliefs about existence, of both instrument and specimen. Modern instrumental techniques do not obediently serve the ruling order of nature but presuppose a conception of substance as a condition for laboratory research. Chemical substance is revealed through the techniques designed to generate experimental evidence.[10] Once selected for investigation, the specimen is subjected to material preparation in which certain properties are isolated and put on display. In this respect, substance is characterized functionally by the properties, capacities, and states that are necessary for the performance of the experiment. This characterization suggests an experimental reduction underlying chemical instrumentation. Again, when modern spectrometers are used, the specimen is reduced to its dynamic properties associated with the transformation of energy states. But no global reduction of chemical substance can be sustained because the characterization of the specimen is highly responsive to techniques of instrumentation.

Notes

A shorter version of this article was read at the Second Meeting of the International Society for the Philosophy of Chemistry, Sidney Sussex College, Cambridge, UK, 1998. I would like to thank the participants of that meeting for many comments and suggestions. Joseph Earley and Eric Scerri were particularly generous and helpful in their recommendations. I also thank the editors of this volume for their meticulous reading and constructive comments of the first draft of this chapter.

1. I first heard the term *epistemic technology* from a discussion with Davis Baird.
2. See Harré (1986, chap. 15) for a discussion of powers.
3. In an illuminating study, Baird and Nordman define a class of instruments by their capacity to generate phenomena through various "striking" actions (Baird & Nordman, 1994).
4. See Bruno Latour's discussion of instruments as inscription devices (Latour, 1987, chap. 2).
5. "In chemistry, for example, I can't find any [thought experiments] at all," writes James Robert Brown (1991, p. 31).
6. See Radder (1996, chaps. 2 and 4) for an exploration of different kinds of reproducibility in science.

7. In this context, I appeal to a proposal by Ewa Zielonacka-Lis: a chemical compound is defined as a causal process (Zielonacka-Lis, 1998).

8. The concept of causation is often included in philosophical discussions of experiments (Cartwright, 1983; Hacking, 1983; Harré, 1986).

9. In fact, the procedure of jury selection in the U.S. system of civil law, a procedure known as *voir dire*, typically leads to the dismissal of those candidates for jury duty who profess an expertise in a technical subject relevant to the case. Such dismissals occur because one of the attorneys, either for the defense or the plaintiff, believes that these candidates are potential threats to the authority of certain expert witnesses who may be called in the case.

10. For Kant, the possibility of science requires that matter be defined dynamically as a summation of the forces of attraction and repulsion. Without succumbing to Kant's a prior categories of the mind, we face the following Kantian task in philosophy of chemistry: what are the metaphysical conditions for the possibility of chemical research (Rothbart & Scherer, 1997).

References

Ackermann, R. 1985. *Data, Instruments and Theory*. Princeton, NJ: Princeton University Press.

Baird, D., & Nordmann, A. 1994. "Facts-Well-Put." *British Journal for the Philosophy of Science*, 45:37–77.

Bogen, J., & Woodward, J. 1988. "Saving the Phenomena." *Philosophical Review*, 97:303–352.

Brown, J. R. 1991. *The Laboratory of the Mind*. New York: Routledge.

Cartwright, N. 1983. *How the Laws of Physics Lie*. Oxford: Clarendon Press.

Coates, J. 1997. "Vibrational Spectroscopy: Instrumentation for Infrared and Raman Spectroscopy." In G. Ewing, ed. *Analytical Instrumentation Handbook* (2nd ed.) (pp. 393–555). New York: Dekker.

Dowe, P. 1992. "Wesley Salmon's Process Theory of Causality and the Conserved Quantity Theory." *Philosophy of Science*, 59:195–216.

Galison, P., & Assmus, A. 1989. "Artificial Clouds and Real Particles." In D. Gooding, T. Pinch, and S. Schaffer, eds. *The Uses of Experiment* (pp. 225–274). Cambridge: Cambridge University Press.

Gooding, D. 1990. *Experiment and the Making of Meaning*. Dordrecht, Kluwer Academic.

Hacking, I. 1983. *Representing and Intervening*. Cambridge: Cambridge University Press.

Harré, R. 1986. *Varieties of Realism*. Oxford: Basil Blackwell.

Latour, B. 1987. *Science in Action*. Cambridge: Harvard University Press.

Latour, B., & Woolgar, S. 1979. *Laboratory Life: The Construction of Scientific Facts*. Princeton, NJ: Princeton University Press.

Lelas, S. 1993. "Science as Technology." *British Journal for the Philosophy of Science*, 44:423–442.

McGucken, W. 1969. *Nineteenth-Century Spectroscopy*. Baltimore: Johns Hopkins University Press.

Radder, H. 1996. *In and About the World: Philosophical Studies of Science and Technology*. Albany, NY: SUNY Press.

Raman, C. V. 1965. "The Molecular Scattering of Light." In *Nobel Lectures: Including Presentation Speeches and Laureates' Biographies: Physics 1922–1941 (pp. 267–275). Amsterdam: Elsevier.*

Rothbart, D., & Scherer, I. 1997. "Kant's *Critique of Judgment* and the Scientific Investigation of Matter." *Hyle: An International Journal for the Philosophy of Chemistry*, 3:65–80.

Rothbart, D. & Slayden, S. 1994. "The Epistemology of a Spectrometer." *Philosophy of Science*, 61:25–38.

Salmon, W. 1994. "Causality without Counterfactuals." *Philosophy of Science*, 61:297–312.

Schweber, W. 1991. *Electronic Communications Systems*. Englewood Cliffs, NJ: Prentice Hall.

Shapin, S., & Schaffer, S. 1985. *Leviathan and the Air-Pump*. Princeton, NJ: Princeton University Press.

Skoog, D., & J. Leary. 1992. *Principles of Instrumental Analysis* (4th ed.). Fort Worth, TX: Saunders.

Wilson, C. T. R. 1965. "On the Cloud Method of Making Visible Ions and the Tracks of Ionizing Particles." In *Nobel Lectures: Including Presentation Speeches and Laureates' Biographies: Physics 1922–1941* (pp. 194–214). Amsterdam: Elsevier.

Zielonacka-Lis. E. 1998. "Some Remarks on the Specificity of Scientific Explanation in Chemistry." Presentation at the Second International Society for the Philosophy of Chemistry, 3–7 August 1998, Sidney Sussex College, U. K.

6

Analytical Instrumentation and Instrumental Objectivity

DAVIS BAIRD

Instrumental Objectivity: Analytical Chemistry and Beyond

The bulk of this chapter is concerned with analytical chemistry during and for the first decade after World War II. At this time analytical chemistry underwent a radical change, which can most easily be characterized as a shift from wet chemistry to instrumental methods (Baird, 1993). Although this transformation of analytical chemistry may interest readers of this volume for a variety of reasons, I focus here on a shift in the concept and practice of a kind of objectivity.

Objectivity is one of those concepts with generally positive connotations, but whose exact characterization proves elusive. A dictionary tells us that, as an adjective, *objective* applies to that which has "actual existence or reality." *Objective observation* is "based on observable phenomena" and "uninfluenced by emotions or personal prejudices" (*American Heritage Dictionary*, 1993, p. 940). Objectivity, it would seem, either sits next to truth or defines the right route to truth. What emerges here, however, is a more complicated concept—a concept with a history that serves various agendas through suitable shades of meaning and marriages of convenience with other concepts.

Instrumental Methods

The development of instrumental methods in analytical chemistry made possible fast, precise, and accurate analyses of a wide variety of important substances. Instrumental methods changed forever the metals industries, medical diagnosis, oil analysis, and forensic analytical chemistry, to mention a few highlights. In a very real sense, these developments in analytical chemistry made contemporary science and technology possible by opening up vast new continents of information about the world, which could be gained relatively easily and applied toward technological and/or scientific ends.

I argue here that these developments in analytical chemistry established a paradigm for one kind of concept of objectivity. Ralph Müller, who will play a central role in this discussion, wrote in the January 1947 issue of the *Analytical Edition of Industrial and Engineering Chemistry* (the journal that subsequently became *Analytical Chemistry*):

> The true instrumental method of analysis requires no reduction of data to normal pressure and temperature, no corrections or computations, no reference to correction factors nor interpolation on nomographic charts. It indicates the desired information directly on a dial or counter and if it is desired to have the answer printed on paper—that can be had for the asking. (Müller, 1947, pp. 23A–24A)

For Müller, objective methods were instrumental methods. Müller sought methods whereby one would insert an unknown in an instrument, push a button, and get the answer. We might best characterize "instrumental objectivity" as "push-button objectivity."

It is important to note at the outset that tying work by analytical chemists specifically to any concept of objectivity is difficult. Their work is in their instruments and methods; objectivity is a meta-concept, which, I claim, plays a role in directing their work, but which need not be articulated by them in the process. Only rarely have I run across claims about objectivity outside of editorials. I analyze in some detail one striking exception to this absence in what follows.

Two Examples Reaching beyond Analytical Chemistry

Objectivity transcends analytical chemistry per se, and, consequently, part of what I am arguing for here is the central role of analytical chemistry in very broad changes in how we relate to our world. Consider two examples.

In 1990 then-President Bush signed legislation requiring the now familiar food labels that appear on nearly all packaged foods.[1] We immediately know how many grams of fat—saturated, unsaturated, and total—each standardized serving contains. This information, brought to us by the chemical analysis of foodstuffs, promotes quick judgments about what to eat: "Seven grams of fat in that cereal? No thanks, I'll have the cereal with three grams of fat." It is as though each package of food was passed through an instrument and "the answer, printed on paper, is ours for the asking."

Here is a different example. One of the first important markets for analytical instrumentation was for medical instrumentation. Consider fetal heart monitors (Hutson & Petrie, 1986; Tallon, 1994; Benfield, 1995). Not too long ago a fetus's progress during birth was followed with a simple stethoscope-like device called a fetoscope. These "low-tech" devices require the physical presence of the nurse, midwife, or doctor to operate. The fetoscope requires concentration, skill, patience, and a quiet environment (Benfield, 1995, p. 6). Now, in many cases a fetus's progress is followed by an electronic fetal heart monitor. These "high-tech" devices employ ultrasound technology to pick up data on the fetus's heart function. This data can be transformed into a variety of outputs, including "faux" heartbeat sounds, CRT representations of heart function, and paper printouts. Such a device can be set up so that it will only output an alarm when it "interprets" the data it picks up as a fetal heart malfunction. As long as the device is properly attached to the pregnant mother, no person needs to be present along with mother and fetus to operate the device. Ultrasound imaging devices can present pictures of the fetus in utero. These, I have been told, are "the gold-standard." "The gold standard" is perhaps too apt, for in addition to capturing the idea that ultrasound provides the best data on the state of the fetus, ultrasound is much more expensive.

Instrumental Objectivity

Ultrasound and food labels share a kind of objectivity. Information is gathered and digested by external devices—instruments—and represented in a way that promotes quick judgments. With ultrasound instrumentation, no human interpretation of fetoscope sounds is required; fetal heart monitors are not subject to misinterpretation because of a lack of skill, time, or a quiet enough room. With food labels we can—and many do—judge the quality of the food we eat from a matrix of 10 or so numbers. The role of detailed or "considered" human judgment is reduced—although human judgments certainly play a significant role in constructing the instruments themselves. I call this "instrumental objectivity," and I argue that the instrumental transformation of analytical chemistry is central to the increased importance of instrumental objectivity.

Objectivity and Accuracy

Although ultrasound and food labels share a kind of objectivity, accuracy is another matter. Ultrasound images do provide valuable information about fetal development. However, there is no evidence that ultrasound heart monitors produce a more accurate account of the condition of a fetus's heart; fetal heart monitors are subject to a variety of systematic sources of error (Benfield, 1995, p. 9; Tallon, 1994, p. 187). Moreover, no statistically significant difference in medical condition, as judged through, for example, APGAR scores, between babies delivered when a fetal heart monitor was in use and those delivered when a fetoscope was in use has been found (Benfield, 1995, p. 11). It clearly is useful to have some key information about food content readily available. There is also no doubt that healthy eating cannot be reduced to 10 numbers. Already there is a major pitched battle over what foods get to be called "organic," presumably another dimension of healthy eating. At the same time, it is clear that many people will use the 10 numbers from the food labels to "eat right."

Objective Ideal

Such examples could be multiplied many fold. Think of measures of pesticides, toxins in the water, or even at a greater distance from analytical chemistry, the numbers we use to assess the quality of teaching. The point is that we now have a model for an ideal kind of objective analysis—be it of steel alloy composition, fetal heart condition, food quality, or even professorial competence: we should be able to subject the object of analysis to some instrument, the operation of which is relatively simple—ideally, push-button simple—and obtain "the answer." Of course, not everything can accommodate such an ideal, but as an ideal it serves to guide us as we develop and critique methods of analysis.

Human Judgment and Money

A variety of elements make up this ideal. I argue that the two most central elements involve minimizing human judgment and cost efficiency. Instrumental objective methods should be simple to perform—requiring minimal human judgment—and the results should be simple to interpret—again, requiring minimal human judgment. Such simplicity typically comes with a cost, the instruments developed typically are expensive. But this expense can be compensated by the ability of the instruments to perform many analyses in a given period of time. This can reduce the cost per analysis, while simultaneously driving out of business laboratories performing small numbers of analyses. Importantly, these two central elements are intertwined for human judgment is relatively expensive: by simplifying analytical operations, allowing less highly trained personnel to perform analyses, objective instrumental methods decrease the labor cost of analysis.

The Feel of Our World

In a nutshell, I argue that the rise of instrumental methods in analytical chemistry helped promote an ideal for what objective analysis could and should be. This ideal, while variously interpreted and achieved in a very wide variety of contexts reaching far beyond chemistry, has insinuated itself deeply into our thinking about and working in the world. Two key interdependent elements of this ideal are cost efficiency and minimizing the role of human judgment.

The result is a profound change in the how the world "feels," in the "texture of the world." (On "texture" and "feel," see Hacking, 1983; 1987, p. 51.) It is a qualitatively different experience to give birth with an array of electronic monitors. It is a qualitatively different experience to teach when student evaluations—"customer satisfaction survey instruments"—are used to evaluate one's teaching. It is a qualitatively different experience to make steel "by the numbers," numbers produced by analytical instrumentation. Push-button instrumental objectivity has changed our world.

Analyzing Objectivity

I write of "a kind" of objectivity, and I write about a "transformation" in the "concept and practice" of a kind of objectivity. My language betrays three key assumptions behind my work. First, I deny there is a single unified concept of objectivity. Second, I deny that there is an eternal meaning for (any kind of) objectivity. Third, I deny that a conceptual analysis alone can be sufficient for understanding (any kind of) objectivity. Each denial has its correlative positive side, which I explore in turn here.

Kinds of Objectivity

At its core, objectivity is supposed to be a guarantor of truth and freedom from ideology. At this most basic level of analysis, objectivity already concerns both results and methods to obtain results. Thus, we may speak of an objective *result* because it is accurate and/or stated in a way we consider to be free of human bias; alternatively, we may speak of an objective *method* because it is specifically designed to avoid human bias in its application. Herein is a general problem. In evaluating a method, be it an ethical rule taken from a rule utilitarian, a statistical method taken from classical Neyman/Pearson statistics, or, indeed, some method of chemical analysis, one confronts cases where the method is good in that on most uses it would produce accurate results, while on rare uses it would produce an inaccurate result.[2] So we can have objective methods that produce (on occasion) results that are not objective. The concept of objectivity is immediately susceptible to contradiction.

The conceptual analyst then requires distinctions: the result is objective in the sense that . . . ; the method is objective in the sense that . . . ; and so on. A special two-volume issue of *Annals of Scholarship* brought together 14 articles by authors from a variety of disciplines to bring light to all the concepts of objectivity (Meg. 11 1991–1992). Allan Megill, in his introduction, analyzed four basic senses to objectivity: absolute, disciplinary, dialectical, and procedural objectivity (1991). Instrumental objectivity does not easily fit into any of Megill's categories, although it comes closest to procedural objectivity:

> Yet the governing metaphor of procedural objectivity is not visual, as in absolute objectivity: it does not offer us a "view." Nor does it stress action, as dialectical objectivity does. Rather, its governing metaphor is tactile, in the negative sense of "hands off!" Its motto might well be "untouched by human hands." (Megill, 1991, p. 310)

Megill references Theodore Porter's work on objectivity in the service of statistics and public administration (Gigerenzer et. al., 1989, chap. 7; Porter, 1992). As becomes evident in the following discussion, there is a close relationship between statistics and instrumental objectivity.

Concept with a History

This key epistemological concept has a history, and I document here a small part of its history. I most closely follow the works of Steve Shapin and Simon Shaffer (1985), Lorraine Daston (1988, 1991, 1992), and particularly Daston's joint work with Peter Galison (1992). These authors document a progressive removal of human judgment from—at least one kind of—objectivity. Shapin and Shaffer show how Robert Boyle sought to "let the air-pump speak;" the air-pump's voice was to be preferred to the voices of people with their various contentious metaphysical interests. Daston and Galison document how, through the nineteenth century, there was progressive removal of human judgment in the production of images used in science; mechanically produced images are more objective than those produced with the aid of human judgment and artistic skill.

I document further developments along this same trajectory. At the middle of the twentieth century, analytical chemists identified objective methods of analysis with instrumental methods. At midcentury, the paradigm for a chemical analysis was to insert a sample of an unknown into a device, press a button, and have the device say of what the sample consisted. Previously, a wide variety of—"subjective"—human judgments were necessary: Has the reaction gone to completion? What are the observable properties of the residue? Does the taper burn more brilliantly in this gas? And so on.

Concept and Practice

Conceptual analysis provides tidy descriptions free from vagueness and contradiction. But people rarely are so careful when they talk about their practices. For this reason, a concept may be used in quite inconsistent ways or in ways that conflate seemingly distinct concepts. In the case at hand, it is vital to recognize that analytical chemists come to their work from a variety of institutional settings with a commensurate variety of concerns. Consequently, it may not be surprising that analytical chemistry, with its ties to industry, provides insight into how the concept of instrumental objectivity *in practice* ties together values frequently analyzed as distinct. These include values discussed under such headings as "de-skilling," "standardization," "black-boxing," and "cost efficiency." Instrumental objectivity is not simply a matter of accuracy and truth. At the most basic level, as already explained, the reduction in the role of human judgment (a component of accuracy) is inextricably tied to reducing the labor costs of analysis by "deskilling" the role of the analyst and "black-boxing" the instrument performing the analysis.

One could argue that, although it is true that the analytical instruments developed during the 1940s and 1950s were more cost efficient, required less skill to operate, and promoted standardization of data collection and presentation, this does not mean that they are more objective; that is a distinct question. Consequently, there is nothing to learn about the concept of objectivity from noting its—accidental—connection with these other values.

Such a point of view, however, misses a central historical point about the changing concept and practice of this kind of objectivity. I am concerned about understanding how the concept of objectivity has changed with instrumentation. These new instrumental devices provide "the gold standard." What kind of devices do this? Those instruments that remove—de-skill—the human judgment do this. Instruments that increase "analytical throughput" do this. Instruments that shift the expense burden away from people to hardware do this. Instruments that standardize the data do this. All of these values—and others, no doubt—must go together for the instrument to be a candidate for the "gold standard." Objective instrumentation, because it is instrumentation produced at a time when standardization and systemic interconnection are important, must accommodate all of these values. In short, it is our place in history that conflates objectivity with these other values. We may conceptually distinguish them, but doing so prevents us from understanding the concept and practice of instrumental objectivity.

Not a Condemnation of Instrumental Objectivity

It could be tempting to read into this discussion a condemnation of instrumentation and the use of instruments to render useful information about the world, instrumental objectivity. From an ivory tower perspective, objectivity should be the straight ideology-free, subjective whim-free conduit for truth about our world. Accuracy, not cost efficiency, should be its standard. I don't dispute this. Yet, concepts in practice frequently have a way of being more complicated and interesting than the ivory tower perspective would have them. Thus, instead of condemning instrumentation, I am, first of all, concerned with describing, not evaluating, changes in one concept of objectivity. Indeed, far from condemning instrumentation, I find many of the changes brought about by developments in instrumentation, even cost efficiency, to be great achievements, a boon for all. The one serious concern I have is this: with the mechanization of objectivity, there has been an associated devaluing of human judgment as "merely subjective." Much human judgment, no doubt, is subjective, and very valuable and important for that. But much human judgment is objective—or should be understood as such—and it would be a great loss to devalue it.

Analytical Chemistry and the "Big" Scientific Instrumentation Revolution, 1930–1960

Before I proceed to some of the details concerning objectivity I put these details into the context of the changes they were part of. More information on these changes is in Baird, (1993). Here, I summarize.

On a cursory analysis, the development of instrumental methods seems simply to augment wet chemical methods; analysts have new better tools in their arsenal, but nothing "revolutionary" has changed in their business. On the contrary, this change in analytical chemistry shook the field to its foundations and, indeed, as I urge in this chapter, had ramifications well beyond analytical chemistry. Within the field, questions arose concerning (1) the fundamental nature of analysis; (2) analytical chemistry's relation to physics, electronics, engineering, and the rest of chemistry; (3) the roles of industry, academics, and government in shaping the field; (4) scale and capital expenditure; and (5) education for the analyst.

Fundamental Nature of Analysis

Now we have no difficulty in recognizing analytical chemistry as the science concerned with the identification and control of substances: "Analytical chemistry deals with methods for the identification of one or more of the components in a sample of matter and the determination of the relative amounts of each" (Skoog & West 1976, p. 1). From our current perspective, a 1929 definition of an analytical chemist as "a chemist who can quantitatively manufacture pure chemicals" (Williams, 1948, p. 2) seems bizarre. The definition is less puzzling when seen in the changing context of analytical practice. Analyzing an unknown substance by wet chemical methods involved breaking the substance down into component parts that could be directly recognized (by color, state, odor, etc.). In the most advanced treatment of the subject at the turn of the nineteenth century, Wilhelm Ostwald discussed the task of identification. He noted that separation typically must precede recognition, and separation, Ostwald noted, "is naturally much the more difficult of the two" (1895, p. 9). Separating substances—manufacturing pure chemicals—was where the work of the analytical chemist was done.

The advent of instrumental methods changed this. Now, for example, a substance with relatively little preparation could be sparked in an emission spectrograph, and its elemental composition could be read off dials. The work of the analytical chemist came in the design of instrumentation to exploit the physical properties of the substances under investigation and with the administration and use of an array of instruments, each exploiting different physical properties for identification.

Thus, at the most basic level of disciplinary identification, the development of instrumental methods challenged the discipline of analytical chemistry. The immediate result of this was a shift in the relation of the analytical chemist to the variety of other disciplines on which his or her work drew.

Analytical Chemistry's Relations

Wet chemistry depends on chemistry. Spectroscopy, by contrast, depends primarily on optics; quantum physics helps to explain the meaning and source of spectral lines, but spectral analysis through the 1940s was largely an empirical affair. Optical equipment requires mechanical stability, and thus spectroscopy also depends on mechanical engineering. Some chemistry is involved in the development of emulsions for the photographic plates used to make spectrograms, but this was not primarily the problem of the analytical chemical spectroscopist; plates with desirable characteristics were bought "off the shelf." The development of direct reading spectrometers, which replace photographic plates with photosensitive electron multiplier tubes, depends on electronics.

The development of instrumental methods brought analytical chemists to question their fundamental ties to chemistry. Beyond chemical reactions, they had to become knowledgeable in optics, electronics, and mechanical engineering. Controversies ensued over the appropriate way for an analytical chemist to spend research energies. Some claimed—as reported by Ralph Müller (see "Objective Instrumentation Is Resisted")—that adding the necessary electronics to make an instrument fully automatic was not a matter of great moment to an analytical chemist. Yet, the drive for "push-button" instrumental objectivity argued against this more insular view.

Shortly after Walter Murphy assumed the editorship of *Analytical Chemistry* (then the *Analytical Edition of Industrial and Engineering Chemistry*), he and his associate editors reconsidered the question of appropriate topics for publication in the journal. They noted that the "expanded interests of the analytical chemist due to the development of new techniques and instruments" (Murphy et al., 1946, p. 218), required reconsideration. The policy they adopted was liberal, taking analytical chemistry away from chemistry proper; among other topics, they solicited articles that "present an improved or new procedure for the analysis or testing of some element, compound, or property, ... The tools used may be chemical or physical, and the procedures may deal with organic, inorganic, physical or biological chemistry. The physical chemistry, in many cases, may approach pure physics" (Murphy et al., 1946, p. 218) They also solicited articles on the application and development of instrumentation, and, indeed, the first regular column they added to the journal was Müller's column on instrumentation. By 1962 H. A. Liebhafsky noted that "like it or not, the chemistry is going out of analytical chemistry" (23A).

Industry, the Academy, and Government

Analytical chemistry has always had strong ties to industry. Determining the nature and composition of materials used in industrial products always has been important to the people making and selling (and buying) the products. Indeed, as noted, the journal, *Analytical Chemistry* is the offshoot of the journal of *Industrial and Engineering Chemistry.*

World War II increased industrial needs for analytical chemistry. No longer could materials be imported from Europe. War demands for various materials resulted in a major effort to find new sources of raw materials or to manufacture synthetic substitutes. All such work demanded analytical determinations of the composition and properties of the new materials (White, 1961, chap. 5). War exigency had two other significant effects: it increased the role of government in analytical-chemical matters, and it demanded the abandoning of wet chemical approaches for instrumental approaches. "When the war came along many control laboratories were caught flat-footed. Time-consuming methods of nineteenth-century chemistry finally had to be dropped in favor of spectrochemistry" (*Fortune*, 1948, p. 133). In his history of the rise of the instrumentation industry, Frederick White concluded that the "search for substitute, synthetic, or new materials was one of the major factors which made scientists aware of their complete dependence on instrumentation" (1961, p. 41). The federal government, because of perceived war needs, largely funded the shift from wet chemistry to instrumental methods. The government itself became a major employer of analytical chemists.

Scale and Capital Expenditure

The government's role in the change from wet chemistry to instrumental methods was important because of the capital expenditure involved; instruments were (and are) expensive. A direct reading spectrometer cost approximately $20,000–$30,000 in 1950 (Chamberlain, 1958, p. 11). Such an instrument could handle an enormous throughput of analytical work; in 1946, the Dow Chemical direct reader ran 7,000 samples per month, roughly 10 per hour 24 hours a day, 7 days a week (Saunderson, 1946, p. 51; Baird, 2000a). Such an instrument was much more expensive than the equipment required for wet chemical determinations. The result of these high up-front equipment costs is that only large-scale operations, performing many analyses a day (large analytical throughput), could bring the cost per analysis low enough to be competitive. Yet, instrumental methods were faster, more precise, and usually more accurate. This made it difficult for smaller scale operations to compete. This is also one of the causes for the exponential growth in the cost of academic science. To do research effectively and quickly—competitively—researchers need to be able to take advantage of the time-savings offered by efficient (but expensive) analytical instruments.

Education

Finally, these changes in analytical chemistry produced major challenges for educating analytical chemists. Editorials in *Analytical Chemistry* struggle with this question. On the one hand, adding new instrumental approaches placed significant demands on limited classroom time. Some time could be saved through decreased emphasis on outmoded wet chemical methods. But these methods were part and parcel to the teaching of chemistry proper. Thus, the question of education brought into focus the question of the *chemical* identity of analytical chemistry. On the other hand, the instrumental approaches did not easily fit into a standard curriculum. Instrumental approaches could be taught as "black-box" affairs—here is a box for the identification of organic compounds; use it like this—with little or no attempt to explain the operation of the black box. This was not satisfactory for several reasons, perhaps the most salient being that ignorance of the principles and operation of the black box would make one insensitive to its limitations. Yet full training in the science of instrumentation required a curriculum that did not exist; a background in electronics, mechanical engineering, optics, and physics was necessary to understand the integration of this knowledge in an instrument.

This, then, is the context in which analytical chemistry developed or promoted the concept of instrumental objectivity. Analytical chemistry long has had important ties with the chemical industry. The development of instrumentation promoted equally important ties to forensic analysis, medical diagnosis, environmental analysis, among other fields. In producing new—better—methods of analysis its goals have had to serve the values of these many masters.

"Modern Objectivity in Analysis"

Perhaps the nicest, albeit indirect, statement of instrumental objectivity can be found in Walter Murphy's March 1948 *Analytical Chemistry* editor's column, "Modern Objectivity in Analysis." In the column, Murphy presents and critiques H. V. Churchill's dinner address to the Third Annual Analytical Symposium of the Division of Analytical Chemistry.

Minuscule Samples

Churchill was concerned about the proportion of material sampled when submitted to an instrumental analysis. At the time, remelting furnaces for the production of aluminum alloys had a capacity of 35,000 pounds. From this, a 60-g sample was taken. When submitted to spectrographic analysis, about 1 mg of material is consumed in the sparking process. Of the total electromagnetic radiation produced, about 30 billionths enters the spectrograph. Thus, Churchill concluded with a note of concern, the ratio of the sample data to the amount of data in the melt is about 1 in 45 quadrillion.

Objective and Subjective

Murphy continued describing Churchill's concerns as follows:

> Commenting on instrumental analysis, Mr. Churchill reminded his audience that most modern objective methods and instruments are methods and instruments for doing faster or in greater volume certain tasks which can be done more slowly and in less volume by classical or traditional methods. Illustrating this point he reviewed a case history of one of the company's plants, where the analysis of aluminum alloys developed from the use of traditional or classical methods to a stage wherein the work was done spectrochemically—that is, by photographic spectroscopy—and finally is being done by the use of direct-reading spectrographs. The speaker reported the relative productivity of workers in these three stages of evolution has been in the ratio of 4 : 20 : 60. This is a 15-fold increase in productivity and speed in changing from subjective methods to those of increasing objectivity and with an increase in both precision and accuracy.

"Is it any wonder," said Mr. Churchill, "that some of us older chemists, who experienced some little difficulty in learning to weight to tenths of milligrams or even as microchemists to weigh to micrograms, are a bit appalled by the brash temerity of these modern-day analytical chemists who go so far into infinitesimals? No wonder we must bolster our faith with the intricate formulas of statistical analysis, and little wonder we have an almost idolatrous faith in the laws of probability." (Murphy, 1948, p. 187)

There is a lot to unpack from this editorial. For starters, it is clear that both Churchill and Murphy identify "modern" objective methods with instrumental methods: "Commenting on instrumental analysis, . . . most modern *objective* methods and instruments." The older wet-chemical methods are subjective—"a 15-fold increase in productivity and speed in changing from *subjective* methods." This usage cannot be seen to be accidental or idiosyncratic. As editor of the primary journal for analytical chemistry, Murphy is bound to be both aware and sensitive to usage in his field. The title of his column, "Modern Objectivity in Analysis," tells us that he is making a point about modern objectivity.

Accuracy and Productivity

We also learn these instrumental objective methods have yielded, in this case anyway, "a 15-fold increase in productivity and speed . . . with an increase in both precision and accuracy." It is immediately apparent that objective methods are tied up with economic ends; they involve the combination of productivity and accuracy. Churchill is worried that modern objective methods may sacrifice accuracy for productivity. The very fact that Churchill can worry that this instrumental notion of objectivity might sacrifice accuracy shows that accuracy is not essential to this concept of objectivity. This is remarkable because separating the idea of accuracy from that of objectivity runs counter to the intuitive, perhaps even analytic, alignment of accuracy and objectivity. These changes tie instrumental objectivity more closely to productivity and loosen its connection with accuracy. While he grudgingly acknowledges the inevitability of the change to instrumental methods, Churchill clearly is nostalgic for older methods, perhaps for an older concept of objectivity that preserves a close connection to accuracy.

Statistics and Standardization

It is also worth pointing out that these new instrumental objective methods rely on "statistical analysis . . . most an almost idolatrous faith in the laws of probability." Indeed, one of the analytical chemists at the University of South Carolina is largely concerned with improving analytical methods through a better use of statistics. Here is a current growth area in analytical chemistry (Deming & Morgan, 1987). This is interesting in its own right. It is more interesting when we reflect on the manner in which developments in statistics have had their own impact on the notion of objectivity (Porter, 1986; 1992; Swijtink, 1987). One of the features of the joining of physicochemical—instrumental—methods with statistical analysis, is the need to standardize. Standardization went hand in hand with the rise of statistical methods during the nineteenth century. Here is another value, prominent in technology studies—which arguably got its start with standardized parts for army weapons during the nineteenth century (Hindle & Lubar, 1986)—inserting itself into this modern notion of instrumental objectivity.

Murphy was a great promoter of instrumental methods. He distanced himself somewhat from Churchill's concerns by quoting Churchill, and while in this way he acknowledged these concerns to an extent, he is impressed with the dramatic increase in sensitivity. His introductory paragraph, stripped of some extraneous material reads, "H. V. Churchill's address . . . illustrated the delicate sensitivity of modern physicochemical devices now available in the field of analytical chemistry" (p. 187). Murphy looked forward to the emerging concept of objectivity: "Modern Objectivity in Analysis."

Ralph Müller as Witness

By the late 1930s, the editors of *Analytical Edition of Industrial and Engineering Chemistry—Analytical Chemistry* sought to bring the contents of the publication more in line with changes in the field—that is, they sought more articles on instrumental methods. The entire October issues for 1939, 1940, and 1941 focused on instrumentation; Müller wrote the entire 1940 and 1941 October issues. In January 1946, Müller's "Instrumentation in Analysis" became the first monthly column in the journal, a column he continued to write until 1968.

Müller's "Instrumentation in Analysis" essays are of interest for many reasons: we can find in them an explicit articulated definition of modern objective instrumental methods; they clearly show how these developments were tied to industrial needs; and, because of this connection, we can see how technological values of blackboxing, standardization, and cost efficiency get tied to this emerging notion of instrumental objectivity. Müller serves as an excellent witness to changes in this notion of objectivity and the tensions that produced them.

"Objectivity" Reserved for Instruments

Müller, with Murphy and the analytical chemical community, understood objective methods to be instrumental methods. While discussing the use of photomultiplier tubes for the relief of eyestrain, Müller wrote, "Thus, whatever gains have been made in the mechanization of this optical procedure, such as objectivity and greater precision, the elimination of fatigue is not one of them" (1946a, p. 29A). Following Murphy's practice, "gains in . . . objectivity" referred to the use of devices external to the human to perform functions previously done by humans. In this case, photomultiplier tubes were used instead of human eyes to sense light.

Instrumentation and Automation

In the same column, Müller made a more general point about the "three Rs" of instrumentation: reading, 'riting, and 'rithmethic—indication, recording, and computation.

In Müller's view, the purpose of instrumentation is (1) to take in information—to read; (2) to make computations as necessary on this information—to perform 'rithmetic; and (3) to present this information to an interested party—to 'rite. Instrumentation—as opposed to instruments—perform all three of these functions in an integrated way (1946a).

Although Müller's "three Rs" provided a memorable way to think of instrumentation, one important feature of instrumentation is left out: the use of servo-mechanisms to provide for feedback and control, both for better data gathering and for the control of materials. Müller devoted an equal portion of his "three Rs" column to servo-mechanisms (1946a, p. 29A). Instrumentation would directly intervene in its own calibration and, ultimately, in the control of the materials on which measurements are being made. To the greatest extent possible, human involvement should be eliminated. Müller's push-button objectivity—see the quote at the beginning of this chapter—required instruments with substitutes for "operator judgment." Instruments had to be able to calibrate themselves and, to as great an extent as possible, interpret their data ('rithmetic!) (1947a, p. 23A).

Objective Instrumentation Is Resisted

Despite strong reasons to fully develop instrumentation, Müller encountered resistance:

> It is strange and difficult to comprehend why the last few steps [in producing fully automatic instrumentation] have not been taken by the analyst in bringing his instruments to this stage of perfection. They are minor details, the absence of which in his motor car, office equipment, or telephone he would not tolerate for a moment. (1947a, p. 23A)

Müller soon found why analysts resisted bringing their instruments "to this stage of perfection." Doing so was not "real" science; at best, it was applied physics:

> Recently we were taken to task by one of America's distinguished chemists for emphasizing these distinctions [between a direct reading instrument and an instrument which provides data requiring further analysis]. 'All a matter of applied physical chemistry,' he explained patiently, 'and therefore not particularly new.' We are obtuse enough to feel that physicochemical techniques bear the same relationship to instrumental analysis as the violent oxidation of hydrocarbons does to the modern motor car. (1947a, p. 23A)

Müller's instrumental objectivity resided in the instrument, not in the principles of applied physical chemistry. The resistance Müller encountered here is tied to the classical notion that, fundamentally, science is concerned with human understanding and not intervening.[3]

> The less sympathetic commentators on instrumentation will insist that the instrument of itself is of little importance, and what is important is the information which it provides and the proper scientific interpretation and application of the results. This is quite proper, in its way, but it has nothing to do with instrumentation. It is merely applying known principles of physics and physical chemistry, and little or nothing is learned about instrumentation. (1947b, p. 26A)

If all that were important was data, then, perhaps, instrumentation would not be an interesting topic on its own—except in the "instrumental sense" in which instruments provide access to data. However, there is a distance between the level of representation (principles of physics and experimental facts) and the level of material devices (instrument). Müller brings a more technologically influenced notion that science is fundamentally concerned with intervening in the world.

Automatic Analysis Is Cost Efficient

Indeed, emphasizing the importance of intervening allowed for the idea of humans eliminating themselves from the loop: automatic analysis. A contemporary text on this subject begins with this sentence: "The partial or complete replacement of human participation in laboratory processes is a growing trend that started in the 1960s and consolidated in the next decade" (Valcárcel & Luque de Castro, 1988, p. 1). Müller anticipated this. His May 1946 column was concerned exclusively with automatic analysis:

> We find numerous examples of distinct "bottlenecks" in analytical or control laboratories which have arisen from changes in manufacturing practice. The use of automatic controls and regulators has speeded up production in many cases to such an extent that the ordinary facilities for analysis or inspection are no longer adequate. . . . In such cases, automatic analysis becomes mandatory, because the cost of conventional methods might well exceed the savings effected by the improved process. A further advantage ultimately arises in this step because the "autoanalyzer" may just as well control the process itself. (1946b, p. 23)

Müller continues by discussing the process of automatic analysis: "Each final step would, of necessity, involve *objective* measurements." As he puts it, "the primary considerations are speed, *adequate* precision, and an equivalent for operator judgment" (1946b, p. 23; emphasis added). Objective instrumentation, as part of systems for automatic analysis and control, perforce requires trade-offs between cost efficiency (in terms of initial costs, throughput, and operator expenses) and accuracy. Adequate precision, not the greatest precision available, is the trick.

Academic "Science of Instrumentation" Needed

Müller sought an amalgam of the technological and scientific. While obviously recognizing—and happily approving—the importance of the economic concerns behind the development of much instrumentation, Müller believed these values needed to be balanced:

> We have long insisted that research in analytical instrumentation of the "useless" variety is urgently needed and that its proper place is the university. Not that this will be conceded in academic surroundings, because there one hears the constant complaint that there are already so many instruments that it is not possible to tell the students about them. This attitude cannot halt the march of progress, but it helps immeasurably. Even where research is being attempted, the approach seems to be from the specific to the general, to illustrate and work with all the instruments which the budget will allow and shrug helplessly for that which must be left undone. The industrial attitude is similar in emphasizing the specific rather than the general. One can name his own price if he guarantees the instrumental solution of a specific headache. (1948, p. 21A)

Left to industry, only immediately commercially viable instruments would be developed. The result would be the predominance of economic values; instrumental objectivity would have to be primarily profit making.

Müller, great promoter of instrumental objectivity, saw problems; a balance was needed between the profit motives of industrial instrument development, which drove toward specific commercially viable instruments and the epistemic motives of academic instrument development, which drove toward a general understanding of instrumentation and of the world through instrumentation. Müller promoted the idea that a university should have a "division of instrumentation." In his January 1947 column, Müller discussed how students could be trained in instrumentation: "The principles of instrumentation have been well defined and classified. As a well organized field of science and technology its rules can be learned in a reasonable time. . . . As we see it, it is simply a matter of a few cardinal principles, a few selected topics in physics, some electronics, and an elementary knowledge of mechanisms" (1947a, p. 24A).

Universities did not follow up. In his September 1947 column he bemoaned the fact that industry had taken the lead in training and developing instrumentation research. At an AAAS instrumentation conference, universities were not well represented—they were only 15% of total attendance: "This situation emphasizes the fact that the initiative and intelligent prosecution of instrumental research have long since passed to industrial research laboratories and a few instrument companies" (1947c, p. 26A). General approaches to instrumentation are fostered in the university setting. The alternative was that only specific, commercially viable, instruments would be developed.

In his November 1949 column, he again brought up the question of a science of instrumentation. Yet he despaired:

> By what type of academic osmosis are the prerequisites for this profession [of instrumentation] to be absorbed from our present curricula? The answer is evident: It can be achieved only by an extensive, detailed curriculum directed precisely to this end. Our guess is that it will require another decade before this is completely recognized and, in the interim, we shall try our best, as individual resources permit, to acquaint our students with the already bewildering array of instruments and techniques. In so doing, the student will become an interim technician, still baffled by the intricacies of "black-box" magic. (1949, p. 23A)

A decade later, departments of instrumentation still were not part of the university and instrument development remained largely in the hands of the instrument-making companies.

Although Müller sought an amalgam of science and technology, the gulf between these two value systems made such a hybrid difficult. The entrenched academic notion that the real goals of research are representation and truth did not easily combine with the more instrument-based goals of intervention and the control of phenomena. But, left to industry, instrumental objectivity had to be tied to industrial values. M. I. T. researcher Eric von Hippel did extensive studies of the sources for instrumental innovation in the 1980s. He found that most innovation (78%) came from the market for the instruments, not the makers themselves, nor universities, except insofar as they provided a market (Von Hippel, 1988).

Cost Efficiency and "Wide Sales"

In November 1959 *Analytical Chemistry* ran an article by Van Zandt Williams, executive vice-president of Perkin-Elmer, a major analytical instrumentation company. Williams was concerned about the lack of cooperation between analytical chemists and instrument manufacturers. He emphasized the urgent need in industry for more efficient methods of analysis: "The lack of chemical analytical instrumentation—particularly automatic, direct concentration readout, chemical analytical instrumentation—may well be a limitation on progress in the chemical industry today" (p. 25A). He devoted a section of his article to the fact that foreign competition was putting pressure on the chemical industries in the United States to be more efficient, by utilizing more automatic means of analysis (pp. 26A–27A). Yet progress was held back by a lack of cooperation between analytical chemist and instrument maker:

> One point of conflict is in the term "wide sales." The major factor limiting an instrument company's growth and profit potential is its development and engineering capacity. Within our own definitions, we aim to get $5 of profit before taxes for each instrument engineering dollar spent. . . . In general we cannot do a profitable business making instruments for only one company's needs, because the number would be too small to warrant the development expense. (p. 31A)

Instrument manufacturers didn't know what analytical chemists needed by way of instrumentation, and analytical chemists didn't know what instrument makers needed by way of profits. Significantly—and obviously—instrument makers had to make profits on their instruments. This required instrument makers to pursue instrument development where a significant market was foreseeable—"wide sales." Müller's "useless" research aimed at increasing our understanding of the world through instrumentation was not the business of instrument makers or industrial analysts with specific analytical needs.

Summary

What then do we learn from Müller's editorials? We learn how, within analytical chemistry, objectivity came to be understood instrumentally and that the best instrumental objective methods were automatic methods. There was a strong economic need in industry for more efficient methods of analysis. Despite Müller's pleas, a science of instrumentation with its own academic departments did not develop in the university setting. In the commercial setting, economic values could not be separated from other desirable values in developing new analytical instruments. In short, instrumental objectivity came to incorporate values from the marketplace.

The Trajectory of Spectrographic Instrumentation

A specific example of instrumental development gives Müller's editorials more life. Here I briefly present some of the history behind the development of emission spectrographic instrumentation. Similar stories could be told for any number of kinds of instrument developed during this period. Müller did indeed have his finger on the pulse of changes in analytical chemistry and instrumentation.

Historical Highlights

Kirchhoff and Bunsen's work in the late 1850s demonstrated the possibility of using spectra to perform chemical analyses. Some significant discoveries were made because of this new method of analysis—notably, the discovery of several new elements (cesium, rubidium, thallium, indium, and gallium), the rare earths, and the rare gases (Ihde, 1984, p. 235). However, for many reasons, spectral analysis did not become a significant tool of the analyst until the late 1930s. At this time, instruments with better resolution, dispersion, and range of operation were brought to market. Several companies were involved, but those principally driving the development and improvement of emission spectrographic instrumentation were Applied Research Laboratories ("ARL"), Baird Associates ("BA"), and Jarrell-Ash ("J-A"). Here, both for specificity and because I am more familiar with the specific chain of events involved, I focus on BA's contributions. This was a competitive market, and consequently, similar stories can be told for both ARL and J-A.[4]

BA developed their first emission spectrograph in 1937; it was sold to the U.S. Bureau of Mines and delivered in 1938. This was a photographic device. Materials were brought to emit their spectra in an electric arc. Wavelengths were spatially separated by a precision grating, and spectral lines were recorded on a photographic plate. To identify the elements present in the sample, the spectrogram produced had to be compared "by hand" with known spectra of suspected elements; to determine the percentage concentrations of the elements involved, the densities of relevant spectral lines (how dark the lines were) had to be measured, again "by hand," with a densitometer. Despite this required "hand" work in interpreting spectrograms, emission spectrographs drastically shortened the amount of time required for many analyses. H. V. Churchill's comments quoted earlier in this article testify to this point.

During the mid-1940s, under intense pressure from World War II to speed up production, researchers at the Dow Chemical Company developed a means to use photomultiplier tubes to record spectra and automatically provide information about percentage concentration. Human "hand work" was no longer necessary. Such "direct-reading" spectrometers had to be configured to provide information on a preselected group of elements of interest. Dow, being in the business of producing magnesium alloy (for aircraft), could easily select the elements of interest.

Dow developed their direct reader for their own in-house use. In 1947 Dow licensed BA to produce and market BA–Dow direct readers. The principal market for these instruments was the steel industry. Virtually simultaneously ARL had developed a technically somewhat different, direct reader (called the quantometer), which was principally sold to the aluminum industry.

On urging from Ford Motor Company, BA undertook a further refinement of the direct reader in the early 1950s. The original BA–Dow direct reader was a moderately large instrument that needed to be housed in a laboratory with a controlled environment. In particular, changes in heat and humidity caused the instrument to malfunction. Ford wanted a small instrument they could put right on the foundry floor. BA's "spectromet," brought to market in 1955, filled this need.

Three features can be readily identified in this trajectory of emission spectrographic instrumentation. First, there has been a steady decrease in the size of the instrument. Second, there has been a steady decrease in the amount of time required for analysis. Third, there has been a steady decrease in the amount of training required of the operators of the instrument.

Size

At late as the mid-1930s, spectrographic analysis at the level of resolution, dispersion, and wavelength coverage required for much analytical work, needed an entire light-tight room. Spectrograms were made by moving plates on steel tracks around the room to capture spectrum lines from different portions of the spectrum. A principle advantage of BA's 3-meter spectrograph was size; the entire instrument, minus some relatively small peripherals, could be fit in a box 2 feet by 2 feet by 12 feet. It could be bought "off the shelf" and delivered to a suitable laboratory. The direct reader was only marginally larger than the spectrograph, and both required an air-conditioned laboratory. Spectromet was smaller—2 feet by 6 feet by 8 feet—and, significantly, did not require a special lab.

Time

The principle advantage to the direct reader was speed, and this was gained by building data interpretation into the instrument, making it an example of "Müller Instrumentation." Photomultiplier tubes collected light at preselected spectral lines; electronics did the 'rithmetic to determine how much light was collected at the various spectral lines and the writing involved in displaying this information on dials. Human interpretation was avoided, and the routine operation of the instrument was deskilled.

Training

Although routine operations could be performed by technicians without special skills, calibrating the instrument—a daily operation—remained important and required skilled personnel. Indeed, by the early 1950s, many of the direct readers sold by BA were not operating properly because the personnel on hand to keep them working did not have the relevant training. In developing spectromet, automatic calibration became a necessity because of the large temperature changes that the instrument would experience on the foundry floor. The significant innovation that made spectromet possible was the development of an "automatic monitor." This device used a mercury lamp, producing a spectrum line of known location and intensity; by using feedback electronics and a servo motor to adjust the optics, this mercury line, and all the rest of the spectral lines, could be kept focused on their photomultiplier tubes. By building automatic calibration into the instrument, it could operate in an environment that would throw it out of alignment and personnel who did not have training to keep optics in alignment could use it.

Spectromet is black-box magic. Foundry workers would take a sample of metal in production taken from "the pots" and, after minimal preparation, put it in the instrument and push the button. Within a minute, information on the concentrations of the elements of interest came up on dials.

You and the Push Button

Traditional Chemistry versus Spectrochemistry

The nicest possible summary of the transformation in instrumentation and objectivity can be read off a 1959 advertisement. In 1959 BA[5] advertised their spectrometers by comparing analytical methods, wet chemistry, spectrographic methods, and direct reading (figure 6.1). An iconic summary was presented for each approach. As with any effective ad, the visual point is made quickly and clearly: wet chemical analysis takes more steps than spectrographic analysis, which itself takes more steps than spectrometric analysis (using a direct reader). Furthermore, the steps involved are easier with spectrographic methods than with wet chemical analysis, and easier still with spectrometric methods.

A reading of the legend behind the icons reveals more. Three icons are involved:— a finger pushing a button, a contented face, and a face of intense concentration. For the finger and button we are told, "indicates an operation of push-button simplicity— human error minimized"; for the contented face, "simple and highly routine human operation—little danger of human error"; for the face of concentration, "Operation which requires skill, care or judgment—subject to human error" (Baird Atomic, 1959).

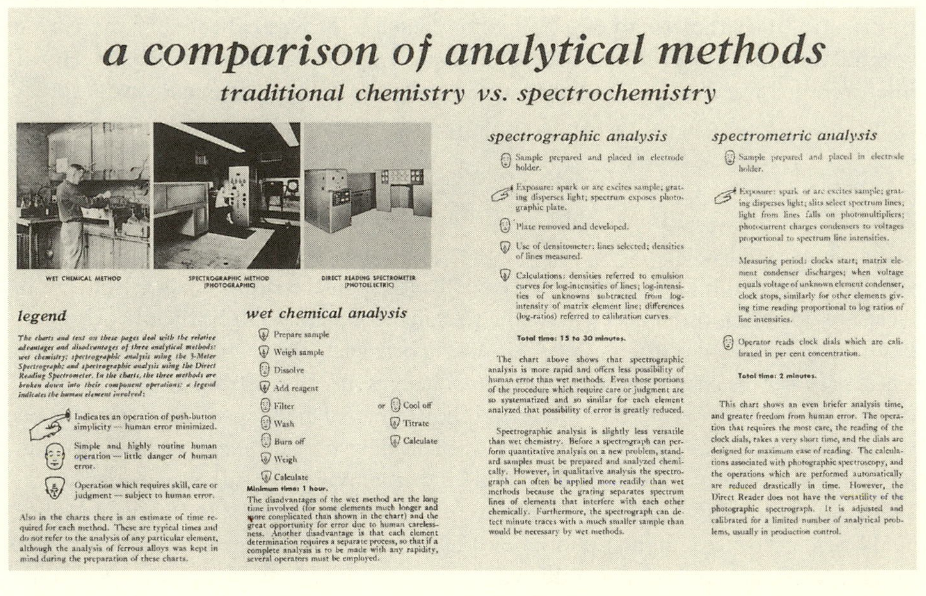

Figure 6.1 Advertising brochure: "A comparison of analytical methods" (Baird Atomic, 1959, pp. 2–3).

The obvious point being, "let an instrument do the work," it will be easier and, what is more important, not subject to human error. The instrument provides the objective ideal.

"For the Cost Conscious"

The brochure is a single fold-out, two pages wide. The comparison of analytical methods runs across the inner two pages. The back page consists of a series of quotes from satisfied customers, all speaking to cost:

> "We saved **2000** man hours per month."
>
> ". . . in using the Direct Reader to cover **8** elements in each of **5** matrices for a little over a year we find we have been able to assign **2** chemists to other laboratory work, reduce the time of analysis **1300%** and increase the quantity of analyses by **200%**."
>
> ". . . actual money saved . . . **$13,800** in laboratory costs alone based on former methods of analysis". (Baird Associates, 1959; emphasis and ellipses in the original)

In the text inside the brochure BA acknowledges that wet chemical methods are more versatile than spectrographic methods, which, in turn, are more versatile than spectrometric methods. But the decrease in versatility—if you can believe advertising—surely is compensated by the savings in money, person power, and time. Another advertising brochure for spectromet notes that one "regularly trained person, probably of shop origin, may operate Spectromet" (Baird Associates, 1956, p. 3).

The Front Page

The front page of the brochure provides, in many ways, the most interesting material (figure 6.2). As an experiment, quickly read the page. What does it say? I have looked over this brochure many times, and I have performed this experiment on several other people.

The quick reading of the first line (below the corporate name) is "you push the button." But it does not say this; it says, "you and the push button." The effect, no doubt, is the result of the choice of type and the placement of the line break, along with the graphic. While it is impossible to say what was intended and what not, it does carry a nice double meaning. On the one hand—the quick reading—we immediately learn that BA analytical instruments are push-button simple. On the other hand—the literal reading—we learn that a comparison is going to be drawn between you (a human) and the push button (the instrument). Given the text of the brochure, it is clear that you are subject to error, while the push button is not. This is BA's "Instrumentation for Better Analysis."

Conclusion

My Claim

I have argued that developments in analytical chemistry during the 1930s, 1940s, and 1950s—the instrumentation revolution—articulated and promoted an ideal for a kind

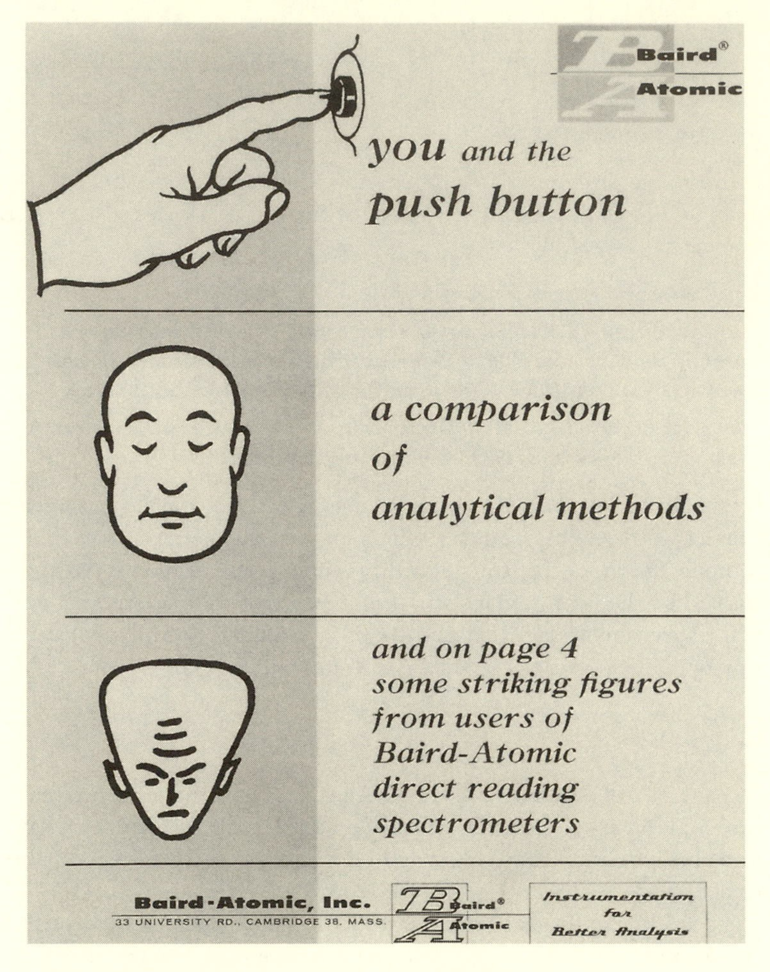

Figure 6.2 Advertising brochure: "A comparison of analytical methods" (Baird Atomic, 1959, p. 1).

of objectivity, "instrumental objectivity." Instrumental objectivity emphasizes the use of instrumentation to obtain data, analyze data, and even initiate interventions in control and production processes. It is objective in the sense that, to the greatest degree possible, human judgment and interpretation is reduced or eliminated, the implicit assumption being that human judgment is subjective and subject to error and/or bias. I have argued that the two central elements to "instrumental objectivity" are cost efficiency and the reduction in the role of human judgment, but not necessarily accuracy and precision, although this is often a consequence of reducing the role of human participation. Finally, I have suggested that this concept of instrumental objectivity has had consequences that reach far beyond analytical chemistry. Analytical chemistry is directly tied to many measurement situations, from food labels to pollution monitors. And beyond those areas directly connected to analytical chemistry, it is clear that the "instrumental ideal" plays a role. Analytical chemistry may not be used to measure teaching effectiveness, but instrumental progress in analytical chemistry has helped to promote an "instrumental approach" to such measurement problems. Analytical chemistry's "instrumental ideal" has changed how we feel about and relate to much of our world.

Notes

I wish to thank audiences at VPI, the International Summer School on the Philosophy of Chemistry in Ilkley England, the South Carolina Society for Philosophy and the editors of this volume for helpful comments and criticisms of earlier drafts. I would also like to thank my student Christine James for her contributions to my thinking about objectivity.

1. I am indebted to the editors of this volume for this example.
2. On rule vs. act utilitarianism, see Smart (1995) and Rawls (1995); on Neyman Pearson statistics, see Baird, (1992, ch. 10) and the references there); on analytical chemical determinations with instruments, see Skoog and West (1976, pp. 45–47) and Lingane (1948, p. 2).
3. For more on the classical—representational—versus this instrumental—interventional—notion of science, see Baird (2000a).
4. Baird Associates was founded by my father, Walter Baird, and two associates John Sterner and Harry Kelly. More detail on the founding of BA and my relationship to BA is in Baird (1991). More detail on the history summarized in this section can be found in Baird 2000a and 2000b.
5. In 1954, Baird Associates, through a merger with Atomic Instruments Inc., changed its name to Baird Atomic—still BA.

References

American Heritage Dictionary. 1993. 3rd ed. Boston: Houghton Mifflin.

Baird Associates. 1956. "Advertising Bulletin #42: Spectromet: Direct Reading Analysis on the Plant Floor." Privately produced advertising brochure, Cambridge, MA: Baird Associates. (In the possession of the author.)

Baird Atomic. 1959. "Advertising Brochure: A Comparison of Analytical." Privately produced advertising brochure, Cambridge, MA: Baird-Atomic. (In the possession of the author.)

Baird, Davis. 1991. "Baird Associates Commercial Three-Meter Grating Spectrograph and the Transformation of Analytical Chemistry." *Rittenhouse* 5(3): 65–80.

Baird, Davis. 1992. *Inductive Logic: Probability and Statistics*. Englewood Cliffs, NJ: Prentice Hall.

Baird, Davis. 1993. "Analytical Chemistry and the 'Big' Scientific Instrumentation Revolution." *Annals of Science* 50: 267–290.

Baird, Davis. 2000a. "Encapsulating Knowledge: The Direct Reading Spectrometer." *Foundations of Chemistry*, 2(1): 5–45.

Baird, Davis. 2000b. "The Thing-y-ness the Things: Materiality and Design, Lessons from Spectrochemical Instrumentation." In P. Kroes and A. Meijers, eds. *Studies in the Philosophy of Technology*. Forthcoming.

Benfield, Rebecca. 1995. "Nursing Science: Considering a Philosophy of Instrumentation." Unpublished manuscript.

Bridgman, Percy. 1927. *The Logic of Modern Physics*. New York: Macmillan.

Chamberlain, Francis. 1958. "Baird-Atomic, Inc. Principal Products Past and Present (1936–1957)." Private corporate document. (In the possession of the author.)

Daston, Lorraine. 1988. "The Factual Sensibility." *Isis* 79: 452–467.

Daston, Lorraine. 1991. "Baconian Facts, Academic Civility, and the Prehistory of Objectivity." *Annals of Scholarship* 8(1): 337–363.

Daston, Lorraine. 1992. "Objectivity and the Escape from Perspective." *Social Studies of Science* 22: 597–618.

Daston, Lorraine. Galison, Peter. 1992. "The Image of Objectivity." *Representations* 40(Fall): 81–128.

Deming, Stanley. & Morgan, Stephen. 1987. *Experimental Design: A Chemometric Approach.* Amsterdam: Elsevier Science.

Fortune. 1948. "Instrument Makers of Cambridge." December, 136–141.

Gigerenzer, G., Swijtink, Z., Porter, T., Beatty, J., Daston, L., & Krüger, L. 1989. *The Empire of Chance: How Probability Changed Science and Everyday Life*. Cambridge: Cambridge University Press.

Hacking, Ian. 1983. "Was There a Probabilistic Revolution 1800–1930?" In M. Heidelberger, L. Krüger, & R. Rheinwald, eds. *Probability since 1800: Interdisciplinary Studies of Scientific Development*. Bielefeld, Germany: B. K. Verlag.

Hacking, Ian. 1987. "Was There a Probabilistic Revolution 1800–1930?" In L. Krüger, L. J. Daston, & M. Heidelberger, eds. *The Probabilistic Revolution. Vol. 1: Ideas in History*. Cambridge, MA: M.I.T. Press. [This is a revised and condensed version of Hacking (1983).]

Hindle, Brook. & Lubar, Steven. 1986. *Engines of Change: The American Industrial Revolution 1790–1860*. Washington DC: Smithsonian Institution Press.

Hutson, J. M., & Petrie, R. H. 1986. "Possible Limitations of Fetal Monitoring." *Clinical Obstetrics and Gynecology* 29(1): 104–113.

Ihde, Aaron J. 1984. *The Development of Modern Chemistry*. New York: Dover.

Liebhafsky, H. A. 1962. "Modern Analytical Chemistry: A Subjective View." *Analytical Chemistry* 34: 23A.

Lingane, J. J. 1948. "Editorial: The Role of the Analyst." *Analytical Chemistry* 20: 2–3.

Megill, Allan. 1991–1992. "Rethinking Objectivity." *Annals of Scholarship* 8(3–4): 301–477; & 9(1–2): 1–153.

Megill, Allan. 1991. "Four Senses of Objectivity." *Annals of Scholarship* 8(1): 301–320.

Müller, R. H. 1946a. "Monthly Column: Instrumentation in Analysis." *Industrial and Engineering Chemistry, Analytical Edition* 18(3): 29A–30A.

Müller, R. H. 1946b. "Monthly Column: Instrumentation in Analysis." *Industrial and Engineering Chemistry, Analytical Edition* 18(5): 23A–24A.

Müller, R. H. 1947a. "Monthly Column: Instrumentation." *Industrial and Engineering Chemistry, Analytical Edition* 19(1): 23A–24A.

Müller, R. H. 1947b. "Monthly Column: Instrumentation." *Industrial and Engineering Chemistry, Analytical Edition* 19(5): 25A–26A.

Müller, R. H. 1947c. "Monthly Column: Instrumentation." *Industrial and Engineering Chemistry, Analytical Edition* 19(9): 26A–27A.

Müller, R. H. 1948. "Monthly Column: Instrumentation." *Analytical Chemistry* 20(6): 21A–22A.

Müller, R. H. 1949. "Monthly Column: Instrumentation." *Analytical Chemistry* 20(11): 22A–23A.

Murphy, W. J. 1948. "Editorial: Modern Objectivity in Analysis." *Analytical Chemistry* 20(3): 187.

Murphy, W. J., Murphy, W. Hallett, L.T. Gordon, G. G. & Anderson, S. 1946. "Editorial Policies: Scope of the Analytical Edition." *Analytical Edition of Industrial and Engineering Chemistry* 18(4): 217–218.

Ostwald, Wilhelm. 1895. *The Scientific Foundations of Analytical Chemistry*. Trans. G. McGowan. London: Macmillan, (First German edition, 1894.)

Porter, Theodore. 1986. *The Rise of Statistical Thinking: 1820–1900*. Princeton, NJ: Princeton University Press.

Porter, Theodore. 1992. "Objectivity as Standardization: The Rhetoric of Impersonality in Measurement, Statistics, and Cost-Benefit Analysis." *Annals of Scholarship* 9(2): 19–59.

Rawls, John. 1995. "Two Concepts of Rules." In S. M. Cahn & J. G. Haber, eds. *Twentieth Century Ethical Theory* (pp. 273–290). Englewood Cliffs, NJ: Prentice Hall.

Saunderson, J. L. 1946. "Spectrochemical Analysis of Metals and Alloys by Direct Intensity Measurement Methods." In *Electronic Methods of Inspection of Metals. A Series of Seven Educational Lectures on Electronic Methods of Inspection of Metals Presented to Members of the A. S. M. During the Twenty-Eighth National Metal Congress and Exposition, Atlantic City, November 18–22, 1946* (pp. 16–53). Cleveland: American Society for Metals.

Shapin, Steven, & Simon, S. 1985. *Leviathan and the Air-Pump*. Princeton, NJ; Princeton University Press.

Skoog, Douglas, & West, Donald. 1976. *Fundamentals of Analytical Chemistry*. New York: Holt Rinehart and Winston.

Smart, J. J. C. 1995. "Extreme and Restricted Utilitarianism." In S. M. Cahn & J. G. Haber, eds. *Twentieth Century Ethical Theory*, (pp. 307–315). Englewood Cliffs, NJ: Prentice Hall. [first published in *Philosophical Quarterly* 6 (1956).]

Swijtink, Zeno. 1987. "The Objectification of Observation: Measurement and Statistical Methods in the Nineteenth Century." In L. Krüger, L. J. Daston, & M. Heidelberger, eds. *The Probabilistic Revolution* Vol. 1: *Ideas in History*. Cambridge, MA: M.I.T. Press.

Tallon, R. W. 1994. "Technology Assessment: Electronic Fetal Monitoring." *Midwives Chronicle and Nursing Notes*, May, pp. 186–188.

von Hippel, Eric. 1988. *The Sources of Innovation*. Oxford: Oxford University Press.

Valcárcel, M., & Luque de Castro, M. D. 1988. *Automatic Methods of Analysis*. Amsterdam: Elsevier Science.

White, F. 1961. *American Industrial Research Laboratories*. Washington, DC: Public Affairs Press.

Williams, C. 1948. "Editorial: The Role of the Analyst." *Analytical Chemistry* 20(11): 2.

Williams, Van Zandt. 1959. "Cooperation between Analytical Chemist and Instrument Maker." *Analytical Chemistry* 31(11): 25A–33A.

Part III

Structure and Identity

7

Realism, Essentialism, and Intrinsic Properties

The Case of Molecular Shape

JEFFRY L. RAMSEY

Figure, or shape, has long been ensconced in modern philosophy as a primary or essential quality of matter. Descartes, Malebranche, Hobbes, and Boyle all apparently endorsed the Lockean claim that shape is "in Bodies whether we perceive them or no" (Locke, [1700] 1975, p. 140). In addition, most seventeenth-century philosophers endorsed the inference that because shape is primary, it is one of the "ultimate, irreducible explanatory principles" (Dijksterhuis, 1961, p. 433; cf. Ihde, 1964, p. 28). Locke has often been read in this way,[1] and in *Origins of Forms and Qualities*, Boyle claims the "sensible qualities . . . are but the effects or consequents of the . . . primary affections of matter," one of which is figure (quoted in Harré, 1964, p. 80).

Little appears to have changed. Most analytic philosophers and realist-minded philosophers of science "would endorse a distinction between primary and secondary qualities" (Smith, 1990, p. 221). Campbell (1972, p. 219) endorses the claim that "shape, size and solidity are generally held to be primary," even though he argues that "the philosophy of primary and secondary qualities" is confused. Mackie (1976, p. 18) discounts solidity but endorses spatial properties and motion as "basic" physical features of matter. Most philosophers also endorse the inference to the explanatory character of the primary qualities. Mackie (1976, p. 25) asserts spatial properties are "starting-points of explanation." Boyd (1989, pp. 10–11) claims "realists agree" that "the factors which govern the behavior . . . of substances are the fundamental properties of the insensible corpuscles of which they are composed." As befits our current situation, explanation purportedly flows from spatial microstructure. A body "possesses a certain potential only because it actually possesses a certain property (e.g., its molecular structure)" (Lange, 1994, pp. 109–110). Even Putnam, who argues all properties are Lockean secondaries, claims powers "have an explanation . . . in the particular microstructure" of matter (Putnam, 1981, p. 58).

In stark contrast to seventeenth-century natural philosophers and modern philosophers, some theoretical chemists have claimed "it is wrong to regard molecular structure as an intrinsic property of a molecule. . . . Molecular shape is not an invariable property of molecules" (Woolley, 1978, p. 1074). As a result, figure is "not an object of belief" (Woolley, 1985, p. 1083) and "not part of the external physical reality" (Amann,

1993, p. 131; cf. Primas, 1982). Thus, invoking molecular shape to explain the behavior of molecules—which chemists do all the time (cf. Pauling, [1970] 1988; Trindle, 1980)—is inappropriate in some way. When one interprets experiments using classical notions of molecular shape, success "is only achieved . . . because one is fitting disposable parameters (bond lengths, force constants, etc.) have no place in the general quantum mechanical theory" of the molecule (Woolley, 1977, p. 397).

I believe everyone is both right and wrong. Philosophers correctly refer to microstructure as explanatory but wrongly claim it is necessarily essential. Primas and Woolley are correct that shape is not essential but wrong that it is not a real property and, thus, not explanatory. Both sides miss an essential (!) point: properties are detected in time or, more precisely, are detected over a span of time. Different properties appear as primary when recorded by different instruments, and more important, different properties are real and explanatory in different time scales of measurement. As Weininger (see chapter 9) notes, we must acknowledge the real-time dynamics of molecules. Recognizing the crucial role of time requires us to both (1) explicitly include the means of detection when distinguishing primary and secondary qualities, and (2) localize the explanatory nature of a given primary quality to a particular measurement context. This is the perspective pursued in this chapter. The upshot is that properties—at least the property of shape—are neither "in bodies whether we perceive them or no" nor mere constructions of the mind.

In the next section I develop the account that says shape takes time to be detected. Then I examine what this implies for our accounts of the primary/secondary distinction and the status of shape as a primary quality. In the conclusion, I speculate how debates about color might inform the debate about shape, and on what the contextual, explanatory account of properties advocated here might look like.

The Physical Account

Shape takes time to be observed. This is the basis for claiming shape is more than mere appearance, but less than intrinsic.[2]

One begins with the Heisenberg uncertainty relation in its time–energy form: $\Delta E \cdot \Delta t \approx h$.[3] If we had infinite time for measuring, we could minimize the energy spread to its theoretical limit. This corresponds to an almost negligible "bump" of the system by the observing instrument. At the limit, we would be measuring eigenstates of a complete molecular Hamiltonian. At any time less than infinity, however, we measure states.

In such states, the molecule does not exhibit or exist in the full symmetry appropriate to the stationary states. *Why* molecules do this is the unanswered question. *That* they do it is simply not at issue. Given that they do, all we have to do is ask whether delocalization takes place on a time scale that rules out the ideas of separated motions and a rigid molecule. This is an empirical matter. For most chemical states, the separability of nuclear and electronic motions is valid.

Take Cl_2 or C_3H_4 as typical molecules. The lifetime for a classical description of these molecules is on the order of 10^{-10} sec (Patsakos, 1976). The chlorine molecule has 70 protons and neutrons; the C_3H_4 has 40. Thus, an isolated molecule of chlorine weighs around 1×10^{-22} g because the bond length in the molecule is about 2 A, the time scale would be reduced to about 5×10^{-10} secs. The C_3H_4 molecule has 40 protons and neutrons, so it weighs around 7×10^{-23} g. It is roughly twice as long as the chlorine molecule, so the time scale is of the same order of magnitude.

In infrared (IR) spectroscopy, the best resolution we can achieve is an energy difference of 0.1 cm^{-1}.[4] The lifetime, Δt, corresponding to this energy difference is 5×10^{-11} sec. Because one needs a process that gives rise to lifetimes of 10^{-11} *or less* to cause broadening—much less eventual separation—of the bands, we will see only one peak corresponding to the classical description (Drago, 1977, pp. 86–87). In other words, a process—here, a vibration—that operates on this time scale, coupled with the limits of detection, causes the two states to appear as one.[5]

If we wish to see the delocalization, we can change the means of detection. For instance, we can trap the molecules in low-temperature beams. This minimizes the energy imparted to the molecule as we "touch" it. In these experimental setups, we can dispense with the separability assumption *if* conditions warrant. However, even in such beams, the assumption is most often justified. Because most rotational levels are separated by about 250 cm^{-1}, we have to trap atoms for minutes at temperatures of milli-Kelvins before the separability assumption fails.[6]

This account of the measuring process makes the properties possessed by a molecule contingent on the way it is detected. Under some measurement conditions, the molecule will "have" a nuclear frame replete with localized nuclei; under others, it will have only a "fuzzed" nuclear frame; and under others still, it will have no nuclear frame at all. For instance, in IR spectroscopy, ammonia (NH_3) "looks" like a trigonal pyramid. If we use a more sensitive spectroscopic technique to, say, distinguish the rotation–vibration levels, the molecule's shape will appear as two overlapping trigonal pyramids, with one pyramid inverted with respect to the other. In the limit of perfect spectral resolution, we would describe the molecule's shape as a sphere.[7] Metaphysically, this may seem quite curious, but the result is quite general: as we "zoom" in using more refined instruments, the shape "changes."[8] In addition, there is some indication (Woolley, 1978) that when molecules react under conditions similar to those in the sensitive spectroscopic measurements, their causal properties are different from those when they react in more "standard" conditions. More on this later.

In some sense, the focus on measurement is arbitrary. If a molecule were to experience conditions in nature like those it experienced in the measuring apparatus, the shape would also vary.

Presumably, this effect makes it difficult to assert that a single shape is universally explanatory. If ammonia engages in a reaction under conditions where it has a trigonal bipyramidal shape, than we probably could not say it reacts in a certain way because it has the shape of a trigonal pyramid. Chemists routinely appeal to shape to explain how a molecule reacts.[9] *Some* shape might be explanatory in every explanation, but the shape will be different in different explanations.

In sum, depending on conditions, the nuclear frame may or may not retain its integrity. There is no "basic" level that determines the truth or falsity of the representations derived under other measuring conditions.[10] Thus, there is no "one" true theoretical description; rather, there are many different appropriate descriptions. Descriptions are explanatory in some conditions, but not in others.

Implications

I have argued elsewhere that the account just given provides good reasons for giving up certain kinds of strict realism and reductionism, the kinds Woolley and Primas see as the source of the "problem" about molecular shape (Ramsey, 1997a, 1997b). What implications does it have for the ideas of primary and secondary properties? Wilson (A) notes, "The traditional concept of a primary quality was . . . very much a product of the assumption of [a] fairly simple relation between (certain) ordinary sensible qualities of perceivable objects, and the nature of the insensible entities which constituted them, providing the basis for explanation of observable interactions in terms of universal laws" (Wilson, 1992, p. 235). In what ways does the account given entail this "simple relation" assumption is incorrect? Let me approach this by considering in turn: (1) how primary and secondary qualities are distinguished; (2) whether shape is "essential" and (3) whether shape is "real."

The physical account satisfies none of the three traditional means for arguing that a quality is secondary or primary. First, the account does not commit us to an error theory of secondary qualities. Woolley and Primas rely on this argument to motivate their case that we detect shape when shape is really not there. For example, Primas argues shape is created only by abstracting from universal EPR correlations (Primas, 1981). For Primas and Woolley, classical shape is not primary because it has not or cannot be shown to be part of the fundamental representation of reality. Woolley claims shape is only a "powerful and illuminating metaphor" (Woolley, 1982, p. 4) rather than an "object of belief" (Woolley, 1985, p. 1083). For them, only eigenstates of full Hamiltonians are invariable, and molecular shapes are not such eigenstates. However, that argument presupposes the distinction between primary and secondary qualities, with eigenstates of full Hamiltonians playing the role of the primary quality and (classical) shape that of the secondary. As the physical account makes clear, there is no such simple relation. A molecule can have any number of appearances. By invoking the error theory argument, Woolley and Primas recapitulate rather than refigure the seventeenth-century relation.

Second, the physical account does not support the argument from perceptual relativity. In fact, one consequence of the physical account pulls in exactly the opposite direction. Instruments with *different* modalities of detection will record the *same* structure as long as they measure on the same time scale. For example, Raman spectroscopy relies on polarizability changes in molecules, and IR spectroscopy relies on absorption of infrared radiation. Nonetheless, both will detect the same vibrational modes of $CHCl_3$.[11]

Third, the account does not necessarily depend on conscious observers. Putnam (1981, p. 61) argues all properties are Lockean secondaries because "nothing at all we say about any object describes the object as it is 'in itself,' independently of its effect on us, on beings with our national natures and our biological constitutions." Our biological nature is surely implicated. Had we evolved at radically different physical scales, we would see different spatial properties for molecules as the "basic" or manifest property. However, we can say things about an object independently of its direct effect on us. We can have instruments do the detection for us; the result does not depend strongly on us as biological beings or as rational agents. If a molecule has shape, it has shape due to its interaction with other things in the world. Sometimes the interaction is with us and our instruments; many times it is not. Shape is not merely an effect on *us*; it is an effect in the world.

In short, Putnam substantiates Wilson's claim that philosophers remain too closely tied to the seventeenth-century conception of primary and secondary qualities. The same applies to A. D. Smith (1990). Smith argues that because the primary properties are not "direct sensibles" but are rather theoretical terms, we should "either ditch the term 'primary property' or advert to what science calls fundamental." But there is reason to say shape is sensible, just not directly so. Because it is recorded by instruments, it can be sensed. "Sensed" need not refer to the five traditional senses. Smith remains too tied to the seventeenth-century definition of a sensible.

Like Putnam, Priest also concludes all properties are secondary, where that term is used to mean "solely as referring to the appearance" (Priest, 1989, p. 32).[12] On Priest's account, seventeenth-century primary properties are produced by "microscopic dispositions plus observers," where the dispositions are "vector properties of quantum states" (p. 36). Because dispositions at the microlevel are quantum vector properties, the old (Newtonian) primary properties—including shape—are really secondary.

In the main, Priest's account meshes with the one given in this chapter. In particular, he relativizes the primary/secondary distinction to a level of description. However, there are some differences in emphasis and at least one important substantive difference between his and my accounts. First, he writes that "registering with an observer is caused (in part) by dispositional properties (or propensities) to be understood *in terms of structure at the quantum level*" (Priest, 1989, p. 36; emphasis added). This suggests Priest is operating with a (passive) observer who simply records the quantum dispositional properties. But on the account I have offered, the structure at the quantum level, while necessary, is not sufficient by itself to account for the appearances. It is the quantum level structure plus the time scale of measurement that provides the relevant account of the properties. The same quantum system will have markedly different properties in different measuring situations. To be sure, Priest writes "in part" in the preceding quote, so perhaps the difference here would disappear if he were to spell out the partial nature of his claim.

But second and more important, Priest continues to endorse the inference that because certain properties are primary, they are explanatory. The behavior of matter "at each level may be explained in terms of the structure of the level below. Thus, the behavior of macroscopic bodies and their properties is explained in terms of the (primary) properties of its microscopic (atomic) parts" (Priest, 1989, p. 36). And the behavior of the microscopic parts, which are the bearers of the primary properties, is "explained in terms of their quantum states and properties" (Priest, 1989, p. 36). On the account I have offered, such talk is problematic. The "bare" quantum structure does not explain—all by itself—the microscopic primary property. Reference to the measurement technique must also be included. But perhaps even more important, as I alluded previously, even the quantum structure is not universal or intrinsic. One and the same hunk of matter can have quantum or classical microscopic properties, depending on the time scale of the measurement. It will be of little explanatory importance to refer to the "quantum states and properties" in most spectroscopic measurement situations. To be sure, quantum corrections can be added in if a finer level of explanation is desired, but the "bare" quantum structure doesn't even exist at common measurement scales.[13] For instance, the bipyramidal quantum structure of ammonia ceases to manifest itself in more energetic measuring situations. Nor are molecules spherical distributions of electrons, neutrons, and protons in most laboratory setups. Thus, although something does remain constant, that which remains constant is not universally explanatory.

In short, philosophers and scientists cannot and should not rely uncritically on shape or spatial properties as "starting points" for explanation. Nor can they rely uncritically on the claim that microstructure is an essential property of substances. This flows from the naturalistic metaphysical stance I have adopted here.[14] Again, on the account given, a particular shape does not "belong" essentially to matter as a "basic" physical feature. Whether a particular "structure" can be imputed to a sample is a contingent matter. That seems enough to conclude it is false that molecular shape is essential. Some other property (such as mere extension) might be essential—that is, invariant and explanatory in all contexts. But shape is not.

The strict essentialist can reply to this in two ways. First, the essentialist can say, "But the matter still has a shape, just not a particular shape." Granted, but this makes shape nonexplanatory. It is supposedly a feature of things that particular shapes explain what we see. Pointed corpuscles explained why acids felt like they did on the tongue; resonances explain why some molecules are colored. Merely possessing "a" shape would not allow recourse to particular molecular structures as producing particular appearances. Also, how is "a" shape different from mere extension in space? Second, the essentialist can reply that "shape" is not the same as "microstructure." The latter term refers only to gross chemical composition, independent of any considerations of topology. This is implausible for two reasons. First, if philosophers who endorse the discovery picture of microstructural essences are aware that one can separate composition and topology, they do not note this fact. Second, and more important, they rely on the standard (at least in traditional chemistry and physics) explanatory connection between structure and properties. That is, because water has a particular topological and compositional structure, it has the properties of being clear, colorless, and liquid. The Twin Earth examples of XYZ having the properties of H_2O trade on the standard topological structure. If water or XYZ were $(OH)^-H^+$ or even H_2O arranged linearly, its properties would be quite different. Philosophers rarely, if ever, mention this qualification, which leads me to believe they must be endorsing a particular topological structure, as well as chemical composition. In short, for most philosophers, the microstructural essence does involve a particular spatial arrangement of atoms in molecules. And again, that is exactly what the given physical account says is contingent.

Finally, let me examine whether we can be realists about primary properties. Traditionally, the primary properties "really exist" and the secondary properties are not "real" in the same sense (Putnam, 1987, pp. 4–5). Philosophers rely on this distinction to claim shape is *always* an object of belief. Woolley and Primas rely on it as well. They presuppose a strict derivational reductionism at the level of theory must be satisfied and, failing to find it, claim that shape is "not an object of belief" (cf. Woolley, 1986). In contrast, the physical account I have given suggests that what is "really" there depends on the time scale of the "looking." A particular shape is sometimes there and sometimes not. Thus, one should be a contextual realist about shape.

By endorsing contextual realism, I take it I am not endorsing Putnam's internal realism. Putnam considers cases where alternate combinations of given individuals, each with its own given properties, are or are not considered to be individuals in their own right (Putnam, 1987). In contrast, the question I have been asking involves whether a given individual has a unique description. The answer is "no" and is such not merely because we decide to describe it differently. It is "no" because how the individual appears to us can legitimately differ. This is a matter decided by the combination of the world and our means of detection. It is not a mereological choice over determinate individuals.

What of other metaphysical realisms, proponents of which reject Putnam's model-theoretic argument and its internal realist conclusion? Lewis (1986, p. 204) remarks, "If we know what shape is, we know that it is a property, not a relation." His treatment of properties "requires things to have or lack them *simpliciter*. . . . I have made no properties that admit of degree . . ." (Lewis, 1986, p. 53). Although this might make sense of the sort of commonsense properties discussed most often by metaphysicians and for classificatory rather than explanatory purposes, it fails on the account of shape I have offered. One can save the analysis by relaxing the concept of "shape" to "spatial extension." Analogously, one can save "color" by opting for "coloredness." But this move makes the concept merely classificatory. It robs the concept of its explanatory content in chemistry and, I expect, its content in attributions of color.[15] As I indicated earlier, it is not merely that molecules are extended in space that accounts for the kinds of reactions they undergo. Rather, it is the particular kind of extension—where the nuclear frame is localized (to some extent) and the electronic motion is also localized (though to a much lesser extent)—that chemists take to be explanatory of chemical structure and reactions. This explanatory notion of shape may be given up in a future science, but it is certainly asking too much to give it up without providing some robust alternative. Current science does not yet have such an alternative, so I would suggest the metaphysics should give way rather than the science. No need to hobble a powerful machine without reason.

Conclusion

I have proposed an account of molecular shape that makes it real and explanatory in most circumstances. I have done so by noting that properties are detected over a span of time. The determinate appearance of a given hunk of matter can change, depending on the length of time it takes to record the property. Thus, reality has different appearances in different situations. We have not been used to thinking in this way because our measuring techniques most often operate over time intervals in which the traditional primary properties are constant. In addition, on the account I have given, whether a given primary property is explanatory must be decided by reference to the context of measurement. Bare primary properties do not—by themselves—explain. They explain in a given context.

As I have noted throughout this chapter, the perspective advanced here does require philosophers to rethink their notions of properties essential to matter irrespective of the means of detection. To be fair, Woolley often stresses this aspect of the situation

(Woolley, 1977, p. 397; 1986, p. 204; 1991), and, thus, his stance is often very much like the one I have adopted. However, he often collapses the correct claim that classical molecular structure is of only limited validity, with the incorrect claim that classical molecular structure cannot be used to interpret and explain the properties of different isomers and isotopically substituted molecules (see Woolley, 1991).

The contextual realism I am advocating meshes nicely with perspectives adumbrated in this chapter and elsewhere. Not the least, it asks us to consider the question of realism by taking into account experimental technique. In addition, it meshes with Grosholz and Hoffmann's claim (see chapter 13) that scientific concepts are multi- rather than univocal. On a contextual realist account, one analysis of molecular shape will not be forthcoming. To the contrary, there will be many. This kind of realism, which foregoes literal, univocal relationships between concept and thing and between different senses of the same concept, will allow us to retain a realistic interpretation of the concept of shape. Different senses of the same concept can easily refer to different causal activities of stuff, and these different senses will be related to each other in a complex fashion. In addition, the account has the consequence that shape is a metaphorical rather than a perfectly literal model. Metaphorical models are potentially dangerous (Bhushan & Rosenfeld, 1995), as are pictures (Luisi & Thomas, 1990), literal scale models,[16] and everyday language. But I think chemists are aware of the danger. In my undergraduate and graduate training as a chemist, as well as in my conversations with chemists, I have yet to encounter one who would claim the various molecular models are absolutely literal, comprehensive representations of reality.[17] If the claim is made that one model is to be taken literally, it is usually done so with the recognition that other important features of the molecule will be suppressed. But the claim of literalness is rarely pressed to its extreme. The language of spectroscopic peaks "corresponding" to the classical description already indicates that the description is not conceived as a strict identity. Perhaps more tellingly, Coulson (1960, p. 174) once remarked that "the concepts of classical chemistry were never completely precise. . . . When we carry these concepts over into the quantum chemistry we must be prepared to discover just the same mathematical unsatisfactoryness." In this sense, modern chemists have moved beyond the strict seventeenth-century account of the microscopic world.

Although the perspective advanced here does get us over some philosophical and scientific hurdles, it is not without its problems in both arenas of inquiry. I have not dealt with the fact that our traditional picture of molecules is only an approximate rendering of what we are detecting. That is, the representations we employ are usually vague. Also, I have been employing "explain" as a synonym for "interpret." This seems questionable. Finally, although endorsing the claim that "shape" is ambiguous, I have said nothing about how one relates classical shape to quantum shape. This is because we do not yet know the answer (Woolley, 1978, p. 1076).

At least with respect to this last issue, debates about color can perhaps help. As the intensity of monochromatic light sources vary, subjects see different colors even though the source is monochromatic (Grandy, 1989, p. 233). Likewise, under varying conditions, we will see different shapes. However, this must be only an analogue as color perception does not—as far as I know—rely on quantum mechanical effects. Another suggestion is to recognize an ambiguity in the terms we use to express properties. There is physical color, and there is phenomenal color. Likewise, perhaps there are various notions of shape, and our job should be to examine the relationships among them. This appears to be the line of argument endorsed by Grosholz and Hoffmann (chapter 13 in this volume). To those who object to the idea of accepting traditional notions of shape, I would reply that, just as we make discriminations among physical objects based on color (Grandy, 1989, p. 230), we make discriminations based on shape. Of course, this will not satisfy the skeptic. We are assuming the existence of two (or more) types of shape or color rather than deriving or inferring one from the other. But then we have moved the game to the plausibility of the skeptic's position, and that is another issue altogether.

Whatever the ultimate connection—if any at all—between shape and color, we do now have two properties that vary under different conditions. One has been traditionally considered as primary, the other as secondary. Whether such contextualization and conditionalization of all properties will be needed is an open question; I think we need not insist on space, time, mass, sound, or taste following the same sort of account pursued here. But whatever the upshot for the other properties, the account does have consequences for certain views about properties and their role in metaphysical systems. I have remarked that Putnam's and Lewis's realisms must be modified to take the contextual nature of shape into account.

So we have two concepts that are contextual or, what amounts to the same thing, perspectival. Importantly, this is not just linguistic contextualism (cf. van Fraassen, 1981). Nor is it a merely representational perspectivalism, in which different "pictures" feature different aspects of a subject. Nor is it merely a matter of the embeddedness of particular properties in given theories (Taylor, 1993). In this sense, the contextualism is not merely epistemological as usually conceived. It is a contextualism tied to how we intervene (cf. Hacking, 1983) and how those interventions reveal or display different causal properties of stuff (cf. Culp, 1995; Wimsatt, 1995). Each experimental (and also theoretical) technique has a particular signature. Different techniques will give different signatures.[18]

At root, the causal, interventionist character of the account defended here is the primary salient difference between the metaphysics of the seventeenth-century and that of the twentieth century. On the account I have provided, shape is not primary, not essential, and not real. But this does not mean shape is only "in the head." That inference itself relies on the seventeenth-century account. Because I have argued it does not follow that shape is only in the world or in the head, we must rethink the seventeenth-century account. Too many philosophers and scientists continue to endorse the seventeenth-century distinction, even when they wish to reject a particular property as primary or secondary or when they try to take into account developments in the sciences. Putnam (1987, p. 17) has noted that "the task of overcoming the seventeenth-century world picture is only begun." Likewise, Wilson has noted that the world picture "rests on assumptions which stand in need of concentrated critical scrutiny" (Wilson, 1992, p. 237). I have attempted to provide such scrutiny in light of modern science.

Notes

Research for this chapter was begun while I was a fellow at the Center for the Philosophy of Science at the University of Minnesota during the 1993–1994 academic year. Portions were completed while I was a fellow at the Oregon State University's Center for the Humanities. A version was read at the Boston Area Colloquium for the History and Philosophy of Science in November 1997. Many thanks to people at these institutions and venues for constructive criticisms and comments. Many thanks also to the editors of this volume for making a number of deep and insightful comments that greatly improved this chapter. Any errors that remain are, of course, my own.

1. Wilson (1992, p. 223) argues Locke did not take this position in contrast to many mechanistic natural philosophers of the period.

2. Very similar versions of this section have appeared in Ramsey (1997a, 1997b).

3. "If the quantum is a wave packet of extension Δx and velocity v, a certain time $\Delta t = \Delta x/y$ is needed for this packet to enter fully into the measuring device. However, because the wave packet is a free particle with energy $E = p^2/2m$, it has an uncertainty ΔE in its energy given by the relation $\Delta E = 2p\Delta p/2m = v\Delta p$. Hence, one has $\Delta E \cdot \Delta t = \Delta p v \Delta t = \Delta p \cdot \Delta x \approx h$, or $\Delta E \cdot \Delta t \approx h$ (Löwdin, 1988, p. 13).

4. The unit of reciprocal centimeters is used for wavenumbers. A wavenumber is defined as the $1/\lambda$, where λ is measured in centimeters. The symbol ν is used for wavenumbers. The relationship between energy and wave number is $\Delta E = h\nu = hc/\lambda = hc\nu$.

5. My thanks to the editors of this volume for suggesting this sentence.

6. With current technology, we can make the lower level lines appear for some light atoms. However, we cannot make them appear for even simple molecules. We can't even yet trap molecules! Even if we could, it would make an observable difference only for very small molecules.

7. We recognize this difference by assigning different molecular symmetry groups to the molecule when it is detected under different conditions. When "seen" as a trigonal pyramid, we say it has C_{3v} symmetry; when "seen" with the more sensitive techniques, we say it has D_3 symmetry (Berry, 1980).

8. This is only partially correct. The scheme of approximation chosen on the theoretical side must also be factored in. Thus, whether a molecule is represented as rigid or floppy is partially determined by the theoretical description we employ. For further discussion, see Berry (1980).

9. Examples here are numerous. Here is one: "In a bimolecular displacement reaction the attacking group attacks at the carbon atom at the rear of the bond to the leaving group. During reaction, the carbon forms a progressively stronger bond to the attacking group, while the bond to the leaving group is being weakened. During this change, the other three bonds to the central carbon progressively flatten out and end up on the other side of the carbon in a manner similar to the spokes of an umbrella inverting in a windstorm" (Streitwieser & Heathcock, 1981, p. 157).

10. This is only partly correct. The occurrent properties depend on how much matter there is and its surrounding environment as well as the manner of detection.

11. The two techniques detect the same modes as long as there is no center of symmetry in the molecule. In general, if there is a center of symmetry, there will be no fundamental lines in common in the two spectra.

12. Rather than biology or rationality, he stresses the role of the (not necessarily conscious) observer in the measurement process.

13. Again, as stated in the previous section, *why* this happens is a fundamental problem as yet unsolved in the science. However, *that* it happens is—it seems to me—without question. We should be able to do the science without having to resolve all the conceptual problems first.

14. LaPorte (1996) has questioned the semantic presuppositions of this position, and Salmon (1981) has argued the account presupposes questionable metaphysical theses of a strictly philosophical variety, i.e., consubstantiality and crossworld identification.

15. The situation seems to be quite parallel to Newton's discussion of the "universal" properties of matter. As McMullin (1978, p. 12) notes, Newton's use in Rule III of the criterion of "extendedness" rather than a particular "extension" "is so broad as to be useless. All the weight

has to be borne by the second criterion specified by the Rule: for a property to be "universal" quality, it has "to be found to belong to all bodies within the reach of our experiments.'" Interestingly, even on this second criterion, shape or figure no longer counts as universal according to the account given in this article.

16. Ask any engineer about scaling factors! Vincenti (1990, p. 139) notes that "data obtained from a model cannot be applied accurately to the full-scale prototype without adjustment." Forget the adjustment, and your model fails miserably.

17. The situation may be entirely different with students of chemistry.

18. Of course, sometimes different techniques will provide the same signature for all intents and purposes. (See the remarks on IR and Raman spectroscopy earlier in the chapter.) Thus, I expect we will get classes of signatures related to each other in quite complex ways.

References

Amann, A. 1993. "The Gestalt Problem in Quantum Theory: Generation of Molecular Shape by the Environment." *Synthese*, 97: 124–156.

Berry, R. S. 1980. "A Generalized Phenomenology for Small Clusters, However Floppy." In R. B. Woolley, ed. *Quantum Dynamics of Molecules*: The *New Experimental Challenge to Theorists* (pp. 143–193). New York: Plenum Press.

Bhushan, N., & Rosenfeld, S. 1995. "Metaphorical Models in Chemistry." *Journal of Chemical Education*, 72: 578–582.

Boyd, R. 1989. "What Realism Implies and What It Does Not." *Dialectica*, 43: 5–29.

Campbell, K. 1972. "Primary and Secondary Qualities." *Canadian Journal of Philosophy*, 2: 219–232.

Coulson, C. 1960. "Present State of Molecular Structure Calculations." *Reviews of Modern Physics*, 32: 170–177.

Culp, S. 1995. "Objectivity in Experimental Inquiry: Breaking Data-Technique Circles." *Philosophy of Science*, 62: 430–450.

Dijksterhuis, E. 1961. *The Mechanization of the World Picture*. Oxford: Clarendon.

Drago, R. 1977. *Physical Methods in Chemistry*. Philadelphia: W. B. Saunders.

Grandy, R. 1989. "A Modern Inquiry into the Physical Reality of Colors." In D. Weissbord, ed. *Mind, Value and Culture: Essays in Honor of E. M. Adams* (pp. 229–245). Atascadero, CA: Ridgeview.

Hacking, I. 1983. *Representing and Intervening*. Cambridge: Cambridge University Press.

Harré, R. 1964. *Matter and Method*. New York: St. Martin's.

Ihde, A. 1964. *The Development of Modern Chemistry*. New York: Dover.

Lange, M. 1994. "Dispositions and Scientific Explanation." *Pacific Philosophical Quarterly*, 75: 108–132.

LaPorte, J. 1996. "Chemical Kind Term Reference and The Discovery of Essence." *Noûs*, 30: 112–132.

Lewis, D. 1986. *On the Plurality of Worlds*. New York: Basil Blackwell.

Lowdin P.-O. 1988. "The Mathematical Definition of a Molecule and Molecular Structure." In J. Maruant, ed. *Molecules in Physics, Chemistry and Biology* (vol. 2, pp. 3–60). Boston: Kluwer Academic.

Locke, J. [1700] 1975. In P. Nidditch, ed. *An Essay Concerning Human Understanding*. Oxford: Oxford University Press.

Luisi, P.-L., & Thomas, R. 1990. "The Pictographic Molecular Paradigm". *Naturwissenschaften*, 77: 67–74.

Mackie, J. 1976. *Problems from Locke*. Oxford: Clarendon.

McMullin, E. 1978. *Newton on Matter and Activity*. Notre Dame, IN: University of Notre Dame Press.

Patsakos, G. 1976. "Classical Particles in Quantum Mechanic." *American Journal of Physics*, 44: 158–166.

Pauling, L. [1970] 1988. *General Chemistry* (3rd ed., altered and correct), New York: Dover.

Priest, G. 1989. "Primary Qualities Are Secondary Qualities Too." *British Journal for the Philosophy of Science*, 40: 29–37.

Primas, H. 1981. *Chemistry, Quantum Mechanics and Reductionism*. New York: Springer-Verlag.

Primas, H. 1982. "Chemistry and Complementarity." *Chimia*, 36: 293–300.

Putnam, H. 1981. *Reason, Truth and History*. Cambridge: Cambridge University Press.

Putnam, H. 1987. *The Many Faces of Realism*. LaSalle, IL: Open Court Press.

Ramsey, J. 1997a. "Molecular Shape, Reduction, Explanation and Approximate Concepts." *Synthese*, 111: 233–251.

Ramsey, J. 1997b. "A Philosopher's Perspective on the 'Problem' of Molecular Shape." In J.-L. Calais, & E. Kryachko, eds. *Conceptual Perspectives in Quantum Chemistry* (pp. 319–336). Dordrecht: Kluwer,

Salmon, N. 1981. *Reference and Essence*. Princeton, NJ: Princeton University Press.

Smith, A. D. 1990. "Of Primary and Secondary Qualities." *Philosophical Review*, 99: 221–254.

Streitwieser, A., & Heathcock, C. 1981. *Introduction to Organic Chemistry* (2nd ed.). New York: Macmillan.

Taylor, B. 1993. "On Natural Properties in Metaphysics." *Mind*, 102: 81–100.

Trindle, C. 1980. "The Quantum Mechanical View of Molecular Structure and the Shapes of Molecules." *Israel Journal of Chemistry*, 19: 47–53.

van Fraassen, B. 1981. "Essences and Laws of Nature." In R. Healey, ed. *Reduction, Time and Reality: Studies in the Philosophy of the Natural Sciences* (pp. 189–200). Cambridge: Cambridge University Press.

Vincenti, W. 1990. *What Engineers Know and How They Know It: Analytical Studies from Aeronautical History*. Baltimore: Johns Hopkins University Press.

Wilson, M. 1992. "History of Philosophy Today; and the Case of the Sensible Qualities." *Philosophical Review*, 101: 191–243.

Wimsatt, W. 1995. "The Ontology of Complex Systems: Levels of Organization, Perspective and Causal Thickets." In M. Matthen and R. X. Ware, eds. *Biology and Society*. Canadian Journal of Philosophy, Suppl Vol 20.

Woolley, R. G. 1977. "Molecular Structure and the Born-Oppenheimer Approximation." *Chemical Physics Letters*, 45: 393–398.

Woolley, R. G. 1978. "Must a Molecule Have a Shape?" *Journal of the American Chemical Society*, 100: 1073–1078.

Woolley, R. G. 1982. "Natural Optical Activity and the Molecular Hypothesis." *Structure and Bonding*, 52: 1–35.

Woolley, R. G. 1985. "The Molecular Structure Conundrum." *Journal of Chemical Education*, 62: 1082–1084.

Woolley, R. G. 1986. "Molecular Shapes and Molecular Structures." *Chemical Physics Letters*, 125: 200–205.

Woolley, R. G. 1991. "Quantum Chemistry Beyond the Born-Oppenheimer Approximation." *Journal of Molecular Structure* (*Theochem*), 230: 17–46.

8

Space and the Chiral Molecule

ROBIN LE POIDEVIN

According to classical stereochemistry, the molecules of some substances have doubles, in the sense of incongruent mirror-image counterparts. This is the phenomenon of *optical isomerism*, first identified 150 years ago by Pasteur. In some cases, the double occurs naturally; in others, it has to be artificially synthesized. These molecules thus share a geometrical feature with such familiar objects as our hands, and, indeed, it is this connection that gives the feature its technical name: *chirality* (from the Greek for hand, *kheir*). Instances of chirality in chemistry are numerous, especially in living things: examples of chiral molecules include adrenaline, glucose, and DNA.

Optical isomerism is interesting, both historically—it played a crucial role in the emergence of structural chemistry and in the attempt to link chemistry with physics—and, I believe, philosophically. I should like to take this opportunity to revisit the scene of an earlier article of mine (Le Poidevin, 1994) in which I examined the implications optical isomerism has for a philosophical debate concerning the nature of space. In that article I argued that chirality in chemistry reinforces a conclusion that Graham Nerlich (1994),[1] in a brilliant reconstruction of a famous argument of Kant's, had derived from more visible instances of chirality: that we should be realists about the geometrical properties of space. I did not, however, want to follow Nerlich (and Kant) in drawing a more radical conclusion: that we should be realists about the *existence* of space. That may sound paradoxical, but it is possible (or so I thought) to regard space as a logical construction from its contents and still think of it, *qua* construction, as possessing certain intrinsic properties that we do not merely impose on it by convention. Since then, I have become more sympathetic to Nerlich's position. Chirality *is* best understood by thinking of space as an entity in its own right. So chemistry has some lessons for the philosophy of space. But the pedagogical relation goes the other way, too: the philosophy of space has interesting implications for chemistry. In particular, it presents an obstacle to microreduction, the view that chemical properties are wholly explicable in terms of intrinsic microphysical properties. Drawing those lessons is the object of this chapter, in which I also try to respond to some of Nerlich's (1995) criticisms of my earlier article.

Before getting down to the business of examining the philosophical issues, however, let us take a brief look at the part played by isomerism in the development of chemistry.

Historical Background

> It is indeed a sign of the times, which suffer from a lack of critical spirit and an aversion to criticism, that two unknown chemists, one from a veterinary school and the other from an institute of agriculture, can decide with such assurance the most difficult problems of chemistry, which no one will be able to resolve—in particular this problem of the arrangement of atoms in space—and present their solution with a nerve that leaves the real scientists flabbergasted. (Hermann Kolbe, quoted in Jacques, 1993, p. 63)

This gloriously snobbish remark appeared in an article entitled "Signs of the Times," published in the *Journal für praktische Chemie* in 1877. Its author was the journal's own editor, Hermann Kolbe, and its objects were Jacobus van't Hoff and F. Hermann, whose crime was to propagate the theory of the asymmetric carbon atom, arrived at a few years earlier, in 1874, by van't Hoff and, independently, by Joseph-Achille Le Bel. Le Bel and van't Hoff's proposals provided a missing link in the theoretical basis of isomerism, the name given in 1830 by Berzelius to the phenomenon in which compositionally identical substances exhibit different properties. The general moral of the theory was that a proper understanding of the properties of a molecule required an appreciation of the relative positions of the atoms within it in the three dimensions of space. Kolbe's resistance to that moral has gained him the reputation of a particularly recalcitrant conservative who used his influence to delay the development of structural chemistry. He lives on, however, in the name of a process for synthesizing alkanes by electrolysis.

Berzelius had been made aware of isomerism by his student, Wöhler, who, at the time, was studying silver cyanate. On analyzing the substance, Wöhler had given it the formula AgCNO—the very same formula, he subsequently discovered, that had been given by Liebig to silver fulminate. Something, it seemed, was wrong because, whereas silver cyanate decomposed when heated, the fulminate exploded violently—the clearest indication that they were different substances. How, then, could they have the same formula? Perhaps a mistake had been made in one of the analyses. Liebig suggested as much to Wöhler, making it clear on whose side he thought the fault lay. But both analyses were correct, as Liebig established for himself. The explanation could only lie in the molecular structures of the two substances, but this was not an idea that readily gained acceptance.

Kolbe's scathing remarks remind us how much resistance to atomism survived technical and theoretical advances in chemistry. Chemists of the early nineteenth-century were careful to distance theories about the proportions in which substances combined from commitment to physical atomism. Atomism, which at the time had much in common with the original Democritean theory, was an account of the ultimate, and unobservable, structure of matter, a theory about what underlay appearances. The law of chemical proportions, by contrast, was simply an empirical generalization to the effect that simple substances combined in proportions expressible as a simple ratio. When, in 1808, the year of Dalton's *New System of Chemical Philosophy*, Gay-Lussac read a paper to the Société Philomatique of Paris, in which he stated that volumes of gases combined in simple proportions (i.e., in ratios of 1 : 1, 1 : 2, or 1 : 3) to form one or two parts by volume of product, he offered a nonatomistic explanation of the law. "Chemical action," he wrote, "is exerted more powerfully when the elements are in simple ratios or in multiple proportions among themselves." To us, the explanatory power of atomism, in contrast to Gay-Lussac's ad hoc explanation, seems obvious, but in the first half of the last century most chemists were suspicious of physical atomism, because it seemed beyond scientific verification. It belonged to what Davy called the "transcendental part of chemistry." The theory of combining proportions was based on experimental fact, whereas the theory of physical atoms had, in Berzelius's words, "only a supposition for its foundation." This refusal to engage with the structure of matter went along with a view of chemistry as essentially to do with observation and experiment, and as long as chemists took an operationalist approach, chemistry remained autonomous.

Isomerism, however, gave great impetus to the attempt to base chemistry on atomistic principles, because it implied that a substance could not adequately be characterized simply by its chemical composition. The phenomenon is also associated with another historical shift in thought. One of the two most famous instances of isomerism concerned urea. It was at one time widely believed that an irreducibly vital element separated organic from inorganic chemistry. Wöhler's synthesis in 1828 of ammonium cyanate, compositionally identical with the urea extracted from a dog's urine, and which is converted to urea on gentle heating, was at one time often cited as a blow to this vitalist conception. In fact, however, the synthesis did not make much impact at the time, probably because at least some of the materials Wöhler used in the synthesis had animal origins.

The other famous instance of isomerism was that of tartaric and racemic acid. Their compositional indistinguishability had already been noted by Berzelius, but it was not until 1848 that their relationship emerged. Louis Pasteur, then fresh from completing his doctorate, discovered that the crystals both of the double salt of tartaric acid (sodium ammonium tartrate) and that of racemic acid were asymmetric, but whereas the crystals of the tartrate all had the same shape, those of racemate had two forms, each the mirror image of the other. This suggested an explanation of a rather puzzling difference between the salts. The difference concerned their optical properties: a solution of the tartrate, like the crystalline form of the salt, rotated a plane of polarized light to the right, whereas a solution of the racemate had no effect on the light.[2] Given that the crystalline structure of the salts would, of course, be destroyed in solution, the optical difference seemed to point to a structural difference at the molecular level, one that would remain when the acid was dissolved and that could, therefore, explain the optical phenomena.[3] The reason the racemic solution had no effect was that the effects of the "right-handed" (or dextrorotatory) molecules were balanced by those of the "left-handed" (or levorotatory) molecules. Sure enough, when in a dramatic experiment Pasteur isolated the "left-handed" crystals from the racemic mixture, he found that a solution of them rotated the plane of polarized light to the left. At this point, he tells us, he rushed from the laboratory, shaking with emotion, to tell the first person he met of his result.

Pasteur clearly saw the molecular implications of his discovery. But what feature, at the molecular level, mimicked the asymmetry of the tartrate crystals? Le Bel and van't Hoff suggested that, because the valency of a carbon atom is 4, it could be represented as occupying the center of a tetrahedron, at the apices of which are other atoms. Where the atoms at the apices are all different, the result is an asymmetric arrangement of which there can be a number of permutations. Different arrangements could give rise to different physical or chemical properties. Previously, the valencies of a carbon atom were conceived of as being equally distributed in a plane, a view that was adequate for the explanation of other forms of isomerism, but not for the explanation of the optical variety. With the theory of the asymmetric carbon atom, chemistry had moved from two dimensions to three.[4]

A Puzzle about Isomerism

The original puzzle of isomerism, of course, was how substances with identical constitutional formulas could have dramatically different chemical and physical properties. That puzzle was solved by considering the arrangement of atoms in space. But the puzzle I want to consider is how isomers can have the *same* properties. What am I talking about? Let me explain.

As indicated, isomerism comes in a number of forms. *Structural* isomers differ in the arrangement of their atoms, and, in many cases, the difference is in their functional groups (i.e., a closely connected group of atoms that is responsible for the characteristic reactions of the compound, such as—OH,—COOH, etc.): that is, either the molecules have different functional groups or the groups are differently positioned. An example of functional group difference is provided by ethanol (or ethyl alcohol) and methoxymethane (or dimethyl ether). Their linear formulae are, respectively, C_2H_5OH (or CH_3CH_2OH) and CH_3OCH_3. The first of these is highly reactive when it is in contact with elemental sodium and (fortunately) liquid at room temperature. The second is relatively unreactive toward sodium and gaseous at room temperature. *Stereoisomers* have the same functional groups, but differ in the position of those groups in space. *Optical isomers* belong to this category, but do not exhaust it. The name derives from the capacity (noted in the preceding discussion) of optical isomers to rotate the plane of polarized light, a capacity that has its basis in the fact that the molecule is asymmetric—that is, has no plane of symmetry. In some cases (for example, that of glucose), only one isomer occurs naturally. Finally, among the class of stereoisomers are the *diastereoisomers* (or diastereomers), which are not mirror images of each other. Maleic and fumaric acids, for example, form a pair of diastereomers.

Thus, tartaric acid occurs in three forms (four, if one includes the racemic mixture): the optically active *dextro* and *levo* forms, which rotate the plane of light to the right and left, respectively, and which are mirror images of each other, and an optically inactive (because symmetrical) *meso* form. *Dextro-* and *levo-*tartaric acid have the same melting point (174°C), but that of *meso-*tartaric acid is much lower (151°C). *m-*Tartaric acid is also less dense, less soluble in water, and a less strong acid than are the *d-* and *l-*isomers (Eliel, 1969). In general, whereas diastereomers typically differ in their chemical and physical properties, optical isomers invariably have identical physical properties, except for their effect on the plane of polarized light. They differ chemically only in reaction with other substances whose molecules are asymmetric. Thus, for example, one way of resolving the racemic mixture of *dextro-* and *levo-*tartaric acids, as Pasteur discovered, is by causing it to react with an optically active base, such as quinine. The result is two diastereomeric salts, whose difference in physical properties thus allows ready resolution—for example, by fractional crystallization. My question is this: *Why* do optical isomers differ not at all physically (optical differences aside), and chemically only in reaction with other optically active substances?

The simple answer, I submit, is that there is no intrinsic difference between the *shapes* of the molecules of two optical isomers, and it is difference in shape that accounts for the different behaviors of other kinds of isomer. This may seem crazy. There must be *some* difference in shape, surely, for what else would explain the difference in optical activity? There is a difference, to be sure, but it is an extrinsic one. It is at this point that we need to start thinking about space itself.

Chirality and Space

Consider a question posed by Kant.[5] In a universe containing only a single human hand, what makes this hand either left or right? It cannot be the internal distances between the various parts of the hand, for these are preserved by reflection, and the mirror image of a right hand is a left hand. Nor can it be the relation of the hand to other objects in the universe, for there are none. What remains? Kant's answer was: space itself. By this he did not mean that it is the relation of the hand to *parts* of space that determines the rightness or leftness of the hand, for these properties do not vary as the hand is moved rigidly through space. Rather, it is the relation of the hand to space *as a whole* that makes the difference between a universe containing only a right hand and one containing only a left hand. Space, then, is something real in its own right. It is not simply a construction from the spatial properties of things. Were it such a construction, it could not provide the ontological basis of handedness.

The obscurity of the suggestion that handedness is a relation between an object and space as a whole may prompt us to look for other solutions. Why should some *global* feature of space determine a local feature of a hand? And what could this global feature be? The simplest alternative suggestion is to say that rightness (or leftness) is an *intrinsic* property of the hand, conceived perhaps as some collection of internal relations. From the fact that the internal distance relations between the parts of a hand are preserved in reflection, it does not follow that *all* relations between them are similarly preserved.

But, argues Graham Nerlich, this postulation of an intrinsic property does not advance things unless it is made clear what the intrinsic property is. John Earman, who, in an early article, is at least prepared to entertain the idea (see Earman, 1971, p. 9), does not offer an analysis, perhaps regarding it as a primitive property. Nerlich offers, instead, a reconstruction of the argument that preserves Kant's main conclusion (that space exists independently of objects), but that employs some decidedly un-Kantian ideas. First, we need to introduce another property: something is *enantiomorphic* if it is incongruent with its mirror image—that is, it cannot be made to coincide with its mirror image by any sequence of rigid motions. It is tempting to say that something is enantiomorphic if and only if it is asymmetric, so our right hand, necessarily, is enantiomorphic. But, argues Nerlich, the two properties are neither identical nor even necessarily coextensive. To see this, it will help to imagine an experiment in two dimensions. Think of a flat object in the shape of an R lying on a table and, next to it, a second object, which is the mirror image of the first. They have no axis of symmetry, and cannot be made to coincide by any series of rigid motions as long as they are confined to the surface of the table. But now imagine these same shapes moving in the two-dimensional space of a Möbius strip: an appropriate sequence of rigid motions will allow them to coincide. Alternatively, imagine them being permitted to move in a third dimension of space: simply lift one of the shapes off the table and turn it over before replacing it. Here, again, the two shapes coincide. So whether or not an asymmetric object is enantiomorphic depends on the geometry of the space in which it is embedded. What obtains in the two-dimensional case can be generalized for three or more dimensions. Thus, in an orientable space of *n*-dimensions, an *n* dimensional-asymmetric object is enantiomorphic. In a nonorientable space of *n*-dimensions, or a space of $n^{+}1$ dimensions, the object is no longer enantiomorphic.[6] But if, in these nonstandard spaces, an asymmetric object and its mirror image can be made to coincide without any alteration of shape, then the shape of the two objects must be the same. Nerlich draws two conclusions. The first is that Kant's intuition about handedness was wrong: a solitary hand need be neither determinately right nor determinately left. The second is that, nevertheless, Kant's conclusion was right: some feature of the hand does depend on space considered as a whole—that is, on a global rather than a local property of space. To be specific, whether a hand is enantiomorphic or not depends on the geometry of space. Hence, space exists in its own right, a position we shall call *realism*. In opposition to realism is *relationism*, which treats space as a logical construction out of, and so having no existence apart from, the spatial properties of and relations between its various contents, such as material objects or electromagnetic fields.

Before proceeding, some further remarks about the realism/relationism issue are in order. The relationist denies that space is an object, persisting through time. What then is it? A system of relations, which of course cannot exist unless there are also *relata*. An analogy with causation may help here. When we say that causation exists in the world, we are not naming a further entity, on a par with trees or electrons. We are asserting the existence of causal relations between things (or, if you prefer, events). We do talk of causes, and these may be objects, but to describe them as such is simply to say that they stand in certain relations to other things. Similarly, for the relationist about space, to say that something is in space is simply to say that it stands in spatial relations to other things. For the realist, however, space is more than a system of relations. What stands in these relations are spatial points, and other items are spatially related by virtue of occupying (a number of) such points. Spatial relations are mediated by these points, in the sense that I stand in certain spatial relations to you by virtue of there being space between us.

Now back to Nerlich's reconstruction of Kant. We have introduced three properties, or, if you prefer, concepts: asymmetry (in the case of three dimensional objects, having no plane of symmetry), handedness (the respect in which the left hand differs from the right), and enantiomorphy (incongruence with one's mirror image). And we have also introduced a distinction between two kinds of property, which we should now make more formal: *F* is an *intrinsic* property of *x* if and only if *x*'s being *F* does not logically depend on the properties, existence, or nonexistence of any object other than *x*. *F* is an *extrinsic* property of *x* if and only if *x*'s being *F* does logically depend on the properties, existence, or nonexistence of any object other than *x*. Being spherical, made of gold, and at a temperature of 62° C are intrinsic; being a great-grandmother, universally admired, and residing in Surrey are, by contrast, extrinsic. The distinction can be characterized in causal, as well as logical, terms: if an intrinsic property of *x* has causal effects, then these must include effects in *x*'s immediate vicinity. Thus, being at a certain temperature may have distant effects, but only via local effects. When *x* becomes a great-grandmother, in contrast, there need be no effects at all in the local vicinity of *x*. Finally, *x* need not currently exist in order currently to exhibit (at least some) extrinsic properties, such as being famous, but it cannot currently exhibit any intrinsic properties without currently existing. Now this distinction, between intrinsic and extrinsic properties, can be applied as much to space as to the objects within it. However, in what sense space can be said to have intrinsic properties, such as geometrical or topological properties, will depend on one's position vis-à-vis the realism/relationism debate. For the realist, the existence of space is logically independent of its contents, so at least some of its properties (and for Nerlich these include geometrical properties) will also be logically independent, and so intrinsic. For the relationist, *any* properties of space will obtain by virtue of the properties of its contents. But this does not make all properties of space extrinsic. Quite the reverse, since space is itself a construction from those contents, what properties it has in virtue of those of its contents will be intrinsic to space.

So, of those three properties—asymmetry, handedness, and enantiomorphy—which is extrinsic, and which intrinsic? Because asymmetry, or lack of it, seems to be determined entirely by internal relations between the parts of an object, this property is surely appropriately regarded as intrinsic. This might be thought to conflict with the experiment just described. For, although an R is asymmetric when confined to a plane, once it is lifted into a third dimension, a plane of symmetry appears (try slicing it horizontally). But this is a cheat. A genuinely two-dimensional R remains asymmetric whatever the dimensionality of the space in which it is embedded: it doesn't have an upper and a lower side that can be separated by slicing the R.

But what of the other properties? Are they intrinsic or extrinsic? And what is the relationship between asymmetry, handedness, and enantiomorphy? Three positions offer themselves for scrutiny (I postpone for the moment the question whether they are all equally coherent):

1. Handedness is intrinsic (to the asymmetric object). Therefore, enantiomorphy is also intrinsic because any rigid motion by definition changes only external relations but leaves intrinsic shape unaltered. On this view, enantiomorphy and asymmetry are necessarily coextensive. Handedness, however, is a further property because being asymmetric does not determine whether an object is right or left handed.
2. Handedness is extrinsic—a matter of an object's relation to other asymmetric objects—but enantiomorphy is intrinsic. Again, on this view, asymmetry and enantiomorphy are necessarily coextensive.
3. Both handedness and enantiomorphy are extrinsic. On this view, all three properties are distinct because asymmetry alone is intrinsic, and handedness and enantiomorphy depend on different things: whether a hand is left or right depends on its relation to other asymmetric objects; whether an object is enantiomorphic or not depends on the global geometrical properties of space. Kant's lone hand, for instance, is neither determinately left nor determinately right, but (given that it is embedded in an orientable three-dimensional space) it is enantiomorphic.[7]

On the face of it, whereas (1) and (2) are compatible with relationism, (3) implies realism (though see the later remarks on Earman).

How are we to choose between these positions? Nerlich recommends position (3) on the basis of its clarity and explanatory power. Treating handedness and/or enantiomorphy as intrinsic is not the provision of an analysis, but the refusal to provide one. Positions (1) and (2) just leave these properties obscure. Position (3), in contrast, offers an explanation of why handedness must be extrinsic: because in certain nonstandard spaces rigid motions of an asymmetrical object *can* be equivalent to reflection. We can readily visualize this in the case of two-dimensional enantiomorphs, as we already saw, and although the corresponding three-dimensional case is beyond our visual imagination, we can at least appeal to analogy.

Against Nerlich, and in defense of (1), one could urge that conceiving of handedness as an intrinsic property does not obscure it, because we can just *see* the difference in shape between our hands. However, I think we can decide the fate of (1) on grounds other than those of explanatory power or epistemological accessibility. Position (1) is false, and I believe that chemistry shows that it is false. Hands and chiral molecules are all part of the same phenomenon. So if being left is an intrinsic feature of a hand, being the *l*-isomer is an intrinsic feature of a sample of, say, lactic acid. Optical isomers, like hands, are enantiomorphic, which is why they are often called *enantiomers*. Now, sooner or later, a difference in the intrinsic properties of two items will show up in the causal interactions of those items: an intrinsic difference makes a causal difference. No intrinsic property of a concrete object is causally impotent. However, as noted in the preceding section, enantiomers have the same melting-point, solubility, and density. In a chemical reaction with symmetric substances, the enantiometers are indistinguishable. In cases where there is an intrinsic difference, as between *d*- and *m*-tartaric acid, for instance, and with other geometrical isomers, the intrinsic difference shows up in a difference of melting point. Thus, enantiomers are intrinsically identical. They do, stand in different relations to asymmetric substances however, and this is why differences emerge in reactions with such substances.

Further, the behavior of isomers tells us something about the geoetrical properties of space, for were space four dimensional, even extrinsic differences between enantiomers would disappear. So, too, would differences (such as melting point) between pure isomers and racemic mixtures.

We are still left with a choice. Or are we? In my earlier article. I argued for the merits of (2), on the grounds that it preserved the insight about handedness (that the property is extrinsic) while avoiding commitment to space as an independent entity because it posited no property that depended on a relation between objects and space. I further suggested that thought experiments involving nonorientable or four-dimensional spaces, insofar as they presupposed that the properties of space were independent of the properties of embedded objects, begged the question. But (2) is not a stable position. Essentially, the debate about chirality and space can be boiled down to one question: Is *the difference between left and right intrinsic or extrinsic?* If the answer is "intrinsic," then we have position (1).[8] If the answer is "extrinsic," then enantiomorphy is extrinsic. For if the difference between a chiral object and its mirror-image counterpart is extrinsic, then, by definition, the difference depends on something outside those objects. If this something is absent, the difference disappears, and the counterparts become congruent. So enantiomorphy is not intrinsic to the asymmetric object. But if it is extrinsic, what could it depend on? If it depends on a property of space itself, then we are led to position (3). Could it depend instead on relations to other objects? This leads us to the following position:

4. Both handedness and enantiomorphy are extrinsic properties of the *hand*, because both depend on relations to external objects. Because of this dependence, enantiomorphy is an *intrinsic relational property* of *pairs* (or other collections) of asymmetric objects.

As I read him, this is the position John Earman (1989) thinks is at least defensible. But it is hard to be sure. What he actually says is strongly suggestive of (3). For instance, his reason for rejecting (1) is that it conflicts with the fact that an asymmetric object is congruent with its mirror image in nonorientable space. But Earman, ultimately, wants to *reject* realism. So the orientability of space (and, indeed, any other geometrical property) must be reducible to a kind of relation between objects. What kind of relation could that be? Earman considers one candidate, that of multiple connectedness (p. 144), but then rejects it as inadequate. There, unfortunately, he leaves the issue of how relationism could account for orientability.[9] This rather confirms Nerlich's assertion that all the relationist can do is gesture toward the property—he knows not what—that grounds enantiomorphy.

I have to confess that my earlier attempt to find a plausible an intelligible *via media* between (1) and (3) was not successful. Number (4) is a consistent position, but, undeveloped, it is an uninteresting one. It explains nothing. As Nerlich says, ontological economy is not the only virtue in a theory: there is also clarity, elegance, intuitive appeal, and explanatory power. All these characterize position (3). Contemplating nonorientable spaces, or spaces of higher dimensions, does help us see why right- and left-handed isomers do not, after all, have different shapes. The puzzle of optical isomerism—namely, the physical indistinguishability of optical isomers—is solved.

Some Objections, and an Observation about Chemical Identity

In this final section, I try to meet some objections to my argument.

The first concerns my appeal, in the last section, to a causal principle: that there are no "silent," or causally impotent, properties. I appealed to that principle in my 1994 article, but managed to give a rather unfortunate statement of it, as follows: "If a property is intrinsic, then it has characteristic causal consequences: an intrinsic difference makes a causal difference" (Le Poidevin, 1994, p. 83). Nerlich, entirely reasonably, objected to this:

> What are the characteristic causal consequences of having a shape, of occupying a region which is a congruent counterpart of a hand? I see no characteristic causal consequences at all. Any consequences depend on what fills the region. . . . A right-handed volume of air doesn't even look different from a left-handed one filled with air, since they'll both be invisible. (Nerlich, 1995, 440.)

I would have been wiser to stick to the weaker principle—that an intrinsic difference makes a causal difference of some kind or other—because that is all I needed for my argument. Why should we accept the weaker principle? Well, first, I challenge anyone to produce a clear counterexample. Second, we might point to the causal characterisation of intrinsic properties offered earlier. According to this, if an intrinsic property of *x* has causal effects, these must include effects in *x*'s immediate vicinity. Now suppose we allow the existence of "silent" properties: ones that could not have any causal effects whatsoever. Then the characterization becomes trivially true, because the antecedent cannot be fulfilled. Any silent property is thus, trivially, an intrinsic property. So the admission of silent properties undermines a useful means of characterizing the intrinsic/extrinsic distinction.

A second objection concerns the relevance or otherwise of enantiomers. Nerlich was skeptical that the appeal to chemistry significantly advanced the argument: "The facts about enantiomers and isomers seem to yield only somewhat more detail than . . . humdrum features of hands" (Nerlich, 1995, p. 440). But I think there is an advance, for two reasons. The first is an epistemological one. Because left and right hands just look different, it is hard to resist the idea that they have different effects, even when interacting with symmetric objects. But when we are dealing with isomers, looks do not enter the picture: the property they share with hands, as indicated earlier, is purely geometrical. We have a number of objective and readily describable means of detecting intrinsic physical differences between substances, and in every case but one, no difference between optical isomers emerges. It is not enough simply to point to the possibility of nonorientable spaces to defeat the idea that chirality is intrinsic, for it would always be open to a particularly recalcitrant relationist to deny the intelligibility of such representations of space. The chemical phenomena provide a much less resistable means of showing that chirality must be extrinsic. The second reason for appealing to chemistry has to do with one rather obvious difference between hands and molecules: that of scale. Because molecules are much smaller than hands, they can indicate, in a way hands cannot, the microproperties of space. Thus, supposing that the curvature of the fourth dimension was so great that ordinary macroobjects were unaffected, the evidence for such a dimension would emerge only at the microlevel. And we can trace chirality to lower levels than that of the molecule. Parity nonconservation among the elementary particles is yet another instance of chirality. Here, though, it has a temporal dimension, which we may illustrate by the example of a rotating cone (Gardner, 1982). As a static object, a cone is symmetric. But once it begins to rotate in a clockwise direction when viewed from above (i.e., with the apex pointing toward one) it is no longer superposable onto its mirror image, a cone rotating anticlockwise. Subatomic analogues of the rotating cone include the beta-decay of cobalt-60.[10]

The third objection is the most serious, and the proximity of this chapter to the ones by Ramsey and Weininger in this volume gives the objection particular urgency. The whole discussion has been situated within the context of classical stereochemistry, where it is axiomatic that the molecule is a thing with a shape. Now we do not have to assume that the molecule is a purely static entity: each atom, and each conglomeration of atoms, is a dynamic system. Not long after van't Hoff's theory became generally accepted, the molecule was represented as a far more nebulous entity than the rigid arrangement of indivisibles pictured by Dalton. But the point, which van't Hoff and others were urging, about the importance of spatial structure survived the revisions in our view of the atom pioneered by Thomson and Rutherford. However, it is far less clear that molecular shape has survived the advent of quantum physics.[11] As Guy Woolley puts it, "Quantum mechanics can predict fairly accurately the way the energy of a molecule may change, but strictly speaking it says *nothing* about the shape of the molecule." (Woolley, 1988, p. 53.) As a consequence, the quantum theory of a molecule "does not give an account of molecular isomerism . . . most striking of all, it fails to explain how certain molecules can exist as optical isomers" (p. 57). Now, one response to this is to say that, because quantum theory does not retain the notion of molecular shape, there is no such thing as the shape of a molecule and, that therefore, stereochemistry is based on a fiction. If this were the end of the matter, then of course the whole argument of this chapter would collapse: we would simply have been drawing pictures in the air.

But there are other responses. One, which seems to be Woolley's own position, is to keep molecular shape as a useful metaphor (Woolley, 1988 p. 2). After all, the centrality of the notion of shape in a chemical understanding of reactions between substances suggests that it cannot simply be dispensed with. Thus, we are led to an instrumentalist interpretation of chemical structure. Theoretical statements within classical stereochemistry are then no longer regarded as descriptions of the world that are either true or false, but rather as useful fictions that generate successful predictions. Another is to question, as Jeffry Ramsey (1997) does, the basis of the inference from "molecular shape does not precisely correspond to any feature at the quantum level" to "molecular shape is not an objective feature of the world." Why should we accept a form of reductionism which insists that any real property is one that can be mapped precisely onto some well-defined quantum state. Ramsey points to an analogous debate over folk psychology in the philosophy of mind. Philosophers who have argued for the elimination of folk psychology (specifically, talk of beliefs and desires) have pointed to the lack of fit between folk psychological concepts and neurological descriptions. A particular mental state does not correspond to a naturally isolable state of the nervous system. But why should we expect it to do so? All that matters is that some complex neurological state should determine the emergence, at the macrolevel, of the mental state. Now, similarly, we have what Ramsey dubs "folk molecular theory," in which molecular shape plays as central a role as beliefs do in folk psychology. Eliminativists point to the absence of any naturally isolable feature at the quantum level that corresponds to shape. But this requirement, as with folk psychology, may be too stringent. What we need, to reconcile stereochemistry with quantum physics, is a different, less rigid, form of reductionism. Yet another position, and one that directly connects with the issues introduced in the last section, is to return a negative answer to a question posed by Steven Weininger: "Is molecular structure an intrinsic molecular property, the existence of which is independent of experimental conditions?" (Weininger, 1984, p. 939). That is, the property of shape may emerge only when certain aspects of the molecule are measured. As Ramsey puts it, "shape is a response property rather than an intrinsic property." (Ramsey, 1997, p. 47.) If this is so, of course, we can no longer assent to the suggestion in the last section that asymmetry is an intrinsic feature of the molecule. Finally, we could simply insist that optical isomerism is a counterexample to the thesis that the world is, in principle, describable wholly in terms of quantum physics, a position as uncompromising toward physics as eliminativism is to chemistry.

Which of these responses to the problem of molecular shape is to be preferred is a matter I shall leave to others. The mere fact that there are alternatives to eliminativism about shape shows that the preceding discussion of the philosophical implications of optical isomerism is not necessarily otiose. More than that, however, there are interesting connections to be drawn between the issue of isomerism and space on the one hand and the debate about molecular shape on the other. The argument of this chapter has been that optical isomerism presents a difficulty for relationism about space—the view that space is simply a construction from spatial relations between things. This would seem to support realism about space, given that realism and relationism exhaust the possibilities. But what if we adopt an instrumentalist view of molecular shape? Then the logical connections between enantiomorphism and space might suggest a corresponding instrumentalism about space itself. On this view, talk of spatial points and of geometrical properties of space is not to be construed as literally descriptive of the world, but neither should this talk be reduced to something else: it is simply a useful fiction. Instrumentalism, or fictionalism, about space is a rival, and a somewhat neglected one, to both realism and relationism. Of course, instrumentalism about molecular shape does not entail instrumentalism about shape per se. Philosophers could still point to more familiar instances of chirality in support of realism. But then the world of macroobjects, and how we conceive of it, is not entirely insulated from developments in physics. It may be that the notion of shape as an intrinsic and objective feature of objects is just as much open to doubt in the context of hands and corkscrews as in the context of the salts of tartaric acid. But now we are drifting into speculation. It is time to take stock.

The results of this meeting between chemistry and the philosophy of space have benefited both sides. Chirality among molecules provides an argument, albeit an incomplete one, for realism about space. In turn, spatial realism has helped us to a deeper understanding of the difference between optical and other forms of isomerism. Are there any further lessons to be drawn? I think there are. One is that a certain form of microreductionism in chemistry is false: this is the thesis that the chemical identity of a substance is wholly determined by the intrinsic microphysical properties of its components. Historically, isomerism led us some way along the path toward microreduction as it showed that a chemistry which entirely ignored the structure of molecules could not provide a satisfactory account of chemical identity. Substances are more than their composition. But enantiomerism I think encourages us to take a step in the other direction. For, as we saw in the last section, a full understanding of the identity of an enantiomer assigns a role to a global property of space. That there is a difference between the *d* and *l* forms of a substance depends on spatial geometry, and so is not wholly determined by the intrinsic microphysical properties of the molecule (or of entities at a deeper level of analysis). This hardly restores chemistry to the position of near-complete autonomy from physics that it enjoyed in the eighteenth-century, but it does suggest that in chemistry one sometimes needs to look further up, rather than further down, the hierarchy of nature.[12]

Notes

1. I refer here to the revised second edition of *The Shape of Space*, though this chapter addressed the original reconstruction in the first edition.

2. Light typically oscillates in all directions perpendicular to the light source. But when it passes through certain materials, such as transparent calcite (Iceland spar), the oscillation is confined to one direction, or plane, and the light is said to be polarized.

3. The suggestion that the optical activity of quartz crystals might be associated with some property of the molecules was made as early as 1824 by Augustin Fresnel.

4. For accounts of the resistance to atomism and the rise of structural chemistry, see Brock (1967, 1992) and Knight (1992). Mason (1976) provides a historical survey of stereochemistry. Some biographical background to Pasteur's researches can be found in Jacques (1993).

5. See Kant (1768). By the time of the *Inaugural Dissertation* (1770), however, he had radically changed his views on the nature of space. By that time, he had rejected both the Leibnizian reductionist view of space and the Newtonian realist account. See Earman (1989, chap. 7) for some interesting suggestions about the changing role incongruent counterparts played in Kant's writing on space.

6. A word about the terms used in this paragraph. A *Möbius strip* can be formed by taking a long strip of paper, giving it a single twist, and then joining the ends. A motion of an object is *rigid* if it preserves all the internal distance relations between the object's parts. A space is *orientable* if any object that exhibits an orientation (including handedness, or the direction in which an object is rotating) preserves that orientation; however, the object is moved rigidly through space. An example of a nonorientable two-dimensional space that has no edges is the surface of a *Klein bottle*: the surface has only one side. The definition of enantiomorphy I have adopted in this chapter is Nerlich's. Earman (1989) proposes a characterization of enantiomorphy that makes it a local property: "An object *O* is an enantiomorph just in case there is a neighborhood *N* of *O* such that *N* is large enough to admit reflections of *O* and that the result of every reflection of *O* in *N* differs from the result of every rigid motion of *O* in *N*" (p. 141).

7. Could an object be handed, in relation to its neighbors, and yet fail to be enantiomorphic—for example, because it was embedded in a four-dimensional space? Not on the definitions we have adopted, for the difference between hands depends on enantiomorphy. However, an object could be globally nonenantiomorphic (i.e., in relation to space as a whole) and locally enantiomorphic (see Earman's definition in note 6). This, suggests Earman, is all that is needed for something to be handed. However, as Nerlich (1995, p. 434) points out, local enantiomorphy is neither more nor less than asymmetry.

8. No wonder Nerlich was puzzled by my earlier article. He could not reconcile what I took to be the lesson of optical isomerism (that chirality is extrinsic) with my insistence that enantiomorphy is intrinsic. In the end, rather than accuse me of incoherence, he charitably assumed that I was arguing for position (1), and took the excursion into chemistry as little more than a distraction.

9. "I will not speculate here" he writes "on how the relationist might respond to this difficulty" (Earman, 1989, p. 145). There is then a footnote to this remark in which he speculates on how the relationist might respond to this difficulty. He considers Sklar's suggestion that the relationist needs to appeal to purely possible objects to ground enantiomorphy (see Sklar, 1974). But, Earman concludes, the controversy over space is independent of the legitimacy or usefulness of *possibilia*.

10. In a famous experiment in 1957, Chien-Shiung Wu subjected a supercooled sample of radioactive cobalt-60 to an electromagnetic field, causing the nuclei to emit electrons in just two directions: north and south (where the north to south direction is defined as the direction with respect to which the nucleus is rotating anticlockwise). She found that more electrons were being emitted from the south end of the nucleus than from the north.

11. See Weininger, (1984), for a useful summary of the debate.

12. I am very grateful indeed to Nalini Bhushan and Stuart Rosenfeld for very detailed and helpful comments on an earlier version of this essay. I am particularly grateful for their drawing my attention to the importance of the debate over whether, and in what sense, molecules can be said to have shape.

References

Brock, W. H., ed. 1967. *The Atomic Debates: Brodie and the Rejection of Atomic Theory*. Leicester: Leicester University Press.

Brock, W. H. 1992. *The Fontana History of Chemistry*. London: Fontana.

Earman, John. 1971. "Kant, Incongruous Counterparts, and the Nature of Space and Space-Time." *Ratio*, 13: 1–18.

Earman, John. 1989. *World Enough and Space-Time*. Cambridge, MA: MIT Press.

Eliel, E. 1967. *Elements of Stereochemistry*. New York: Wiley.

Gardner, Martin. 1982. *The Ambidextrous Universe* (2nd ed.). London: Penguin.

Jacques, Jean. 1993. *The Molecule and Its Double* (trans. L. Scanlon). New York: McGraw-Hill.

Kant, Immanuel. 1768. "*Concerning* the Ultimate Foundation of the Differentiation of Regions in Space." In Kerferd and Walford, 1968, 36–43.

Kant, Immanuel. 1770. *The Inaugural Dissertation*. In Kerferd and Walford, 1968, 45–92.

Kerferd, G. B., & Walford, D. E., eds. 1968. *Kant: Selected Pre-Critical Writings*. Manchester: Manchester University Press.

Knight, David. 1992. *Ideas in Chemistry: A History of the Science*. London: Athlone.

Le Poidevin, Robin. 1994. "The Chemistry of Space." *Australasian Journal of Philosophy*, 72: 77–88.

Mason, Stephen F. 1976. "The Foundations of Classical Stereochemistry." *Topics in Stereochemistry*, 9: 1–34.

Nerlich, Graham. 1994. *The Shape of Space* (2nd ed.). Cambridge: Cambridge University Press.

Nerlich, Graham. 1995. "On the One Hand: Reflections on Enantiomorphy." *Australasian Journal of Philosophy*, 73: 432–43.

Ramsey, Jeffry. 1997. "Molecular Shape, Reduction, Explanation and Approximate Concepts." *Synthese*, 111: 233–251.

Sklar, Lawrence. 1974. "Incongruous Counterparts, Intrinsic Features, and the Substantivality of Space." *Journal of Philosophy*, 71: 227–290.

Weininger, Steven. 1984. "The Molecular Structure Conundrum: Can Classical Chemistry Be Reduced to Quantum Chemistry?" *Journal of Chemical Education*, 61: 939–944.

Woolley, R. G. 1982. "Natural Optical Activity and the Molecular Hypothesis." *Structure and Bonding*, 52: 1–35.

Woolley, R. G. 1988. "Must a Molecule Have Shape?" *New Sc entist* (22 October): 53–58.

9

Butlerov's Vision

The Timeless, the Transient, and the Representation of Chemical Structure

STEPHEN J. WEININGER

In a now famous article, Alexander Mikhailovich Butlerov declared that

> there will be possible only ONE such rational formula for each substance. If then the general laws will have been derived which govern the dependence of the chemical characteristics of the substances on their structure, such a formula will express all these characteristics. . . . Time and experience will teach us best how the new formulas will have to appear if they are to express chemical structure. (Butlerov, [1861], 1971, p. 291)

Butlerov's statement contains a multiplicity of claims, implicit and explicit, with respect to ontology, epistemology, and representation. He clearly believes that each substance has a single structure that makes it possible to represent it by means of "one . . . rational formula." Butlerov further implies that all chemical characteristics of a substance are knowable, as well as being representable, by means of a single structural formula. The boldness of Butlerov's assertion is perhaps related to its having been advanced at a particular historical moment. In the decade or so preceding Butlerov's article, chemists' ability to deduce chemical structure had been greatly aided by the reformation of atomic weights and the rise of the valency concept (Brock, 1993, chaps. 6 and 7). At the same time, chemical synthesis was still in its infancy, and the number and type of compounds whose chemical characteristics had to be accounted for had not yet reached astronomical proportions.

Butlerov's prophecy has been taken as heralding the first step toward the creation of molecular representations with which all chemical properties of a substance could be rationalized. Some imagine these to be iconic formulas of a "photographic" fidelity. Others assert that the ultimate representation of a chemical substance is a mathematical equation—the molecular wavefunction—that would permit not only the rationalization but even the prediction of the material's chemical behavior (Mosini, 1994). One of my theses is that no such single "rational formula" exists or could exist. However, exploring why Butlerov's vision is unrealizable can shed light on what entities chemists see as "fundamental" and the role of time within chemistry.

At present, quantum mechanics is the most comprehensive theory we have to explain "the chemical characteristics of substances." It is a holistic theory that tells us (1) that all particles in the universe are "entangled" with one another and (2) that it is we who divide the world into object and context:

> We can no longer adhere to the old dream of the existence of a single frame of reference that eliminates the pluralism of physical, chemical and biological theories. Every description of nature depends on the division of the world into a part whose effects are to be considered, and a part whose effects are to be ignored. . . . What counts as a phenomenon depends on the breaking of the holistic symmetry of the world by division and abstraction. Adopting different partitions, we will in general observe different phenomena. (Primas, 1980, 41)

There are pragmatically sound reasons for making the divisions we do. When we boil a liquid, it is converted into a vapor that seems to contain microscopic particles capable of penetrating barriers impervious to the bulk liquid. Furthermore, the vapor may be recondensed to the self-same liquid. In this instance, it makes good sense for us to say that the substance is "composed of" these microscopic entities, which we call molecules. In other cases, molecular explanations for substantial properties are less effective; in these instances, we might decide to invoke submolecular entities such as electrons or supermolecular entities that could vary in size from the oligomolecular (of the dimensions of a few molecules) to the macroscopic.

The chemist's criteria for distinguishing between object and context are usually based on energy differences. Thus, in the example just given, the fact that molecules survive at temperatures sufficient to dissipate liquids contributes to our belief that the molecule is a "fundamental" constituent of matter. Because of the fundamental constitutive status accorded molecules, molecular *structure* is ipso facto seen as a "fundamental" property of substances. And many chemists believe there to be a "best" molecular structure, the one of lowest energy, which is associated with a quantum mechanical stationary state.

However, molecules and supramolecular aggregates are not static. Even when they are confined to a narrow average temperature range, they exhibit a wide variety of structures because they experience a wide variety of internal motions, which operate on several different time scales. Furthermore, we explore these structures by means of chemical and physical probes that each interrogate only a portion of the wide time domain within which chemical phenomena occur. For instance, infrared spectroscopy shows us a range of values for a particular absorption frequency, reflecting the range of environments experienced by the molecules in the sample. By contrast, nuclear magnetic resonance spectroscopy, which operates on a slower time scale, often gives us only average values. Which "picture" is "correct?"

The foregoing discussion illustrates the considerable difficulties attending the representation of supermolecular aggregates whose composition varies with time; even more severe are the problems of calculating their properties.

In this Chapter I claim that the temporal aspect of chemical structure has been neglected relative to the energetic one and offer an historically based explanation for this neglect. I then discuss recent work that enriches our conception of molecular and supermolecular structure by attending more closely to their temporal dimension. The final pages contain my assessment of why this story in particular and chemistry in general might merit the attention of philosophers.

How Might Chemical Structure Be Expressed?

We can approach the question of how chemical structure should be expressed by asking what Butlerov himself had in mind.[1] He explicitly denied being able to draw any conclusions "at the present time" about "the atomic arrangement in the interior of the molecule" or the "relationship that exists between the chemical influence and the mutual mechanical arrangement of the atoms." Furthermore, after claiming that the "chemical nature" of a molecule depends on the number and arrangement of its constituent atoms, Butlerov tells us that he is "far from alleging that such a law would be absolutely valid and cover everything; . . . a further follow-up . . . would show how far this law . . . is valid and what is left out" (Butlerov, [1861] 1971, p. 290; see also Rocke, 1981, pp. 35–38).

The statements of Butlerov and his contemporaries allow us to infer reliably what meaning he attached to "molecular structure." For him it denotes the grouping of atoms within the molecule as determined by the valencies of those atoms, but not their actual spatial positions. (Distinguishing between "chemical atoms"—the smallest entities conserved in chemical processes—and "physical atoms" was standard in Butlerov's time.) It is worth remembering that in the nineteenth-century virtually all information about chemical structure was derived from chemical transformations. Given their inherently dynamic character, it seemed doubtful to contemporary chemists that chemical reactions could yield reliable information about the physical ("mechanical") structure of molecules (Rocke, 1981, pp. 30–38). Even more abstruse was the relationship between the "mechanical" and "chemical" structures of molecules.

Under these circumstances there were many possible "rational formula[s] . . . to express chemical structure." To illustrate the nature of their relationships I'll use an analagon and assume the object being studied is a power supply. In one approach, our representation would consist of nothing more than a list of settings for the various switches and dials set beside the different outputs each would produce. There might also be one or more algorithms useful for predicting the best combination of settings to produce a desired output, even a simple heuristic model of the device. There might be little or nothing to show regarding the internal "structure" of the power supply because it might be permanently sealed and/or we might consider it *irrelevant* to *our principal interest*, which is getting the outputs we want when we want them. A representation of this type would presumably appeal to chemists whose views matched those of Charles Gerhardt, the nineteenth-century chemist who famously declared that there could be "as many rational formulas as there are reactions" (quoted in Crosland, 1962, p. 331).

At the other end of the spectrum, our goal might be a set of differential equations that describe the behavior of each of the components and their interactions. These would likely attract anyone who aspired to reduce chemical phenomena to mechanics, whether classical (in Butlerov's time) or quantum (in our own). Butlerov explicitly rejected Gerhardt's option while expressing considerable doubt about when, if ever, the second option would be attainable. As it happened, the subsequent history of chemical structure has seen the development of a number of intermediate representations.

What might these intermediate representations look like? In the case of the power supply, one could use a photograph of the inner structure of the device, which would tell us accurately how it looks but rather less about how it functions. Alternatively, we could aim to produce a circuit diagram, which could convey a good deal of information about the device's function without much resembling it. Analogously, chemical signs could aim at representing composition, or reactivity under a particular set of conditions, or potential synthetic derivations, or transportability through a lipid membrane, and so on, without limit. Note that the category of "chemical characteristics" was not defined by Butlerov, and it is not clear that it would be definable, in large part because chemistry is the science that keeps *creating* new objects of study which exhibit previously unanticipated "chemical characteristics." So it is hardly surprising that chemists have entertained a range of representations for any single substance because of differing conclusions over what was known and knowable, over what best served their needs and interests, *and what was most appropriate to the particular communicative situation.*[2]

Among the chemical characteristics that virtually any chemist would want a representation to convey are a substance's dynamic properties, particularly its reactivity under a variety of conditions. It goes without saying that any attempt to portray dynamic properties in a static medium is handicapped from the outset and entails substantial compromise and loss (Lawrence, 1971). In chemistry, this problem has been complicated, and to some degree concealed, by the elision of two meanings of the term *reactivity*. One meaning refers to the likelihood of a substance being transformed in a certain way under a specific set of conditions; a second meaning refers to the actual dynamic process that accomplishes this transformation. In chemical terms, the first meaning is oriented around rationalizing and predicting the structures of the products to be expected from a particular reaction, and the second alludes to depictions of the pathways and calculations of the rates of those reactions.

I and others have raised doubts about the extent to which molecular structure can be reduced to quantum mechanics (Woolley, 1978; Primas, 1980; Weininger, 1984; for a contrary view, see Mosini, 1994). Furthermore, by the time quantum mechanics found systematic application to chemical problems, chemists had developed a system of structural formulas that is astonishingly effective at guiding the prediction of reaction products ("reactivity" in the first sense). For example, an able undergraduate using this system could correctly anticipate the outcome of a number of organic reactions he or she had never seen before. This same system of representation also plays a central role in our attempts to analyze and anticipate the actual rates of chemical processes ("reactivity" in the second sense). I would argue that its use in this second context was symptomatic of the relative neglect of dynamic aspects of chemical processes, and to the relegation of time to a subordinate status within chemical thought.

Time and Chemical Structure—Architecture versus Dynamics

It is a curious fact that although chemistry is first and foremost the science of transformation, discussions of time in the chemical literature are quite meager compared to discussions of energy. There are innumerable works addressing the energetics of chemical reactions and structures, yet a far smaller number are devoted to examining the temporal aspects of these subjects. I do not mean that the rates of chemical reactions are slighted but, rather, that differences in *time scale* among various physicochemical events, and the consequences of those differences, receive rather little notice.

Processes that generally interest chemists require energies between about 10^{-12} eV (the energy of a nuclear transition in a typical NMR spectrometer; eV stands for electron-volts, a unit of energy) and 10 eV (energy sufficient to break several bonds in the same molecule)—a span of 13 orders of magnitude. By contrast, the *time* span of chemically significant processes covers at least 20 orders of magnitude—from about 10^{-15} seconds for an electronic transition in a photoexcited molecule to about 10^{5} seconds, the half-life for a slow but measurable first-order reaction. Yet far more attention is paid to interpreting small differences in energy than in thinking about the consequences of having multiple processes take place simultaneously on very different time scales or about the relationship between the phenomenological kinetics of reactions under ordinary conditions and the rates of those molecular events that presumably underlie them.

How should we account for this relative neglect of the dynamic dimension? Ted Benfey admits to being "tempt[ed] to characterize the history of chemistry as a continual attempt to avoid the explicit incorporation of time into chemical theories" (Benfey, 1963, p. 574). This was not a universal tendency throughout history—the first half of the nineteenth-century saw a general preoccupation with chemical transformations (Kim, 1992), and chemists such as Alexander Williamson took the very ubiquity of such transformations as prima facie evidence for the dynamic character of molecules. Williamson also believed, in common with many of his contemporaries, that the study of change is what makes chemistry a unique science and, in particular, differentiates it from physics (Williamson, [1851], 1963, pp. 69–70). Furthermore, the uncanny character of many chemical transformations—apparently instantaneous and often dramatic—attracted many future chemists to their vocation: the adjectives "mysterious" and "marvelous" are applied to these transformations by such serious thinkers as Lothar Meyer, Walther Nernst, and J. H. van't Hoff. However, the latter half of the nineteenth-century saw an increasing preoccupation with chemical architecture and a concurrent abridgement of the scope of thinking about chemical dynamics.

In her comprehensive study of the rise of physical organic chemistry, Mary Jo Nye says that only a minority of his contemporaries and immediate successors agreed with Butlerov about the possibility of writing "true rational formulas" (Nye, 1993, p. 102). One very significant member of that minority, van't Hoff, asserted in 1884 that chemistry was somewhat belatedly experiencing the second of two phases through which all sciences pass, in which phase it becomes "rational or philosophic." Furthermore, according to van't Hoff, rational studies in chemistry are characterized by a movement to connect the constitutional formula with the properties of the substance: "As all the properties of a body arise from the intimate relations of its constituents, one can easily foresee that in the future the constitutional formula will, after further development, be able to tell us *exactly* and *completely* the properties of the bodies they represent" (Hoff, 1884, p. 2; emphasis added).

As is well known, van't Hoff played a central role in convincing chemists that the three-dimensional structure of molecules could be both apprehended and accurately represented, and it would be easy to categorize him as a naive realist with respect to molecular structures. That would do him a serious injustice—as Jeffry Ramsey correctly observes, "van't Hoff was [not] a naive ball-and-stick guy" (Ramsey, 1997, p. 248). He saw the two weak points of chemical theory as its failure to resolve the position of the atoms in a molecule and its failure to account for their intramolecular motion. In fact, van't Hoff criticized contemporaneous rational formulas precisely for their inability to depict this dynamic character. How he envisaged changing them to remedy this defect he left unsaid, however, perhaps because "his proposal for a tetrahedral carbon solved the first problem at the expense of the second" (Nye, 1993, p. 112).

An examination of van't Hoff's writings tells us much about the obstacles that hindered the integration of the dynamic dimension into chemistry. When van't Hoff's article on the "Extension into Space of the Structural Formulas at Present Used in Chemistry" first appeared (Hoff, [1874], 1975), it met a variety of objections, both theoretical and practical. One of the theoretical objections was that it did not take into account the "rotatory and vibratory movements by which are 'animated' every atom in particular and every group of atoms in the molecule" (Berthelot, 1875, quoted in Snelders, 1975, p. 56). The substance of this objection was not in dispute, but it failed to derail van't Hoff's hypothesis because his "suggestion" was used to make sense of and further predict a large number of experimental measurements, and it represented a major step toward solving one of the crucial problems of nineteenth-century chemistry—geometrical isomerism. As we shall see, these successes effectively smothered Berthelot's objection without disposing of the issue that provoked it.

Van't Hoff's proposal came within the context of growing confidence in and exploitation of the structural theory, which emphasized the stable motifs of chemical architecture. In the earlier decades of the nineteenth-century, chemists' increasing mastery of techniques for isolating and synthesizing organic compounds began to outstrip their abilities to relate and classify them. Many chemists felt that they verged on being overwhelmed by "the thousands of organic forms . . . so prolifically strewn about us" (Sanders, 1860, pp. 7–8). Some practitioners characterized organic chemistry of this period as a "dark forest" and a "labyrinth" (Brock, 1993, pp. 210–211). While remaining sensitive to the dynamic character of molecules, chemists desperately needed to discover what was conserved in chemical processes if they were ever to find a way through the "labyrinth." Structural theory provided the necessary illumination, not least because of its compatibility with the visual culture of chemistry (Weininger, 1998). G. N. Lewis, who was as thoroughly conversant with contemporary developments in physics as he was with the leading issues in chemistry, confidently asserted:

> No generalization of science, even if we include those capable of exact mathematical statement, has ever achieved a greater success in assembling in simple form a multitude of heterogeneous observations than this group of ideas which we call structural theory. The graphical formula is far more than a mere theory of atomic arrangements; it has become a remarkable short-hand method of representing a great variety of chemical knowledge. (Lewis, 1923, pp. 20–21)

One of the preeminent creators of structural theory was August Kekulé, probably best known for advancing the hexagonal structure for benzene, C_6H_6. Any structural formula for benzene had ultimately to take account of the following facts: the well-established quadrivalency of carbon; the complete equivalence of benzene's six

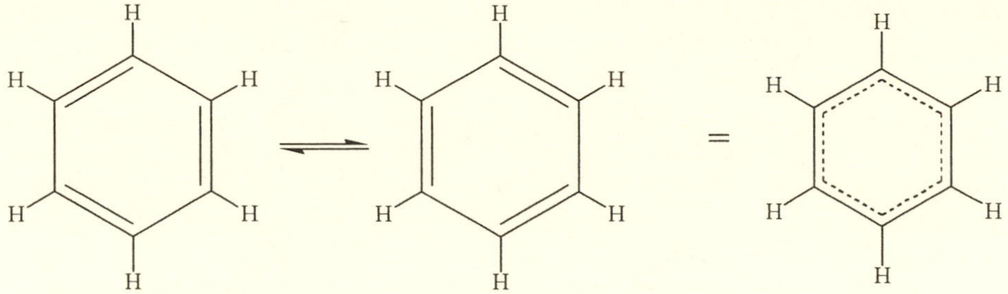

Figure 9.1 Rapid interconversion of the two threefold-symmetric forms of benzene to produce apparent sixfold symmetry.

hydrogen atoms (and equally the six carbon atoms); and the compound's remarkable lack of reactivity. What is noteworthy about the molecular formula is that there are only six hydrogens per molecule, while six carbon atoms could accommodate as many as fourteen hydrogen atoms if the carbons were connected solely by single bonds. The inevitable conclusion was that the carbon atoms in benzene must be connected by multiple bonds, which were known to confer high reactivity on compounds. Benzene's very low reactivity thus stood as a major challenge to established reactivity rules in general and structural theory in particular.

Rocke has shown how Kekulé struggled over a period of years to devise a formula that would satisfy all the above-mentioned conditions (Rocke, 1985, pp. 373–376). Although his efforts fully justify the praise they garnered, neither Kekulé nor anyone else could devise a formula that was in complete harmony with all the criteria. One of Kekulé's most original suggestions for solving this conundrum was a dynamic one—he proposed in 1872 that two equivalent threefold symmetric structures interconverted so rapidly that the carbon–carbon bonds alternated ceaselessly between being single and double, allowing the molecule to attain effective sixfold symmetry (figure 9.1).

Although this "Oscillation Hypothesis" appeared to solve the problem, it was far from decisive; discussions of the "benzene question" continued for decades (Henrich, 1922, pp. 175–236). Given that the rates of very few ordinary chemical reactions had been measured by the end of the century (Walker, 1899, p. 277), the presumably much more rapid benzene "oscillation" would have been totally beyond experimental reach. A thorough review of the still-unsettled benzene question 50 years after publication of the "Oscillation Hypothesis" concluded that this imaginative resort to dynamics had "long since been discredited" (Henrich, 1922, p. 175; for an extensive discussion of the Oscillation Hypothesis, see Brush, 1999).

Van't Hoff and Kekulé were not alone in recognizing the inherently dynamic character of molecules—relatively few of their contemporaries were "naive ball-and-stick guy[s]." The problem was how to use that awareness to solve chemical problems. On the one hand, chemists were not able to derive very much of practical value from treating molecules as dynamic objects; for example, the kinetic theory of gases wasn't even capable of predicting correct heat capacity ratios for polyatomic molecules (Sackur, 1917, pp. 154–166). On the other hand, a large body of previously confusing experimental data readily made sense if one treated molecules as more or less rigid objects.

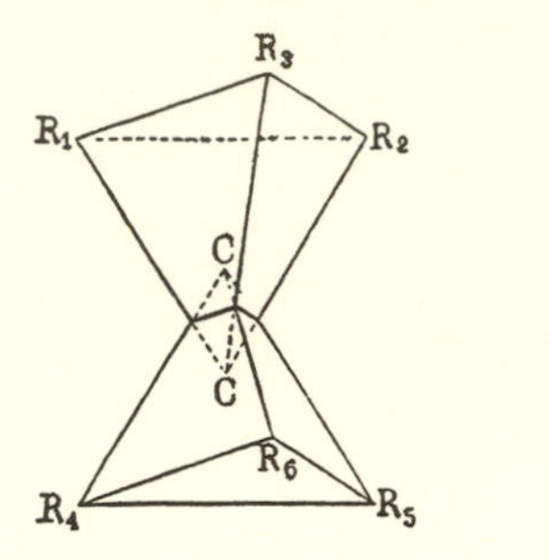

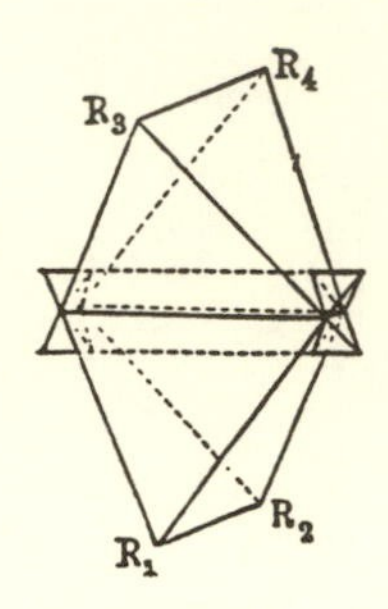

Figure 9.2 Molecular models showing carbon atoms, represented as tetrahedra sharing (*left*) vertices, forming a single bond, and (*right*) edges, forming a double bond (Hoff, 1899, pp. 116, 122).

Once again, we can turn to van't Hoff for a glimpse of the tensions between these opposing poles. Speaking about the study of constitution (molecular structure), he says that "in it the molecule is supposed motionless, so that the theory could at most represent the facts accurately at the absolute zero" (Hoff, 1900, p. 83). Yet if we look at the stereochemical *illustrations* accompanying van't Hoff's writings, we see molecules assembled from solid tetrahedra, not only motionless, but seemingly incapable of motion—durable and enduring (figure 9.2; Hoff, 1900, pp. 116, 122). Van't Hoff and his contemporaries would only be able to deal with the paradoxes presented by molecular dynamics under the aegis of thermodynamics.

Dynamics and Thermodynamics

Conventionally, the origins of thermodynamics are dated to about the mid-nineteenth-century with the work of Mayer, Joule, Clausius, Thomson, and Helmholtz (Laidler, 1993, pp. 83–130). In the main, thermodynamics was initially developed by physicists and did not attract much of a following among chemists. The "translation" of thermodynamics into a form accessible and useful to chemists we owe to van't Hoff, whose commanding role in this tale can hardly be overestimated (Laidler, 1993, pp. 114–122; Kragh & Weininger, 1996, pp. 101–103). The arrival of thermodynamics in this more "user-friendly" form was welcomed by a variety of chemists for a multitude of reasons.

Pragmatically, it had immediate and important applications in both industrial and academic laboratories, confirming the centrality of chemistry for other sciences and for society (Hoff, 1903). There were ideological benefits as well. The claim to fundamental status of thermodynamics rested in part on its asserted independence from any hypotheses about the structure of matter. The terms *hypothetical* and *metaphysical* were often terms of distaste and disapproval among nineteenth-century chemists (Brock, 1993, pp. 167–172). Thus, the use of thermodynamic arguments not only provided new tools for attacking chemical problems, but also bolstered chemistry's claim to being a proper science and the equal of physics, a subject on the minds of more than a few chemists (Nye, 1992). A similar boon accrued from the (relatively) sophisticated mathematical content of thermodynamics.

The generally warm embrace of thermodynamics by chemists did not extend to all its components, however; the Cinderella among these was the concept of entropy. Significantly, entropy is *the* thermodynamic function that confers directionality on time—it is "time's arrow." By focusing their attention on closed systems, chemists were able to approximate an idealized condition of equilibrium, "a static and time invariant state of a system where no spontaneous processes take place and all macroscopic quantities remain unchanged" (Denbigh, 1951, p. 1). Although one criterion for the attainment of equilibrium is the maximization of entropy, it is possible to carry out thermodynamic calculation on closed systems without reference to entropy. In fact, it was van't Hoff who showed how this could be done, and entropy did not become a standard part of chemists' vocabularies until after World War I (Kragh & Weininger, 1996, pp. 126–130).

Van't Hoff noted that up until about 1890, "dynamics, that is the study of reactions and of equilibrium, took a secondary place. But . . . since the study of chemical equilibrium has been related to thermodynamics, and so has gained a broader and safer foundation, it has come into the foreground of the chemical system, and seems more and more to belong there" (Hoff, 1899, p. 9). Indeed, van't Hoff himself had managed to bring some light to "one of the darkest chapters in chemical mechanics," the temperature dependence of reaction rates, by starting from the expression for the temperature dependence of the equilibrium constant. Svante Arrhenius took van't Hoff's treatment and derived from it several important insights: that an energy barrier lies between reactants and products, that only activated molecules can surmount the barrier, and that these activated molecules are in equilibrium with the unactivated majority (Laidler, 1993, pp. 238–242). Undergirded by and closely tied to the "broader and safer foundation" of *equilibrium* thermodynamics, chemical kinetics was to play a major role in reawakening the "elusive dream" of a chemical mechanics (Nye, 1993, p. 71).

Transition-State Theory: Explaining the Temporal via the Atemporal

Organic chemistry provided the arena in which this renaissance flowered. It required the reconciliation of two strange bedfellows, thermodynamics and structural theory—the first supposedly independent of any hypotheses about the structure of matter, the second inescapably committed to the atomic–molecular hypothesis. Their meeting ground would be transition-state theory.

Although attempts to model reaction rates mathematically extended back to 1850 and even before (Farber, 1961), the thermodynamically grounded work of van't Hoff and Arrhenius was the indispensable starting point for transition-state theory. Enriched by contributions from complementary approaches (Laidler & King, 1983, pp. 2658–2662), transition-state theory emerged in 1935 in a form that, at least in principle, allowed the calculation of reaction rate constants (Glasstone et al., 1941). It elaborated on Arrhenius's proposed equilibrium between unactivated and activated molecules, in which only the activated ones could cross the energy barrier. The architects of the theory—Evans, Polanyi, and especially Eyring—laid great stress on the state of the system at the apogee of this barrier (the transition state) and on the configuration of the molecule passing over it (the activated complex).

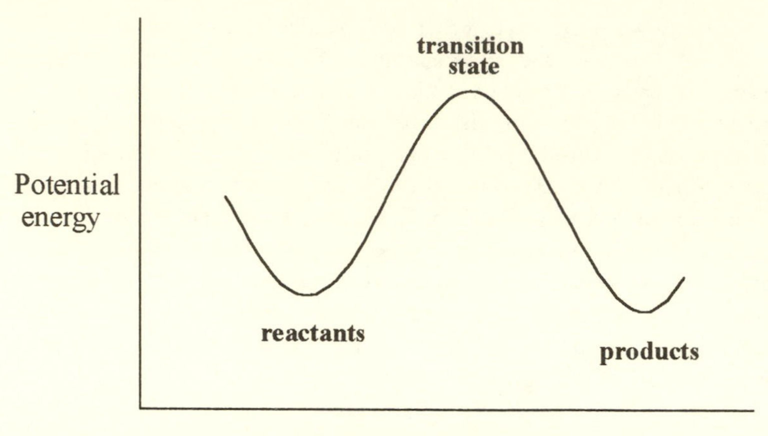

Figure 9.3 A generalized plot showing the relative potential energies of the reactant, transition state, and product. For a simple reaction ($A + BC \rightarrow AB + C$), the reaction coordinate depends on all the A–B and B–C distances.

During its passage from some equilibrium reactant state through the transition state to an equilibrium product state, a molecule will undergo a large number of internal rotations and vibrations. There are, in fact, $3N–6$ independent internal modes of motion, where N is the number of atoms in the molecule. Thus, keeping track of the molecule's internal energy changes requires a plot in ($3N–5$)-dimensional space. For molecules with more than two atoms, such a plot cannot be constructed physically, and for moderately larger ones, an even approximate visualization is impossible. The fact that these motions take place on different time scales only exacerbates the conceptual and representational problems.

It was Eyring's "very important and novel contribution in appreciating the great significance of the reaction coordinate" (Laidler & King, 1983, p. 2662). The reaction coordinate represents the *one* specific motion out of the plethora of possible motions that carries the reacting system over the col, or "mountain pass," that lies on the lowest energy path between reactants and products: "the 'reaction coordinate' . . . is virtually the reaction path drawn in one plane" (Glasstone et al., 1941, p. 98; figure 9.3). Furthermore, the configuration of the activated complex located at the col is decisive in determining the reaction rate—it is "down hill" after that, and any further changes that ensue were assumed to have no effect on the rate or on the final distribution of products.

Reviewing the transition-state theory 50 years after its creation, Laidler and King accorded it the "highest marks" as "a valuable working tool for those who are not concerned with calculation of absolute rates but are helped in gaining some insight into chemical and physical processes" (Laidler & King, 1983, p. 2664). I believe that few chemists would dispute this claim; an article on "Direct Observation of the Transition State" was included in a recent journal issue devoted to "Holy Grails in Chemistry" (Polanyi & Zewail, 1995). The wide utility of transition-state theory rests in large part on its ability to enable chemists to *visualize* reaction dynamics. Chemistry has always

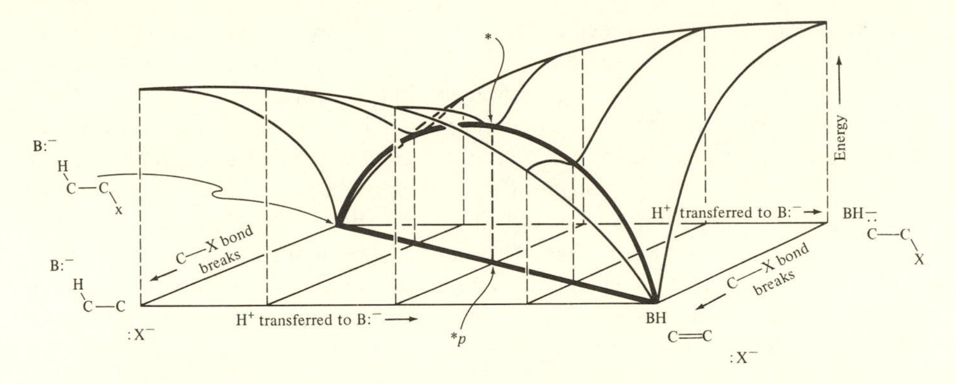

Figure 9.4 Representation of a three-dimensional approximation of a (3N−5) dimensional reaction surface in the vicinity of the transition state, whose position is marked by *. The chemical structures depict different possible intermediates that would be found along different reaction trajectories, and the arc in boldface shows the favored reaction trajectory, as illustrated in Figure 9.3. From Mechanism and Theory in Organic Chemistry, 2nd ed., by Thomas H. Lowry & Kathleen Schueller; Copyright © 1981, by Harper and Row Publishers. Reprinted by permission of Addison-Wesley Educational Publishers.

been a highly visual science, and the last few decades have seen a growing effort to make explicit the implications and consequences of this visuality (Hoffmann & Laszlo, 1991; Nye, 1993). As chemical transformations themselves grew more difficult to "read," their representations grew increasingly "realistic" and gradually assumed the role of *surrogates* for the processes they represented (Klein, 1997; Weininger, 1998). By treating the activated complex as a normal molecule except for the motion that becomes the reaction coordinate, transition-state theory made it possible to use structural drawings to depict molecules undergoing dynamic change.

Transition-state theory also employed another type of representation that had been put to effective use decades before by van der Waals. "Constructibility and visualizability became the dominant criteria" in van der Waals's approach to thermodynamics, which included the building of three-dimensional surfaces to represent thermodynamic functions. "It should also be noted that in the case of mixtures, one has to do with the depiction of *properties* and not *entities*" (Gavroglu, 1997, p. 295). Transition-state theory represented molecular properties by means of potential energy surfaces (Glasstone et al., 1941, chap. 3). Figure 9.4 is a two-dimensional representation of a three-dimensional approximation to a (3N−5)-dimensional plot for a reaction, in the vicinity of the transition state. Reaction coordinate diagrams like figure 9.3 are obtained by slicing through such multidimensional surfaces along the lowest energy path. By combining representations that visualize *both* the structure of evolving entities and their potential energies, as in figure 9.4, transition-state theory allows the chemist in a sense to visualize *time itself*.

But what kind of time is being visualized? First, it is a "time" that is completely tethered to space (Lawrence, 1971). Note that the reaction coordinate does not even have the *units* of time—it is a more or less complex function of internuclear distances. Second, this "time" has no *direction*. The system is completely isolated from the rest of the world and so could proceed equally well in either direction. Implicit in this formulation is a notion of time designated by the themodyamicist G.S. Denbigh as the time of theoretical physics. Denbigh characterizes theoretical physics as an "Eliatic" science that "tends to regard only those things which are constant and changeless as providing an adequate basis for the understanding of nature" (Denbigh, 1981, pp. 6, 96–101). It is this form of time—an undifferentiated, isotropic continuum without past, present, or future—that is implicit in transition-state theory. The theory rests on a conception of dynamics that Koyré characterized as "a motion unrelated to time or, more strangely, a motion proceeds in an intemporal time—a notion as paradoxical as that of a change without change" (quoted in Prigogine, 1980, p. 2). And its limits are becoming increasingly apparent.

The Organization of Chemical Structures in Time

A close look at the place of time within chemistry raises questions about that science's fundamental conceptual and explanatory entities. Put very simply, what is chemistry *about*? A conventional narrative depicts chemistry, in its youth a science of substances, as reaching maturity when it metamorphosed into a science of molecules. The development of transition-state theory certainly conforms to and reinforces that narrative because the theory's successes can be ascribed to its "reduction of the dynamics problem to the consideration of a single structure" (Truhlar et al., 1983, p. 2665). Yet questions have been raised recently as to whether molecular explanations are adequate to account for all chemical phenomena (Woolley, 1978; Weininger, 1984), and the view that substances are still the primary subject matter of chemistry has by no means disappeared (van Brakel, 1997). I suggest that chemists can call on a *variety* of explanatory entities that are intermediate between the molecule and the substance, and these entities need not have the permanence of either molecules or substances.

The conventional hierarchy of the sciences places chemistry between physics and biology, and throughout its history chemistry has had a rich and complex interchange with both. In the course of the nineteenth-century, physics provided extensive methodological guidance for chemistry, and in the twentieth-century, new conceptual paradigms as well. However, for several decades now, chemists have felt comfortable reclaiming their traditional roots in medicine and agriculture and, therefore, in biology. This trend has encouraged a reconsideration of what constitutes a fundamental entity in chemistry, along with the role of time in chemical processes.

Roughly 25 years ago *supramolecular chemistry* began to gain recognition as a coherent subdiscipline. The biological inspiration behind its creation was emphasized by one of its principal founders:

> Chemistry is the science of matter and its transformations, and life is its highest expression. [The objects of study in this new field are] *supermolecules*, well-defined, discrete *oligo*molecular species that result from the intermolecular association of a few components; [and] *supramolecular* assemblies, *poly*molecular entities that result from the spontaneous association of a large undefined number of components into a specific phase having more or less well-defined microscopic organization and macroscopic characteristics. (Lehn, 1995, pp. 1, 7)

Life itself is evoked by the use of metaphorical phrases like "self-assembly," "self-organization," and "self-recognition" (Lehn, 1995, chap. 9).

Although supramolecular chemistry is concerned with "soft bonds" that make its entities "thermodynamically less stable, kinetically more labile and dynamically more flexible than molecules," its viewpoint is still mainly architectural. "A chemical species is defined by its components, by the nature of the bonds that link them together and by the resulting spatial (geometrical, topological) features" (Lehn, 1995, p. 6). The thermodynamic paradigm that reigns here is still that of equilibrium.[3] However, it has long been recognized that there are other thermodynamic paradigms, for which the ideal state,

> the steady state, is the corresponding time invariant condition of a system which is open to its environment. In a sense, the [steady state] is more comprehensive than the [equilibrium state], for equilibrium is simply the limiting case of the stationary state when the flux from the environment approaches zero. Open systems are, of course, very familiar in biology and in the continuous reaction processes of the chemical industry. (Denbigh, 1951, pp. 1–2)

The study of these *constructive* and *dissipative* systems has given rise to a new thermodynamics applicable to "irreversible processes" or "the steady state."

Within such systems, new forms of organization can appear. Understanding their behavior requires that we attend to both the energy *and* time scales that govern the internal motions of their constituents:

> There are two very useful views of vibrationally excited molecules, two views which at first sight are in a very real conflict. At lower energies . . . the motion of the atoms in the excited molecule can be described by normal modes [that] retain their identity and hence their energy content for a long time. The atoms can vibrate in a given mode for more periods than the earth has so far revolved around the sun. . . . The other point of view, which has been developed to describe the process of bond breaking . . . is just as old. In this picture the energy is delocalized and moves from one mode of motion to another on a time scale short compared to the scale of the chemical event of interest. . . . The two paradigms, the spectroscopic and kinetic one, cannot be simultaneously right. . . . Of course, the answer is that both descriptions have limited ranges of validity. . . . Whether or not the products of a reaction obey the spectroscopic or kinetic paradigms is dependent on the *lifetime* of the transition state, measured in vibrational periods. Short lifetimes favor the former, long lifetimes the latter. (Remacle & Levine, 1996, pp. 1–2, 4)

In the regions intermediate between these limiting cases, normal modes of vibration "erode" at different rates and product distributions become sensitive to the precise conditions of the experiment. Intramolecular motions in different product molecules may remain coupled by "long-range forces even as the products are already otherwise quite separated" (Remacle & Levine, 1996, p. 51). These circumstances make possible a kind of *temporal* supramolecular chemistry. Its fundamental entities are "mobile structures that exist within certain temporal, energetic and concentration limits." When subjected to perturbations, these systems exhibit restorative behavior, as do traditional molecules, but unlike those molecules there is no single reference state—a single molecular structure, for example—for these systems. What we observe instead is a series of states that recur cyclically. "Crystals have extension because unit cells combine to fill *space*; networks of interaction that define [dissipative structures] fill *time* in a quite analogous way. In these . . . cases sequences of states periodically recur in time, just as specific distributions of atoms periodically recur in a spatial arrangement of a crystal" (Earley, 1993, pp. 279, 282). Under these conditions, the "time" that is being "filled" is *not* the isotropic time that Denbigh ascribes to theoretical physics. It is, rather, what he denotes as thermodynamic/biological time that has a definite direction and thus allows for genuinely new entities to come into existence.

Chemistry, Plurality, and Representation

What conclusions should one draw from this overview? One option is to see it as a kind of Passover story: Chemistry, whose originary god is Transformation, is enslaved by the Pharaoh of Structure but eventually freed by a Moses in the guise of Non-equilibrium Thermodynamics. This interpretation may have some elements of truth, but it is far too Whiggish. And it obscures a more judicious interpretation that focuses on the irreducible plurality of chemical methods and theories.

For example, coincident with the growth of interest in spatial and temporal supramolecular phenomena has been a very successful research program aimed at probing the dynamics of individual molecules over very short time intervals. The combined application of molecular beams and ultra fast lasers has given birth to *femtochemistry* (Zewail, 1994). Because one femtosecond (10^{-15} second) is a period considerably shorter than the time required for a single molecular vibration, this technique allows investigators access to "the very act of molecular motion that brings about chemistry" (Zewail, 1988, p. 1645). Although femtochemical experiments are often carried out in molecular beams under conditions of high dilution, their results are still interpretable within the framework of transition-state theory. This interpretation produces "snapshots" of the individual molecules as they metamorphose in real time (Zewail, 1988, p. 1651). One of the pioneers of fast dynamics, George Porter, sees this technique as carrying us to the limits of dynamic investigations in chemistry: "The study of chemical events that occur in the femtosecond time scale is the ultimate achievement in half a century of development and, although many future events will be run over the same course, chemists are near the end of the race against time." (quoted in Zewail, 1994, p. xi).

With due respect to Lord Porter, I do not believe that femtochemistry will give us the last word about chemical dynamics, nor that femtochemical experiments are more fundamental than supramolecular ones. The theories that interpret these different sets of experiments have substantial but finite ranges of validity, like the spectroscopic and kinetic theories of intramolecular energy distribution. Analogously, I do not believe that one has to choose the molecule, the substance, or the supramolecular aggregate as the foundational unit of chemical analysis and explanation. Joseph Earley speaks for many chemists when he maintains that we "*do not hold* that there is a single fundamental level of description—that there is any *one* scale of time or size (or single group or classes of entities) in terms of which *all* interesting questions can be answered" (Earley, 1993, p. 281).

Nor does chemistry encompass one *conception* of time. We could adapt Denbigh's notion of different concepts of time and designate two of them "molecule time" and "substance time." The first is isotropic and directionless; the second is inherently asymmetric. They each have their domain of validity, and neither is "truer" than the other. Based on past history, we can anticipate some very interesting phenomena in those border lands where systems are not completely under the sway of either.

The plurality of chemical theories makes apparent why Butlerov's vision was chimerical. No single structural formula, no matter how rich, could represent the entire panoply of chemical properties (whatever "entire" could ever mean). At one extreme we have such classical properties as melting and boiling points that are best correlated by graph-theoretical abstractions that consider only connectivity patterns within molecules (Hansen & Jurs, 1988). At the other extreme, we seek to explain photochemical processes with the aid of traditional structural pictures loaded down with the quantum chemical paraphernalia of orbitals, nodes, and charge densities (Michl & Bonačić-Koutecký, 1990). The molecular structure diagram has so far managed to accommodate them all. Nonetheless, the increasing recognition accorded chemical entities structured in time is one development that forces us to ask whether molecular structures and molecular explanations are adequate in every chemical situation.

Thus, I have reservations about Luisi and Thomas's claim that "the realm of chemistry is, then, rather sharply delineated. It is the domain in which the chemical structural formula is the main operational tool for talking science. Cross its boundaries and this form of graphical representation loses its validity, in one direction due to the structural complexity of living systems and in the other because one cannot represent intelligently the reality of elementary particles with a drawing." My doubts arise from the precise flaw that they themselves identify: "There is no adequate visual method for the portrayal of the real-time dynamics of chemical systems which are so often crucial to their understanding. Thus, by the constant use of cartoons in chemistry and biochemistry one may be brought to forget or at least neglect the dynamic aspects" (Luisi & Thomas, 1990, pp. 72, 74). Rather than curtail our growing appreciation of the dynamic range and complexity of chemical phenomena, I think it better to abandon the cherished notion that one mode of representation is sufficient to account for the totality of chemical experience—even if that mode has proved itself uncannily effective and resilient.

Lehn characterizes chemistry as "a science of transfers, a communication centre and a relay between the simple and the complex, between the laws of physics and the rules of life, between the basic and applied" (Lehn, 1995, p. 1). As for chemists themselves, they are, according to Gavroglu, "obliged to proceed to ontological commitments which are unambiguous and clearly articulated" while working with "rules [that] form a framework where it becomes possible to accommodate more than one theory" (Gavroglu, 1997, pp. 284, 287). The untidiness of chemistry (with its fuzzy boundaries, its plural, and sometimes contradictory theories, and, almost vanishing demarcation between "pure" and "applied") seems not to have impressed philosophers as worthy of much attention. This neglect was often justified on the grounds that any philosophical question that could be answered by looking at chemistry could be more clearly and cleanly answered by studying physics (Theobald, 1976, p. 204). The narrow outlook betrayed by such sentiments was often founded upon a narrow conception of the acceptable styles of explanation and theory formation in science (Hofmann, 1990, and references therein). In actuality, chemists deploy a variety of theoretical and explanatory styles, and that fact alone should argue for a careful philosophical scrutiny of chemistry, "an area in which rationalizing, pattern explanations rub shoulders with Hempelian, deductive explanations to a greater extent perhaps than in any other science" (Theobald, 1976, p. 213).

There is now a growing appreciation for a new model of the world, one "that is *hierarchically* layered into quasi-autonomous domains, with the ontology and dynamics of each layer essentially quasi-stable and virtually immune to whatever happens in other layers" (Schweber, 1997, p. 172). To evaluate such a model, one wants to know how the sedimentation of these layers takes place and how their boundaries are maintained. How porous are those boundaries, and what types of exchange take place through them? As a test case for the assessment of this new vision, chemistry should be an area of continuing interest for the philosophical community.

Notes

I am grateful to HarperCollins for permission to reproduce figure 9.4. Earlier versions of this chapter were presented to the Division of the History of Chemistry, American Chemical Society, and the Boston Colloquium for the Philosophy of Science, Boston University. I thank the organizers of these sessions for their invitations and the participants for their comments. Most of this chapter was written while I was on sabbatial leave at the Beckman Institute and the Division of Humanities and Social Sciences, California Institute of Technology. The boundless hospitality of my colleagues there is gratefully acknowledged.

1. The answers themselves have been the subject of extended debate among historians. Butlerov's beliefs and his role in the development of structural theory have even become ammunition in twentieth-century scientific disputes with marked political and ideological content (Graham, 1964, pp. 25–27). The "mutual influence" (to borrow Butlerov's phrase) of history, historiography, and ideology in the Butlerov controversy has been painstakingly examined by Rocke (1981).

2. Discussions of representation in the scientific literature devote too little attention to their communicative functions. Yet "representation" never takes place in a vacuum, and the issue of representation for what end is always relevant, even if masked. The context shapes the character and usage of the representations, as well as highlighting the social character of the scientific enterprise. An important exception to this general neglect is the intensive study of chemical language, particularly its symbolic components, by Mestrallet (1980; see also Hoffmann & Laszlo, 1991; Laszlo, 1993; Weininger, 1998).

3. Theobald characterizes a number of chemical concepts, including molecule, reactivity, equilibrium and transition state as "static, organizing, descriptive." He sees them as "more like concepts in biology than the dynamic causal concepts of so much of physics" (Theobald, 1976, p. 209). Theobald's view of physics seems to be diametrically opposed to that of Denbigh (1981, p. 6). This disjunction raises interesting questions about whether any science is inherently "static" or "dynamic," and the extent to which these (admittedly vague) attributes are context dependent.

References

Benfey, O. T. 1963. "Concepts of Time in Chemistry." *Journal of Chemical Education*, 40: 574–77.

Brock, William H. 1993. *The Norton History of Chemistry*. New York: Norton.

Brush, Stephen. 1999. "Dynamics of Theory Change in Chemistry: Part I. The Benzene Problem 1865–1945." *Studies in the History and Philosophy of Science*, 30: 21–79.

Butlerov, A. M. [1861] 1971. "On the Chemical Structure of Substances." Trans. Frank F. Kluge and David Larder. *Journal of Chemical Education*, 48: 289–291.

Crosland, Maurice. 1962. *Historical Studies in the Language of Chemistry*. Cambridge: Harvard University Press.

Denbigh, K. G. 1951. *The Thermodynamics of the Steady State*. London: Methuen.

Denbigh, Kenneth G. 1981. *Three Concepts of Time*. Berlin: Springer-Verlag.

Earley, Joseph E. 1993. "The Nature of Chemical Existence." In P. A. Bogaard & G. Treaty eds. *Metaphysics as Foundation: Essays in Honor of Ivor Leclerc* (pp. 272–284). Albany: State University of New York Press.

Farber, Eduard. 1961. "Early Studies Concerning Time in Chemical Reactions." *Chymia*, 7: 135–48.

Gavroglu, Kostas. 1997. "Philosophical Issues in the History of Chemistry." *Synthese*, 111: 283–304.

Glasstone, Samuel, Laidler, Keith J., & Eyring, Henry 1941. *The Theory of Rate Processes*. New York: McGraw-Hill.

Graham, Loren. 1964. "A Soviet Marxist View of Structural Chemistry: The Theory of Resonance Controversy." *Isis*, 55: 20–31.

Hansen, Peter J., & Jurs, Peter C. 1988. "Chemical Applications of Graph Theory. Part I: Fundamentals and Topological Indices." *Journal of Chemical Education*, 65: 574–580.

Henrich, Ferdinand. 1922. *Theories of Organic Chemistry*. (Trans. and enl. T. B. Johnson & D. A. Hann) New York: Wiley.

Hoff, J. H. van't. [1874] 1975. "A Suggestion Looking to the Extension into Space of the Structural Formulas at Present Used in Chemistry." In O. B. Ramsey, ed. *Van't Hoff-Le Bel Centennial* (pp. 37–46). Washington, DC: American Chemical Society.

Hoff, J. H. van't. 1884. *Etudes de Dynamique Chimique*. Amsterdam: Frederik Muller.

Hoff, J. H. van't. 1899. *Lectures on Theoretical and Physical Chemistry: Part 2—Chemical Statics*. Trans. R. A. Lehfeldt. London: Edward Arnold.

Hoff, J. H. van't. 1903. *Physical Chemistry in the Service of the Sciences*. Chicago: University of Chicago Press.

Hoffmann, Roald, & Laszlo, P. 1991. "Representation in Chemistry." *Angewandte Chemie International Edition in English*, 30: 1–16.

Hofmann, James R. 1990. "How the Models of Chemistry Vie." *Philosophy of Science Association*, 1: 405–419.

Kim, Mi Gyung. 1992. "The Layers of Chemical Language, I: Constitution of Bodies *v.* Structure of Matter." *History of Science*, 30: 69–96.

Klein, Ursula. 1997. "Paper-Tools and Techniques of Modelling in Nineteenth-Century Chemistry." In *Nineteenth-Century Chemistry: Its Experiments, Paper-Tools, and Epistemological Characteristics*, (Preprint 56, pp. 27–47). Berlin: Max-Planck-Institut für Wissenschaftsgeschichte.

Kragh, Helge, & Weininger, Stephen J. 1996. "Sooner Silence Than Confusion: The Tortuous Entry of Entropy into Chemistry." *Historical Studies in the Physical and Biological Sciences*, 27: 91–130.

Laidler, Keith J., & King, M. Christine 1983. "The Development of Transition-State Theory." *Journal of Physical Chemistry*, 87: 2657–2664.

Laidler, Keith J. 1993. *The World of Physical Chemistry*. Oxford: Oxford University Press.

Laszlo, Pierre. 1993. *La parole des choses, ou, Le langage de la chimie*. Paris: Hermann.

Lawrence, Nathaniel. 1971. "Time Represented as Space." In E. Freeman & W. Sellars, eds. *Basic Issues in the Philosophy of Time*, (pp. 123–132). La Salle IL: Open Court.

Lehn, Jean-Marie. 1995. *Supramolecular Chemistry: Concepts and Perspectives*. Weinheim: VCH Publishers.

Lewis, Gilbert Newton. 1923. *Valence and the Structure of Atoms and Molecules*. New York: Chemical Catalog.

Lowry, Thomas H, & Schueller Richardson, Kathleen 1981. *Mechanism and Theory in Organic Chemistry*, (2nd ed.). New York: Harper and Row.

Luisi, Pier-Luigi, & Thomas, Richard M. 1990. "The Pictographic Molecular Paradigm: Pictorial Communication in the Chemical and Biological Sciences." *Naturwissenschaften*, 77: 67–74.

Mestrallet Guerre, Renée. 1980. "Communication, linguistique et sémiologie: Contribution ã l'étude de la sémiologie. Études sémiologique des systèmes de signes de la chimie." Doctoral thesis, Faculty of Letters, Universitat Autònoma de Barcelona.

Michl, Josef, & Bonačić-Koutecký, Vlasta 1990. Electronic Aspects of Organic Photochemistry. New York: Wiley.

Mosini, Valeria. 1994. "Some Considerations on the Reducibility of Chemistry to Physics." *Epistemologia*, 17: 205–224.

Nye, Mary Jo. 1992. "Physics and Chemistry: Commensurate or Incommensurate Sciences?" In M. J. Nye, J. L. Richards, & R. H. Stuewer, eds. *The Invention of Physical Science: Intersections of Mathematics, Theology and Natural Philosophy since the Seventeenth Century*, (pp. 205–224). Dordrecht: Kluwer Academic.

Nye, Mary Jo. 1993. *From Chemical Philosophy to Theoretical Chemistry: Dynamics of Matter and Dynamics of Disciplines, 1800–1950*. Berkeley: University of California Press.

Polanyi, John C., & Zewail, Ahmed H. 1995. "Direct Observation of the Transition State." *Accounts of Chemical Research*, 28: 119–132.

Prigogine, Ilya. 1980. *From Being to Becoming*. San Francisco: W. H. Freeman.

Primas, Hans. 1980. "Foundations of Theoretical Chemistry." In R. G. Woolley, ed. *Quantum Dynamics of Molecules: The New Experimental Challenge to Theorists*, (pp. 39–113). New York: Plenum.

Ramsey, Jeffry L. 1997. "Molecular Shape, Reduction, Explanation and Approximate Concepts." *Synthese*, 111: 233–251.

Remacle, Françoise, & Levine, Raphael D. 1996. "Spectra, Rates, and Intramolecular Dynamics." In R. E. Wyatt & J. Z. H. Zhang, eds. *Dynamics of Molecules and Chemical Reactions*, (pp. 1–58). New York: Marcel Dekker

Rocke, A. J. 1981. "Kekulé, Butlerov, and the Historiography of the Theory of Chemical Structure." *British Journal for the History of Science*, 14: 27–57.

Rocke, A. J. 1985. "Hypothesis and Experiment in the Early Development of Kekulé's Benzene Theory." *Annals of Science*, 42: 355–381.

Sackur, O. 1917. *A Textbook of Thermo-chemistry and Thermodynamics*. Trans. and rev. G. E. Gibson. London: Macmillan.

Sanders, J. Milton. 1860. "Introduction." In J. M. Sanders, ed. *William Gregory, Handbook of Organic Chemistry; For the Use of Students* (4th ed., pp. 7–9). New York: A. S. Barnes and Burr.

Schweber, S. S. 1997. "The Metaphysics of Science at the End of a Heroic Age." In R. S. Cohen, M. Horne, & J. Stachel, eds. *Experimental Metaphysics: Quantum Mechanical Studies for Abner Shimony* (vol. 1 pp. 171–198). Dordrecht: Kluwer Academic.

Snelders, H. A. M. 1975. "Practical and Theoretical Objections to J. H. Van't Hoff's 1874–Stereochemical Ideas." In O. B. Ramsay, ed. *Van't Hoff—Le Bel Centennial* (pp. 55–65). Washington, DC: American Chemical Society.

Theobald, D. W. 1976. "Some Considerations on the Philosophy of Chemistry." *Chemical Society Reviews*, 5: 203–213.

Truhlar, Donald G., Hase, W. L., & Hynes, J. T. 1983. "The Current State of Transition-State Theory." *Journal of Physical Chemistry*, 87: 2664–2682.

Van Brakel, J. 1997. "Chemistry as the Science of the Transformation of Substances." *Synthese*, 111: 253–282.

Walker, James. 1899. *Introduction to Physical Chemistry*. London: Macmillan.

Weininger, Stephen J. 1984. "The Molecular Structure Conundrum: Can Classical Chemistry Be Reduced to Quantum Chemistry?" *Journal of Chemical Education*, 61: 939–944.

Weininger, S. J. 1998. "Contemplating the Finger: Visuality and the Semiotics of Chemistry." *Hyle*, 4(1): 3–25. Also available on the Internet; http://www.uni-karlsruhe.de/ ~philosophie/ hyle.html.

Williamson, Alexander W. [1851] 1963. "On the Constitution of Salts." In O. T. Benfey, ed. *Classics in the Theory of Chemical Combination* (pp. 69–75). New York: Dover.

Woolley, R. G. 1978. "Must a Molecule Have a Shape?" *Journal of the American Chemical Society*, 100: 1073–1078.

Zewail, Ahmed H. 1988. "Laser Femtochemistry." *Science*, 242: 1645–1653.

Zewail, Ahmed H. 1994. *Femtochemistry: Ultrafast Dynamics of the Chemical Bond*. Singapore: World Scientific.

10

The Nature of Chemical Substances

JAAP VAN BRAKEL

The Manifest and the Scientific Image

Professor Hare, delivering the presidential address to the Aristotelian Society in Oxford in 1984 (p. 12), said:[1] "It is commonly said that the property of being water supervenes on the chemical (or ultimately on the physical) property of being H_2O. As it stands this view seems to me to be obviously false." In terminology, that will become clearer as we proceed, Hare defended the *manifest* image—in this case, ordinary liquid water against elimination by the scientific image (which reduces "being water" to "being H_2O"). Hare used the verb *to supervene* instead of *to be reducible*, but the difference between the two is slight (as we shall see in a later section).

A more common view among philosophers and scientists is expressed in the following citation from Kim (1990, p. 14): "Chemical kinds and their microphysical compositions (at least, at one level of description) seem to strongly covary with each other, and yet it is true, presumably, that natural kinds are asymmetrically dependent on microphysical structures." Kim takes the view that manifest objects are "appearances" of a reality constituted by systems of imperceptible particles. Such a view takes for granted that the macroscopic, manifest world is dependent on the microstructure of the world in such a way that it is underlying things that are more real and determine appearances. In crude jargon: science uncovers the *Dinge-an-sich* that explain the phenomena we see. I chose the quotations of Hare and Kim because *both* point to, though fail to address, the philosophical issue I discuss in this chapter, viz. the tension between manifest and scientific image, focusing on chemistry.

"Manifest" versus "scientific" imagery talk stems from Sellars. The manifest image refers to things like water, milk-lapping cats, injustice-angry people, as well as sophisticated interpretations of "people in the world." The scientific image is concerned with things like neurons, DNA, quarks, and the Schrödinger equation, again including sophisticated reflection and a promise of more to come. I use "manifest image" with a different inflection from Sellars, avoiding associations with sense data (which was an important part of his concern), associating it rather with forms of life. That is, "manifest image" is short for "manifest form(s) of life, understood interculturally," and that is how the term will be used in the sequel.[2] The manifest image is not a unified whole—it is full of contradictions, contestations, undefined notions, incommensurable views, and so on. Still, it is more complete and more fundamental that any unified, end-of-ideal-science scientific picture. The scientific image makes a strict separation between fact and value, subjective and objective, cognitive and noncognitive. The manifest image (understood interculturally) doesn't do that and, therefore, is more homogeneous and complete than the scientific image. Moreover, the scientific image is dependent on manifest image language and intuitions: the notions of cause and chance are grounded in the manifest image, as are the epistemic virtues; even logic or quantum mechanics only exist relative to a final grounding in the manifest image.[3] None of this implies that the manifest image is static or "one thing," or that the scientific image wouldn't have an immense effect on the manifest image, or that manifest image views are by definition better than scientific image views. But it does imply that, when the question of priority or primacy *is* raised, then (but only then) it is the manifest image that can claim priority over the scientific image, and not the other way around.

Although manifest can often be identified in this chapter with macroscopic and scientific with microscopic (and this tends to be Sellars's view), these terms should not be taken as synonyms. Assume a nonmanifest scientific theory that takes certain macroscopic objects as basic objects that are not understood in terms of manifest concepts (e.g., supramacroscopic objects or middle-sized unobservables). Such a theory would be a straightforward part of the scientific image. By contrast, lots of unobservables are postulated in the manifest image.[4]

According to Sellars (1963, p. 4) "the philosopher is confronted by *two* pictures of essentially the same order of complexity, each of which purports to be a complete picture of man-in-the-world, and which, after separate scrutiny, he must fuse into one vision: the manifest and the scientific image." He then suggests that there are no independent grounds to adjudicate between the two rivals. Any arguments given are always arguments from within one or the other image. I consider this wrong. The issue of rivalry is an external question (in the sense of Carnap),[5] which to the extent it is meaningful, is a question *within* the manifest image.

Sellars (1963, pp. 27–31) claimed primacy for the scientific image because he found the two images incompatible, and because of the incompleteness of the manifest image. The question that arises immediately, is: From which perspective does Sellars make this claim? Like Bernstein (1966), van Fraassen (1998) argues that that is not clear, and concludes he is telling it from nowhere. But it is obvious which stance Sellars adopts or where nowhere can be found: although Sellars' writings sometimes suggest a third stance,[6] he is clearly telling the story from the scientific bench. The lacunae or incompleteness Sellars sees in the manifest image are only there from the perspective of the scientific image. Similarly, the manifest image is only "a sort of naive-scientific image" when it is seen from the scientific image. Sellars is equally mistaken in thinking that within the manifest image the mental cannot live happily with the physical. Combining the mental with the physical is only a problem for the scientific image. Therefore, Sellars's arguments for the primacy of the scientific image do not touch the manifest image—they are offered from the scientific perspective, and, even by scientific standards, they are not valid unless assumptions not part of the scientific image are used.[7]

Against the background of the tension between the manifest and scientific image, my question will be: Where does chemistry fit in? Does it belong to the manifest or the scientific image? Sellars, and most other philosophers of whatever bent would probably reply: the scientific image. My intention is to raise doubts about that apparent self-evidence. A number of approaches to the question of how chemistry fits in are possible. One is to "reconstruct" the scientific image starting from the manifest image. This is the approach of protoscience, or, in this case, protochemistry.[8] Although sympathetic to this perspective, I'll analyze the manifest/scientific issue in the context of more central concerns in (general) philosophy of science on the one hand and focus on what is most specific to *chemistry* on the other. Therefore I'll address two issues: (1) that of interdiscourse relations, of which reduction is the most well known, and focus on the relation of macro- to microchemistry and of microchemistry to microphysics; and (2) the notion of chemical species (also: chemical substance, compound, or kind): Should it have a macro- or microdefinition? Is it reducible to physics?

The next Section is a brief introduction to what can be considered the main motivation for this chapter: the prejudice that chemistry is reducible to physics. The following section critically reviews a number of proposals of how to understand interdiscourse relations (epistemic and/or ontological relations between different theories or disciplines), and settles down on the metaphysical model of anomalous monism (never before applied to chemistry and physics). Then I show in what way the (chemical) notion of pure substance is methodologically, epistemologically, and ontologically *independent* of microchemistry and quantum mechanics and *dependent* on the manifest intuition of pure substance.

Is Chemistry Reducible to Physics?

According to Primas (1991, p. 163), "the philosophical literature on reductionism is teeming with scientific nonsense," and he quotes, among others, Kemeny and Oppenheim (1956), who said: "a great part of classical chemistry has been reduced to atomic physics."[9] Perhaps it was not philosophers who invented this story after all. Almost certainly, Oppenheim and other philosophers of science at the time were familiar with the influential statements of Dirac, Heisenberg, Reichenbach, and Jordan on this issue.[10] Notoriously, the physicist Dirac (1929, p. 721) said, 'the underlying laws necessary for the mathematical theory of a large part of physics and the whole of chemistry are thus completely known, and the difficulty is only that exact applications of these laws lead to equations which are too complicated to be soluble." Less famously, the philosopher of science Reichenbach (1978, p. 129) reiterated that "the problem of physics and chemistry appears finally to have been resolved: today it is possible to say that chemistry is part of physics, just as much as thermodynamics or the theory of electricity." These views clearly stuck. For example, in a recent review of quantum electrodynamics (QED), to which Dirac made important contributions, the historian of science Schweber (1997, p. 177) says, "the laws of physics encompass in principle the phenomena and the laws of chemistry."

A more correct way of putting the issue is that of the physicist Gell-Mann:

> When Dirac remarked that his formula explained most of physics and the whole of chemistry of course he was exaggerating. In principle, a theoretical physicist using QED can

> calculate the behaviour of any chemical system in which the detailed internal structure of atomic nuclei is not important. [But:] in order to derive chemical properties from fundamental physical theory, it is necessary, so to speak, to ask chemical questions of the theory. (1994, p. 109–110)

Even Gell-Mann (pp. 110–111) tends to suppress the "chemical questions to be asked," or at least invites misquotation when he says, "the laws of chemistry can in principle be derived from QED, provided the additional information describing suitable chemical conditions is fed into the equation."

Still, the general consensus among chemists and philosophers sticks to Dirac's view as reported by Reichenbach, Schweber, and others. Let me give an example from chemistry. Bader (1990) claims his work in quantum chemistry is firmly grounded in standard quantum mechanics, and he is explicit about his presuppositions. In a coauthored article that he says (Bader et al., 1994, p. 620) "a scientific discipline begins with the empirical classification of observations. It becomes exact in a predictive sense, when these observations are classified in such a way as to reflect the physics governing its behaviour." From such a perspective, the fact that so called ab initio methods in quantum chemistry are approximate and guided by observations will be no problem;[11] it is all part of getting closer to the truth, the essence of which has been specified a priori. It is not something that needs empirical evidence (except for hand waving toward progress in physics). Bader (1990, p. 249) admits that "the total electron density distribution in a molecule shows no indication of discrete bonding or non-bonding electron pairs." But this merely raises the question (p. 252): "Where then to look for the Lewis model, a model which in the light of its ubiquitous and constant use throughout chemistry must most certainly be rooted in the physics governing a molecular system." The answer is that (p. 248) "the Lewis model of the electron pair does find a more abstract but no less real physical expression in the topological properties of the Laplacian of the charge density." Were Bader's account of electron pairs to be criticized, it would be an in-house technical discussion and could never undermine the reductionistic program. The question to which the (fallible) answer is a reply, presupposes that reduction is a necessary characteristic of science. Moreover, Bader starts from premises like "the present theory evolves from the unfolding of a single assumption: that atoms are objects in real space." He refers to the physicist Feynman, who had said that if in some cataclysm all scientific knowledge were to be destroyed and only one sentence could be passed on to the next generation of creatures, it should be the atomic hypothesis. That might be a good suggestion, but the atomic hypothesis is not one of the postulates of quantum mechanics.

I'll touch briefly on these issues in a later section on molecular structure and quantum mechanics. To set the scene, however, let me stress that Primas (1991) is absolutely right when he says that if reduction is used in the sense of "higher-level theory together with its interpretation can be deduced from the basic theory" then it is a *brute fact* that:[12]

- Chemistry has not been reduced to physics: "How are the nonlinear differential equations of chemical kinetics derived from linear quantum mechanics?"
- Chemical purity is not a molecular concept: "Pure water does *not* consist of H_2O molecules."
- The theory of heat has not been reduced to statistical mechanics: "How can the zeroth law of thermodynamics be derived from statistical mechanics?"

Reduction and Beyond

Interdiscourse Relations

The question "Can chemistry be reduced to physics?" is unclear. First, it is not clear how one would delineate and separate chemistry and physics—what, for example, about chemical physics or mechanical and physical separation methods in chemistry and chemical engineering? Second, the question of reduction has to be made much more concrete. For example:

- Can chemical thermodynamics be reduced to statistical mechanics?
- Can the (chemical) Lewis model be reduced to quantum mechanics?
- Can "being water" be reduced to "being H_2O"?

Third, many more than two discourses (apart from chemical and physical) need to be distinguished: there is manifest water, the physical, physico-chemical, chemical, and biochemical *macro*properties of water, the (chemical) molecular *micro*discourse, and the (physical) quantum mechanical discourse, and that's only the beginning. Fourth, the notion of reduction has to be spelled out: (a) What is it that is being reduced (theories, concepts, properties, natural kinds, laws, explanations)? (b) How does it differ from related notions such as replacement, elimination, integration, supervenience, and emergence? (c) What *sort* of relation is it (dependence, identity, . . .)?

Before addressing these issues I'll first make a few general remarks on what I'll refer to as "interdiscourse relations." For a start, there is nothing wrong in understanding the phrase as "interlevel relations," "linking propositions," "theoretical identities," or "bridge laws." Examples of different discourses (or levels or domains or disciplines or theories) are physical, chemical, behavioral, brain, genetic, cultural, moral, and so on. In this chapter I'm primarily concerned with specific discourses, for example, that of thermodynamics, molecules, or quantum mechanics.[13]

The expression in scientific English that most often exemplifies referring to interdiscourse relations is "to underlie" and its cognates. Its use can be illustrated with the following examples: A particular molecular structure underlies the property of being brittle. A conjunction of states representing the kinetic translational energy of the constituent molecules underlies the temperature of an ideal gas. Physical kinds underlie chemical kinds. In general, "B" (the "base") underlies "S" (the supervening facts, properties, kinds, or whatever). It is usually unclear what saying that B underlies S means, as there are many different ways to paraphrase this statement. In unreflective scientific language use, talk of B underlying S seems to make no distinction between things being identical; being somehow related; one thing constituting, causing, controlling, or determining another; one thing making some sort of contribution to another; one thing being *ceteris paribus* statistically relevant for another; and so on. Again, words in the last sentence suffer from the same sort of ambiguities. I suggest the vagueness exemplifies a kind of covering up avoid properly addressing the question of interdiscourse relations. Still, what seems common to all occurrences is that "to underlie" implies a form of hierarchy or asymmetry between the different discourses that are related (cf. the quotation from Kim in the first section). If B underlies S, apparently it is *not* the case that S underlies B.

Reduction

"Reduction" is related to a variety of other issues in the philosophy of science, for example the (dis)unity of science, models for the development of science, and for the change of theories, the analysis of theoretical terms, the purpose of science, and types of scientific realism. Moreover, in one guise or another, reduction is pervasive throughout the history of philosophy. Not only has it played a central role in the history of analytic philosophy and philosophy of science, but also the philosophies of Hegel, Marx, Nietzsche, Foucault, and Derrida, to name but a few, can be read as reductive programs.

The classical notion of reduction in the philosophy of science, which originated with the Unified Science program of the logical positivists, had the following characteristics (Lévy, 1979):

- It is an asymmetrical, nonreflexive, and transitive relation.
- The relation is linear in that it holds between not more than two theories.
- Its first aim is an explanatory one, usually that of connecting successor theories, in particular to provide an explanation of the continuous process of scientific growth, rationalization, and formalization. Such "synchronic" or "intralevel" reduction will not further concern us, however.[14]
- Its second aim is the unification of scientific theories, either from the same or different disciplines.

The classical view, summarized by Nagel (1961), proposed two criteria for reduction: the theoretical terms of S should be definable in terms of the lexicon of B, and the laws of S should be deducible from the statements of B. In a famous article, Oppenheim and Putnam (1958) proposed the following "package deal": (a) microreduction of the sort described by Kemeny and Oppenheim (1956); (b) levels of reality exist with lower levels being more fundamental and composed of simpler elements; (c) unity of science; and (d) cosmic evolution.[15] Criticism of these older proposals tended to parallel criticisms of the Vienna Circle's unity of science program: too much rational reconstruction toward the ideal of a universal language and ignoring scientific practice. In reply to such criticisms, various modifications were proposed—though, notwithstanding their greater sophistication and rigor, they were not fundamentally different from earlier proposals.[16]

Without dwelling on detail, two aspects of reduction *must* be made precise, lest any statements about it lose their meaning. First, the question of what sort of thing it is that is being reduced must be posed. Three types of reduction are often distinguished:

- *Constitutional reduction* concerns the question of whether the domains B and S are ontologically identical—that is, whether the S-entities are constituents of the same elementary substrates with the same elementary interactions as B-entities.
- *Epistemological reduction* concerns whether the concepts (properties, natural kinds) necessary for the description of S can be redefined in an extensionally equivalent way by the concepts of B and whether the laws governing S can be derived from those of B.
- *Explanatory reduction* concerns the question of whether for every event or process in S there is some mechanism belonging to B which (causally) explains the event or process.[17]

I'll concentrate on epistemological reduction, because the other two don't make sense without some sort of cognitive connection between the S- and B-domains. The second, more crucial issue is hidden in the phrase "supplemented by suitable supplementary assumptions," which I omitted from the definition of epistemological reduction but which has to be added to make any sense of it (acknowledged in Nagel, 1961, 1974). Without this addition, the reduction won't work, undermining the idea that what is at issue is *reduction*. There are mainly two kinds of "supplementary assumptions": (a) bridge laws and (b) initial and boundary conditions. Bridge laws are usually taken to be nomological biconditionals that connect one term of the reducing theory with a corresponding term of the reduced theory or of equations that do the same job for a whole family of terms; either they translate kind-predicates in one science into those of a more basic one, or they specify a metaphysical relation, like *being identical to* or *being a necessary and sufficient condition for*, between the kinds of one science and those of the reducing science.

The obvious question to ask is: Aren't the bridge laws themselves in need of some explanation in terms of the reducing theory *alone*? The prototypical example of a bridge law since Nagel (1961) is "temperature is mean kinetic molecular energy." Another notorious example is "water is H_2O." The defender of reductionism will say that temperature is *nothing but* mean kinetic energy of molecules. However, putting aside the assumptions that underlie the notion of averaging,[18] this bridge law doesn't apply to all (or even any) occurrences of temperature. In general, temperature is *not* the same as average mean molecular kinetic energy. This seems to be so for "model" gases (i.e., "perfect" gases of idealized "billiard-ball" molecules in random motion) but not for solids, plasmas, or a vacuum. Moreover, if for the sake of the argument, we pass over the distinction between a gas and a model of a gas, for a dense gas at low temperature, quantum mechanics implies that the kinetic energy does not depend just on the temperature. Hence, the connection between temperature and mean kinetic energy is nonlinear, its value depending on other characteristics of the system. Therefore, temperature is more accurately described as a functional property, having to do with the mechanism of heat transfer between different bodies (in accordance with the zeroth law of thermodynamics).[19]

Nagel (1961, p. 372) was well aware that reduction was not possible "unless a postulate is added relating the term 'temperature' to the expression 'mean kinetic energy of molecules.'"[20] Later writers underplayed this problematic aspect of bridge laws. However, even if bridge laws were understood as identities that justify elimination or replacement of the "reduced theory," there would remain the rarely addressed question of initial and boundary conditions. One thing is clear: they are not part of the reducing theory. This is where Gell-Mann's "chemical questions to be asked" fit in, or where Bader's *chemical* intuitions steer his work in quantum chemistry.

Supervenience

Since the 1980s talk of reduction has been replaced by talk of supervenience, not just in the philosophy of mind but also in the philosophy of science. The most common claim is that one level or domain or description or discourse, S, supervenes on another level, B, *if*, in some sense of *necessary*,

[S1] It is necessary that if an S-truth changes, some B-truth changes.

Or

[S2] Necessarily, if two situations (objects, states, events) are identical in their B-properties they are identical in their S-properties.

The connection with older ideas about reduction will be obvious: [S1] and [S2] are the generalized successors of the "bridge laws" or "linking propositions" that limit the relation to a token–token instead of a type–type relationship. The hope was that supervenience relations would have all the advantages of reduction with no disadvantages. Reduction has the advantage that it explains why events are not causally overdetermined. The disadvantage of this picture is that it seems to undermine the autonomy of the S-discourse unduly. The literature on supervenience is dominated by this tension between determinacy and autonomy.

Much of the literature on supervenience is on such an abstract level that one wonders whether anything at all is being said (van Brakel, 1996a). For example, most definitions of supervenience relations refer to objects or worlds or other things being indiscernible or identical. But what does it mean to say that two objects are exactly alike? The problem passed over here is that in the *actual* world there are no two objects that are exactly alike.[21] At the very least, there are no manifest objects that are exactly alike: identical objects occupy different places, so their contexts at least are different, if only slightly so. Lewis (1983, p. 355) gives as an example the output of a perfect photocopying machine: "Copy and original would be alike in size and shape and chemical composition of the ink marks and the paper, alike in temperature and magnetic alignment and electrostatic charge, alike even in the exact arrangement of their electrons and quarks." But it's unclear what this means, because the arrangement of electrons and quarks is (a) in constant flux and (b) not well determined while not being measured.

The most noteworthy characteristic of almost all supervenience definitions that have been proposed is that it is *not* an asymmetrical relation, contrary to microreductionistic expectations (such as the quotation of Kim in the first section. If one likes the word "supervenience," and one believes that water is H_2O, one can say that being water supervenes on being H_2O, but it is then equally true that being H_2O supervenes on being water. Although most writers on supervenience *suggest* that there is a dependence relation, it has never been specified what kind of dependence this is; the implied asymmetry is missing from the formal definitions.[22]

Supervenience is best seen as a sophisticated version of the mirror of nature paradigm. The picture is sophisticated because more than one mirror is allowed. Each mirror gives a different autonomous picture of (part of) the world, but one mirror—the ideal physical one—mirrors reality as it *is* (ontologically speaking). All other mirrors supervene on (part of) the ideal mirror. One could perhaps say that, on the one hand, the supervening mirrors have somehow emerged, but, on the other, they picture mere appearances, without cosmic significance. In all this, it remains an open question how far supervenience extends; that is, what is to be included in the ideal mirror, and what supervenes on it?

Emergence

After the initial hype about supervenience being *the* answer to a form of nonreductive physicalism, it was realized that it delivered no goods, and in the past few years interest has shifted fast to emergence. This notion goes back to the 1920s, being the predecessor of the notions of reduction and supervenience just discussed.[23] At the time, the terms *supervene* and *emerge* were used interchangeably. For example, Pepper (1926, p. 241) said, "an emergence [is] a change in which certain characteristics supervene upon other characteristics." Roughly, emergent properties can be defined as properties of a whole, which are not possessed by its component parts. For Nagel, there wasn't much to find wrong with this, as long as such views were not connected to a metaphysical commitment to a variant of creative evolution.

The following set of definitions, with P a property of a system x, is representative of contemporary discussions (Spencer-Smith, 1995):[24]

- P is emergent in x: P is novel in x, and no physical theory of the components of x can explain or predict P (radical emergence).
- P is emergent in x, relative to T: T is a theory of the components of x and P is novel in x, but T can neither predict nor explain$x$ (epistemic emergence).
- P is emergent in x: P is novel in x, and P is explained by interactions between the components of x (interactional emergence).

Most scientists in modern disciplines such as nonlinear dynamics, connectionist modeling, chaos theory, and artificial life have, like Nagel (1961), no problem combining reductionism and the weak form of interactional emergence.[25] A chemical example would be the spatial configuration of a molecule: it is a property of the whole molecule and of none of its atoms. But, it could be argued, it is not emergent in a radical sense, because (in many cases) the spatial configuration of a molecule can be derived from knowledge about the electronic configuration in the constituting atoms. However, the fact that some weakly emergent properties can be given a reductive explanation doesn't presuppose that whole theories or disciplines can be reduced. A stronger emergent property might be the tetravalence of carbon (Schröder, 1998). To deduce this from underlying interactions, the hybridization of orbitals has to be postulated.[26] The latter is a principle that does not belong to quantum mechanics. This is a case of an interaction, not of the components of the system, but of discourses at different levels.[27]

Today's radical emergentism and reductionism might agree that (a) the ultimate base is physical, (b) systems have *systemic* properties, and (c) a form of mereological supervenience applies, but they differ on the question of whether the systemic properties are *resultant* (i.e., reducible to the properties and relations of the parts of the system) or *emergent* (not reducible, not even to *relational* properties of the parts). The difference between resultant and emergent can be fought out over Broad's (1925) traditional example of sodium chloride, the tetravalence of carbon just mentioned, or more complicated cases like fermentation, where an unexpected mode of organization at the chemical level can explain that fermentation is more than an *ordinary* chemical process (Bechtel & Richardson, 1992).

As will become apparent in a later section, if the notion of emergence is to make sense when considering interdiscourse relations between talk of pure substances, chemical molecules, and quantum mechanical calculations, then it is more a case of backward emergentism. It is the details of molecular structure or the introduction of quantum mechanics that is novel, not the properties of pure substances. In contrast, if one takes "novel" in such a way that the most recent theory is the place from which everything is looked at, "backward emergence" turns into "radical emergence". Then "macrochemistry is not reducible to microchemistry" and "microchemistry is not reducible to quantum mechanics" can be restated in terminology like "the property of being pure water is an emergent property relative to molecular chemistry" and "the property of being H_2O is an emergent property relative to quantum mechanics."

Anomalous Monism

Summarizing the discussion on reduction, supervenience, and emergence, it would seem that there are three intuitions, motivations, or requirements that underlie contemporary discussions (Kim, 1990): (1) all interdiscourse relations rest on an ultimate physical base (i.e., dependence); (2) somehow interdiscourse relations must be explained in terms of things indiscernible from the B-view being also indiscernible from the S-view (i.e., covariance); and (3) each S-discourse should be autonomous (i.e., nonreducible).

Contemplating the intuitive meaning of dependence, covariance, and nonreducible suggests that the task set is not an easy one (if coherent at all). The solution I favor is that of anomalous monism (Davidson, 1970, 1993). It differs from each of the proposals for interdiscourse relations discussed so far on the following points (van Brakel, 1999a):

- No strict separation between ontology and ideology, i.e., between things and, properties;[28] no strict distinction between ontological, epistemological, and explanatory reduction, and no strict distinction between *ontological* reduction and *conceptual* autonomy
- Interdiscourse relations (bridge laws, supervenience, or emergence relations) are empirical ceteris paribus regularities, not metaphysical necessities—the only circumstances in which they might apply strictly, are model situations, completely isolated from the rest of the universe
- Interdiscourse relations are symmetrical, leaving the autonomy of both sides intact (and without preventing forms of explanatory interaction or extension by borrowing, synthesis, criteria of overall coherence, and so on)

Events[29] cause one another independently of how they are described, even independently of how they are identified. The same event can have a chemical and a physical description, or a mental and a neurophysiological description, or a moral and a physical description, or a macroscopic and a microscopic physical description, and so on. Of course, the only way to talk about causes is under some description or other, but no privileged description exists, independent of particular interests and the particular surroundings of the event discussed. In a physical discourse, events can be identified as physical events; in a chemical discourse, events can be identified as chemical events. But because chemical and physical predicates are not made for one another—after all, that is why the whole discussion about reduction or emergence arises—it is not possible to say whether an event identified under some physical description is exactly the same event (or not) as the one described under a chemical (or a psychological, or a moral, etc.) description. By assuming that each event that can be given a chemical description also has a physical description, some insight is gained in the autonomy of the chemical *and* the physical, while keeping both in the same world.

Substances and Molecules

Pure Substance

The stereotypical meaning of substance is "material from which something is made and to which it owes many of its characteristic qualities."[30] But substances can also be transformed into other substances. Such transformation occurs naturally or intentionally. Furthermore, depending on properties of interest to user or contemplator, substances are classified. I will argue that chemistry is not primarily the science of molecules, but of substances. This goes against the grain of all common sense knowledge about chemistry. For example, the philosopher Forbes says (1985, p. 199) that "we expect the superficial and easily detectable differences between pieces of stuff to reflect fundamental differences which explain the superficial ones, and it is the fundamental differences which have the final say in classification; so someone who refuses to classify samples in this way may fairly be said not to understand what a substance is."

My question would be, What sort of criteria are to be used to assess whether Forbes understands what a substance is? Note that the point of discussion is not whether successful correlations can be found between microscopic and macroscopic descriptions. The question is what the right way is to explain what (pure) substances are, or, what the "first" meaning of "pure substance" might be. It might be suggested that Forbes's suggestion also characterizes the intuitions of the working chemist. However, in the practice of chemistry, the intuition that observable (manifest) properties "emerge" from (are caused by) unobservable microstructures is on a par with the intuition that (macroscopic) knowledge about chemical reactivities permits certain conclusions to be drawn about microstructural models.

I suggest that the right way to approach the question is from the manifest side, and, more specifically, from the observation that many substances can occur in solid, liquid, and vapor form. The phase rule[31] and the theory of chemical thermodynamics provided Gibbs with the theoretical background for the concept of phase (the state of aggregation of a substance such as solid, liquid, vapor)[32] and hence, made it possible to give precise definitions of "solution," "compound," "pure substance," and "element," independent of any atomic hypothesis (Timmermans, 1963). A bubble of air, a piece of sugar, a drop of salt water, a fragment of glass—each is a phase. A tiger, milk, and most paints are polyphasic aggregates. In principle, polyphasic aggregates can be separated into different phases by mechanical methods. For example, paints left to themselves long enough will separate into a solid phase and a liquid phase on top, which would not have happened if it would have been one phase.

Mechanical methods, using mechanical forces (e.g., filtration, centrifugation, and grinding) can be used to separate heterogeneous and homogeneous materials.[33] Physical methods using thermal energy and hydrostatic pressure (e.g., distillation, crystallizing, and melting) can be used to divide mixed and pure materials.[34] Thermodynamic methods (using energy or pressure at higher levels) can be used to divide compounds and elementary materials. A pure substance can be defined as a substance of which the properties, such as temperature, density, and electric conductivity, do not change during a phase conversion (as in boiling a liquid or melting a solid). This definition is independent of one's beliefs in atoms, in an atomistic hidden variable interpretation of quantum mechanics, or in any other microphysical story.[35] Such a pure substance persists as a phase of constant composition when the conditions of temperature, pressure, and composition of the other present phases undergo continuous alteration within certain limits (i.e., the limits of the existence of this pure substance).[36] Or, in short, a pure substance is a body that forms hylotropic phases in a finite range of temperature and pressure.[37] Then chemical elements are substances that *never* form other than hylotropic phases.

That is to say, a material is pure if it is perfectly homogeneous after being subjected to successive modes of fractioning that are as different as possible (when attempts at further purification produce no further change in properties). Ideally, separation techniques have to be applied an infinite number of times; moreover, techniques that depend on the phase rule are subject to a whole range of pitfalls. Later refinements of separation techniques may show that what was once thought to be a pure substance is, after all, not pure. Different separation techniques (crystallization, electrophoresis, and so on) set different standards of purity.[38] Moreover, the thermodynamic notion of pure substance is an idealization. Water, after all, is *not* a pure substance—that is so only under "ordinary condition." At high temperatures, it changes into a mixture: above 500°C, its vapor partially dissociates and its two gaseous constituents can be separated. Similarly, at low temperatures, a mixture of acetylene (C_2H_2) and benzene (C_6H_6) is a solution, but at high temperatures, it behaves as a single substance (in particular, in the presence of porcelain).

Categories that cause problems for this definition of chemical substance include (1) enantiomers (species containing equal amounts of two optical isomers, like *l*- and *d*-tartaric acid); (2) azeotropic mixtures; (3) dissociative compounds in equilibrium; (4) certain types of mixed crystals or other polymorphic compounds (e.g,*d*- and *l*-camphoroxime); (5) synthetic polymers; (6) many biochemical compounds;[39] (7) systems that are not in "pure" thermodynamic equilibrium; and (8) isotopes. In each case, pragmatic decisions have to be made, as the notion of pure substance cannot be essentialized. There are no competing definitions of "pure substance" that can avoid the need for "inspired adhoccery" to deal with difficult cases.

This definition of a pure chemical kind, species, or substance goes back to Wald (1897) and Ostwald (1902, 1904, 1907). Ostwald, following Hume and Mach, had two aims: first, to distinguish between what is given in experience and what is postulated by the mind (nothing *compels* us to affirm that mercury oxide contains mercury and oxygen), second, to show that energy is the most general concept of the physical sciences. This led to an approach to chemistry that can be associated with the conventionalism of Poincaré, the monism of Mach, the later positivism of the Vienna Circle, and the operationalism of Bridgman. Such a view implies that thermodynamics is the most basic physical science (and not statistical or quantum mechanics).[40] As late as 1901 Ostwald vehemently argued in his lectures against the "dead track" of appeals to molecular structure or forces between molecules (Ostwald, 1902). In his Faraday Lecture of 1904, he said that "chemical dynamics [i.e., his approach], has, therefore, made the atomic hypothesis unnecessary for this purpose and has put the theory of stoichiometrical laws on more secure ground than that furnished by a mere hypothesis"

(Ostwald 1904, p. 187f). Instead of atomic mass, Ostwald (1909, p. 326) uses "relative combining weight," for example, by saying, "It is possible to ascribe to each element a certain relative combining weight in such a way that every combination between the elements can be expressed by these weights or their multiples." Also Gay-Lussac's gas law is expressed without appeal to the hypothesis of Avogadro (Ostwald, 1909).

Neither Timmermans (1963) and Prélat (1947) nor later the more philosophical approach of protochemistry add much to the program as laid out by Ostwald.[41] However, an important, though overlooked observation of van der Waals (1927) should be added.[42] Although he addresses the issue only briefly, his rigor is way above that of Ostwald or later writers. To define the notion of pure substance in macroscopic terms (as presented here), (Gibbsean) thermodynamics is drawn upon, which *itself* presupposes an undefined concept of pure substance.[43] In terms of the scientific/manifest issue introduced in an earlier section, notions like temperature, density, phase conversion, and so forth are used to give a definition of pure substance *within* the scientific image,[44] against the background of thermodynamics. Any talk of atoms, molecules, and valences will be relative to this macroscopic scientific definition of pure substance. Moreover, this scientific *macroscopic* definition *presupposes* a notion of pure substance that can only be justified by an appeal to its "vague" meaning in the manifest image.

Problems for the Concept of Molecule

A pure substance is often considered a collection of molecules of the same type. For example: "Substance *x* is chemically pure if and only if *x* is composed exclusively of either atoms or molecules of a single species or kind" (Bunge, 1985, p. 222), where, presumably, lest the definition becomes circular, "molecules of a single species or kind" has to be understood as "molecules having the same composition and structure". However, this definition only applies in rare cases—pure water or pure acetic acid not consisting exclusively of one type of molecules all having the same "structure."

The definition doesn't work for metals, salts, electrolytes, or dissociating liquids. It also breaks down for enzymes, antibodies, viruses, or, more generally, isochemical compounds and homeomers (Pirie, 1952).[45] It sounds good to say something like "a molecule is the smallest particle of a definite compound which still has the same properties." But "smallest particle" only makes sense for ideal gases and a few liquids, not for water, carbon (diamond, soot, buckminsterfullerence), salt crystals, proteins, or cellulose. Then there is the problem of definite and indefinite compounds—a problem already recognized by Mendeleyev.[46] In fact, even in the simplest cases, and approaching the issue from the molecular point of view, the majority of pure materials are tautomers:[47] they do not consist of identical molecules, but of an intimate mixture of different species of molecules (metamers or polymers) in statistical equilibrium with one another and inseparable under ordinary experimental conditions.

The concept of molecular structure seems to derive its meaning more from the way molecules are represented in models than from anything else. Molecule is an indispensable, but thoroughly theoretical concept; it is part of theories that are impressively empirically adequate, but without giving a clear idea of what entities are thought to exist. Moreover, to the extent it makes sense to say things like "a pure substance consists of identical molecules," any empirical evidence will depend on a prior understanding of what a pure substance is.

It has been suggested that Ostwald was forced to accept the atomic hypothesis because the issue of isomeric substances forced it upon him. Isomeric substances can be defined as substances having the same composition but different energy content, but as this says nothing about which chemical reactions to expect, Ostwald was allegedly forced to admit to a kind of "grainedness" in the physico-chemical world, on pain of having "to admit alternative structures of *nothing in particular*!" (Bradley, 1955; italics and exclamation mark in original).[48] As we are brought up in a world of molecules, strengthened further by the dominance of DNA-talk, this sounds utterly plausible, but the appeal to the nonsensicality of "structures of *nothing in particular*" breaks down under minimal scrutiny. Consider the benzene molecule: it is impossible to specify exactly where single and double bonds are. The "real" microsituation is a kind of mixture of a number of (logically) possible fixed arrangements of nuclei (of atoms) and electrons. The structure of the benzene molecule has been represented on different occasions as "intermediate" between two, three, and five structures. There is no end to the extent of this kind of hybridity: to explain the known reactions of anthracene more than 400 different diagrams have been proposed. If there is any "it" to which these 400 stories apply, it would seem to be anthracene and not "molecule which depending on circumstances can have any of 400+ structures." Bullvalene ($C_{10}H_{10}$), is said to have 1,260,000 electronic tautomers, each of which is separated from the other by an energy barrier of 12 kcal/mol.

The citation of Bunge at the beginning of this section can be seen as an informal statement of a bridge law. In more recent literature, bridge laws have returned as theoretical identity statements, which, according to influential publications of Kripke (1972) and Putnam (1975), give the essence of natural kinds, for example, the essence of water is that it is H_2O—the view Hare (quoted in an earlier section) reacted against. Although this is incorrect, virtually all philosophers believe that water consists of H_2O molecules. "Essentialistic realism," therefore, doesn't work either. The underlying microscopic essences vary as much with context or circumstance as nominal essences (van Brakel, 1986).

If one is drawn to microreductive "psychological essentialism" (Medin, 1989), one ends up eliminating all substance, being left only with quarks, strings, or whatever final physics settles on (van Brakel, 1992).[49] We started out being interested in the "essence" of water, but end up with the "essence" of matter—which itself evaporates on further scientific scrutiny. The whole notion of a *particular* and *pure* substance has been eliminated in the process of reduction. It is taken for granted that pure substance *must* be a welldefined category in terms of mathematical or atomic primitives, and, if not, it is simply eliminated during the reductive process.

Molecular Structure and Quantum Mechanics

Many phenomena are of interest to the chemist which can only be explained in terms of quantum mechanics. From this it doesn't follow that chemistry can be reduced to quantum mechanics. For example, chemical molecules have a shape, but the latter concept cannot be found in quantum mechanics. For simple molecules, outstanding agreement has been obtained between calculated and measured data of certain parameters. Yet, the concept of, say, a chemical bond has not been found anywhere. A chemical bond is an entity with a length, an orientation, a dissociation energy, a contribution to the enthalpy of atomization, an electric dipole moment, and other properties. But it does not "appear" in the Schrödinger equation.[50]

The general point against the suggestion that molecular chemistry can be reduced to quantum mechanics is that the decision when and where to suppress the interaction with the environment is not something that can be derived from quantum mechanics—this is where Gell-Mann's "chemical questions being asked" (mentioned earlier) enter the discussion. But it is these decisions that, as it were, abstract objects out of the quantum mechanical formalism. Quantum mechanics describes the material world, in principle, as one whole. Within quantum mechanics an object can only be defined in terms of its relations to its environment. To separate out objects from this whole requires a justification that lies outside the principles of quantum mechanics. Because "the" environment consists of the rest of the universe, it can never be given a precise description and must therefore be replaced by a model environment that mimics aspects of the real situation.

Woolley's article "Must a Molecule Have a Shape?" (1978) has been referred to in many subsequent publications, though, perhaps due to its provocative title, it has not always been understood correctly. The ambiguity is in the term *molecule*.[51] For example, although Bunge (1985) is skeptical about the reduction of chemistry to physics, nonetheless García-Sucre and Bunge (1981, p. 91) conclude that "quantum chemistry makes room for a concept of molecular shape, recently criticised by Woolley." But what they mean by that is that the quantum system they call "molecule" "will extend over the entire space and its shape will be a closed surface infinitely far removed from the origin of the reference frame" (p. 91) and this system "lacks a definite size" (p. 91). Woolley wouldn't deny that. His point, after all, was that the *classical* concept of molecule cannot be derived *ab initio* from quantum mechanics. The classical concept of molecule does not allow each molecule to "extend over the entire space." So García-Sucre and Bunge did not show that the classical definition of molecular structure carries over smoothly into the quantum realm. Of course, quantum mechanics has structure, but Woolley (1986) is surely correct to say that "quantum structure" is *not* synonymous with the classical terms *shape* and *size*. The reason such considerations are easily dismissed as contentious is, I believe, due to the fact that microreductionism is presupposed as the goal of science and requires no further discussion (cf. the earlier example of Bader).

Another way of putting the point is that "quantum chemistry regards the existence of molecules as beyond any doubt and doesn't ask what quantum mechanics has to say about it" (Primas, 1983, p. 292). Quantum chemistry *borrows* the notion of molecular structure from classical chemistry. Even if an isolated molecule is considered, there is no room for concepts like molecular shape because of the Pauli principle and the superposition principle.[52] Relative to quantum mechanics, classical observables are truly *emergent* quantities: they can only be "constructed" from quantum mechanical principles by *selecting* one of the infinitely many inequivalent representations of the formal system. The selection is governed by irreducible chemical knowledge, including intuition grounded in experience with chemical practice.

Concluding Remarks

If much of the preceding discussion sounds incredible, it is because of the way the history of the scientific image has been written. First, such history has tended to distort the relation between chemistry and physics or, more particularly, the relation between chemistry and atomism. Second, the scene has been set by Kant's influential remark that chemistry is not a proper science.[53] From the perspective of physics, it is not too far fetched to suggest that with the advance of quantum mechanics, atomism finally dug its own grave. Modern physics, therefore, rather supports the views of Mach and Ostwald, who opposed the reality of atoms as material objects. But there is no reason to draw the conclusion that, therefore, chemical substances or chemical molecules are not "real." To draw that conclusion is exactly the eliminative version of the scientific image in which everything disappears, except quarks or strings or something else nobody, not even a few hundred theoretical physicists, can really grasp. That the notion of "(chemical) pure substance" is methodologically prior to notions like atom or molecule is indisputable. To see the notion of pure substance as merely a ladder to reach the true reality of atoms and molecules might, therefore, be called the essentialistic fallacy—and similarly for the ladder from molecules to quantum mechanics.

Reduction either of chemistry to microphysics or, more generally, of the manifest to the scientific image doesn't work. Similarly, within chemistry (and physics) it is not the case that the microdescription will, in principle, always give a more complete and true description. If quantum mechanics would turn out to be wrong, it would not affect all (or even any) chemical knowledge about molecules (bonding, structure, valence, and so on). If molecular chemistry were to turn out to be wrong, it wouldn't disqualify all (or even any) knowledge about, say, water. Interdiscourse relations between macrochemistry and microchemistry or between microchemistry and quantum mechanics are best seen as symmetrical relations, locally valid under well-described ceteris paribus boundary conditions—where the latter are governed by top-down explanatory interests.

In a way, the primacy of the manifest image is rather trivial. The manifest image is the highest meta-level where questions of justification stop and appeal is made to certainties not grounded in the scientific image. All discourses are related via the manifest image, because they are grounded in it. Chemistry partakes in *both* images as defined by Sellars. As the science of the transformation of substances however, it is finally grounded in the manifest image, not the scientific image.

Notes

1. In citations ellipsis dots are omitted and interpunction and capitalization is changed to fit the surrounding text.

2. In the sense here intended, the manifest image roughly corresponds with Wittgenstein's "form of life," Husserl's *Lebenswelt*, and Heidegger's *Dasein*, as well as expressions like *praxis*, "background," *Vorverständnis*, Moorean corpus, and "natural realism of the common man." For an elaboration, see van Brakel (1994, 1998s). The manifest image in my sense should not be associated with "basic level" in the sense of Rosch and Lakoff—cf. van Brakel (1991).

3. For details and references, see van Brakel (1996b); cf. also note 28.

4. Folk psychology draws on an abundance of unobservables. Not only gods, but also beliefs, desires, and meanings are unobservables. As to the physical world, very young children invoke "tiny invisible particles" in explanatory strategies, for example in understanding why water tastes like sugar when sugar has "disappeared" in the water (Au et al., 1993).

5. See Carnap (1956, 1963, p. 929) and DeVidi and Solomon (1994).

6. As when he says that the task of the philosopher is to fuse the manifest and scientific perspective into one "stereoscopic" or "synoptic" view or that the critique he is engaged in is "one which compares this [manifest] image unfavourably with a *more* intelligible account of what there is" (Sellars, 1963 pp. 4, 18).

7. For a fuller account and criticism of Sellars' views, see van Brakel (1996b).

8. Most of the literature on protochemistry is in German (Janich 1994; Psarros, 1994, 1995a, 1995b: Gutman & Hanekamp, 1996); for a brief characterization in English, see van Brakel (1999b), and here note 42.

9. cf. Oppenheim and Putnam (1958, p. 5) on: "the possibility that all science may one day be reduced to microphysics (in the sense in which chemistry seems today to be reduced to it)."

10. Reichenbach (1978, p. 129), Dirac (1929), Heisenberg (1972, p. 112), Jordan (1957, p. 19).

11. For more on *ab initio* methods, see Scerri (1991), van Brakel (1997).

12. Primas (1983, 1985a, 1985b, 1991).

13. For the more general account, see van Brakel (1996a).

14. For the difference, see Nickles (1973).

15. For an early critique of Oppenheim and Putnam's "sweeping" statements, see Schlesinger (1961).

16. See Lévy (1979), Sarkar (92), Endicott (1998), Hooker (1981).

17. Note that this is a much stronger requirement than the observation that there are many phenomenological facts that can be given a plausible explanation in terms of a microtheory.

18. And a host of other problems connected to the suggestion that thermodynamics can be reduced to statistical mechanics (Sklar, 1993).

19. cf. Primas (1985b), Sklar (1993). There is also talk of "negative, infinite and hotter than infinite temperatures" (Ehrlich, 1982). Even if temperature could be reduced, it doesn't yet follow that "boiling point" can be reduced (Hooker, 1981 pp. 497–500).

20. cf. Nagel (1974): "the theory T is not derivable from (and hence not reducible to) the theory T′, although T may be derivable from T′ when the latter is conjoined with an appropriate set of bridge laws."

21. Would it help to loosen the identity requirement by talking about "similar" or "roughly identical" S-properties overlaying "roughly identical" B-properties? No, not in general; what is roughly similar and what is not, depends on the context and interest at hand.

22. Armstrong (1989, p. 104), who makes symmetry an explicit part of the definition of supervenience, is a minority view. The dominant view is that the underlying discourse should be somehow prior to the overlaying discourse (Menzies, 1988).

23. Nagel (1961, pp. 366–380) gives an excellent survey of the early literature. For the history of emergence and reductionism, see also Stöckler (1991).

24. The discussion about emergence is traditionally linked to that of downward causation. This notion will change in meaning with the specification of the kind of emergence. In the minimal sense of interactional emergence downward causation is the influence the relatedness of the parts of a system has on the behavior of the parts (not the influence of a macroproperty itself).

25. As Wimsatt (1996) puts it: "a reductive explanation of a behaviour or a property of a system is one showing it to be mechanically explicable in terms of the properties of and interactions among the parts of the system." See also Humphreys (1996).

26. Note that hybridization undermines microreductionistic pictures of even the simplest molecules. Consider water. On the basis of the Pauling principle, one would expect oxygen to

form covalent bonds with two hydrogen atoms yielding a structure with an HOH bond angle of 90°. Experimentally the value is 104.45°, which is used to calculate the appropriate amount of hybridization. Hybridization is part of the modern conception of molecular structure, but it is grounded in a whole range of discourse levels.

27. In connection with Mendelian and molecular genetics, Kitcher (1984) called this "extended explanation" and Needham (1999b) in his discussion of Hanna's (1966) *Quantum Mechanics in Chemistry*, speaks of "overall coherence." It is one thing (while appealing to an inference to the best explanation) to say that a theoretical "picture" plays a unifying role, quite another to suggest that some isolated phenomenon can, in some sense, be derived from it.

28. Which individuals there are is relative to the language used to discuss the world; the world does not come to us presorted (Carnap, 1956: 'Lowe, 1989': Quine, 1993). This is one way in which the scientific image depends on the manifest image for its basic intuitions. The labeling "ontology" (referents of names or values of variables) versus "ideology" (referents of predicates) stems from Quine.

29. This is a metaphilosophical term, but still close to the manifest image term "even." It is not a term like "strings" (or "superego"), but more like "boiling point." Justification of the proposal of anomalous monism (based on the primitive notion of "event") is first grounded in philosophical discourse, and this discourse is grounded, in the end, in the manifest image.

30. It should perhaps be stressed that the notion of chemical substances is completely different from the metaphysical concept of substance that figures in the philosophical tradition since Aristotle (Witt, 1989; Hoffman & Rosenkrantz, 1994). As will appear, the chemical notion of substance is, at least as presented in this chapter, an empirical concept, whose referent is wholly defined in terms of laboratory procedures, and cannot be given an essentialist definition.

31. According to Gibb's phase rule, a system of c components in p phases has $c - p + 2$ degrees of freedom. It is a law completely free from all hypothetical assumptions as to the molecular condition of the substances involved.

32. A phase is a macroscopic continuum which, when in a state of thermodynamic equilibrium, has constant and uniform properties throughout. Here "properties" refers to properties such as density, electric conductivity, magnetic susceptibility, and so on—the so-called physical constants.

33. A material is homogeneous if random samples taken from the material have the same properties—in case of a solid, it should first be ground. In other words: a material is homogeneous if it cannot be separated into different materials by mechanical methods (more precise: if it cannot be separated by external or capillary forces).

34. Mixed materials can be further divided into solutions (of the gas/gas, liquid/liquid, and solid/solid sort), addition compounds (such as hydrates) and aggregates (emulsions, conglomerates, colloids, smokes, etc.).

35. It shouldn't be underestimated how theoretical the notion of "pure substance" already is, as can clearly be seen from the history of this concept (Timmermans, 1963, pp. 10–16, 67–73); cf. main text to note 44.

36. Note that, depending on its history, the same pure compound may display different secondary qualities: the precipitate of mercury oxide is yellow; as a product of calcination, it is red (the color difference is due to a difference in grain size).

37. Ostwald (1907, p. 166): "The mode of phase change in which the newly formed phases have at every moment the same properties and the same total composition as the original system is called a *hylotropic* transition." Ostwald (1907, p. 170): "An element is a substance which cannot be transformed into another non-hylotropic substance within the entire range of attainable energy influence."

38. Spectroscopic methods for purity tests always remain secondary to traditional purity tests in terms of the inseparability by any separation technique (Schummer, 1997): (i) there is

no pure "standard" unless it has been made by conventional laboratory procedures (or in rare cases of "naturally" pure products, has been checked); (ii) no spectroscopic method can be used unless one already knows what the spectra of a particular pure substance look like. Although it is possible to make predictions of the electromagnetic properties of substances, there is no theory that provides a general criterion to distinguish, on the basis of spectrographic data alone, spectra of pure substances, and mixtures.

39. In biochemistry, the concept of purity, with the connotations it has in the chemistry of small molecules, starts to break down (Pirie, 1952).

40. This, too, was the view of Duhem (Needham, 1996). There are many similarities in Ostwald's and Duhem's understanding of chemical structure. For Duhem's views, see Needham (1998, pp. 50–55). See, also, Needham (1999a) on the ontological implications of the macroscopic view represented by thermodynamics.

41. However, at least two wider issues were introduced by protochemistry. First, it is argued that chemical laws should be understood as norms, not as natural laws. For example, the laws of constant and multiple proportions contain the undefined terms "compound" and "part" or "element." The first of these laws, in fact, represents several norms and definitions: (i) the sum of the weights of the compounds that enter into a (chemical) transformation is equal to that of the products that are formed. (ii) It must be possible to isolate the products of a transformation as pure chemical substances. (iii) A transformation that fulfills the previous two criteria is called a chemical reaction. (iv) Chemical reactions have to be carried out in such a way that its products are chemically pure substances with constant composition. The first of these norms is not the *law* of conservation of mass discovered by Lavoisier. The last norm is not the *law* of constant proportions (Psarros, 1994). A second important theme protochemistry introduced is the distinction between "ordinary" chemical language and "reflective" chemical terms, which are part of a meta-language. The prime example of such a reflective term is "stuff" (*hyle*, chemical substance). Only *after* the chemist has distinguished "chemical stuff-properties" from other properties, does it make sense to introduce talk of atoms and molecules (Janich, 1994). A similar argument can be applied to physical notions like force, mass, or energy.

42. He doesn't mention Ostwald, and gives credit to Wald for an earlier rigorous approach (van de Waals, 1927, p. 229n2). Wald (1897) credits Ostwald for motivating his research.

43. "Wir setzen also bis auf weiteres den Begriff einer 'chemisch reinen Substanz' als bekannt voraus" (p. 14); "die Betrachtungen stützen sich auf die Annahme, daß wir wissen, was wir unter diesen Worten ['chemische reinen Substanz' and 'chemischen Individuums'] zu verstehen haben" (p. 227).

44. Note also that the definition of a pure substance in terms of a combustion analysis ("elemental analysis," going back to von Liebig) is a macroscopic definition, based on weighing macroscopic quantities of material.

45. Isochemical substances have the same gross composition, but with different average size of aggregates. Homeomeres are substances with identical chemical activity.

46. The variation can be considerable, for example, $Na_xWO_3 (0.93 > x > 0.32)$ or Li_xWO_3 $(0.57 > x > 0.31)$. Of course, such cases present problems for any definition of pure substance.

47. The term is used here in a wide sense. In a narrow sense it may refer merely to compounds of which the keto- and enol-form of an isomer are in equilibrium.

48. In the preface of Ostwald (1907) he says: "Three years ago, on the occasion of the Faraday lecture [Ostwald, 1904], I made an attempt to arouse the interest of chemists in these matters, but the result was not very encouraging. I know from personal experience that patient and continued labour can accomplish wonders even when the case seems hopeless. One must wait for the right time, and I am convinced that the time for this matter has arrived." But by the end of the book, when addressing metamerism, polymerism, and the role of valence in structure theory, he has to acknowledge that, when compared with "structure theory" his "purely empiri-

cal" account in terms of differences in energy content "predicts, however, nothing whatever about the chemical reactions which are to be expected, and is therefore not applicable as an aid to building up a system. Energy in the sense in which the word is used here is expressible by a mere number; it has no further properties, and is therefore not of value in expressing the qualitative differences belonging to chemical reactions" (p. 326). And he further acknowledges "the introduction of a new factor, one involving differences in the spatial arrangement of elements, and in a few cases even this assumption appears insufficient"—though its status for Ostwald doesn't rise above that of "a very important aid" (p. 329).

49. cf. Quine (1992). "My tentative ontology continues to consist of quarks and their compounds, also classes of such things, classes of such classes, and so on, pending evidence to the contrary."

50. For this and the next paragraph, see Primas (1983,1985a), Amann (1990), Del Re (1996), and also the discussion between Scerri (1997, 1998) and Needham (1999b).

51. For a start, a chemical molecule is not a physical molecule. For a chemist, electrons are individuals, but not for the physicist.

52. Although Pauli's principle was rescued by the new quantum theory, the notion of individual quantum numbers for each electron was lost. The concept of electronic configurations cannot be derived from quantum mechanics. It represents an approximation and a bookkeeping scheme for finding the number of outer electrons in an atom, but does not necessarily provide information as to the inner electron shells (Scerri, 1991). On the superposition principle, see Amann (1990) and Woolley (1991).

53. For sources and discussion on Kant on chemistry, see van Brakel (1999b); cf. quotation of Bader in text.

References

Amann, A. 1990. "Chirality: A Superselection Rule Generated by the Molecular Environment?" *Journal of Mathematical Chemistry*, 6: 1–15.

Armstrong, D. M. 1989. *A Combinatorial Theory of Possibility*. Cambridge: Cambridge University Press.

Au T. K., Sidle, A. L. & Rollins, K. B., 1993. "Developing an Intuitive Understanding of Conservation and Contamination: Invisible Particles as a Plausible Mechanism." *Development Psychology*, 29: 286–299.

Bader, R. F. W. 1990. *Atoms in Molecules: A Quantum Theory*. Oxford: Clarendon.

Bader, R. F. W., Popelier, P. L. A., & Keith, T. A. 1994. "Theoretical Definition of a Functional Group and the Molecular Orbital Paradigm." *Angewandte Chemie*, 106: 647–659. *Int. Ed. Engl.*, 33: 620–631.

Bechtel. W., & Richardson, R. C. 1992. "Emergent Phenomena and Complex Systems." In A. Beckermann, H. Flohr, & J. Kim, eds. *Emergence or Reduction? Essays on the Prospects of Nonreductive Physicalism* (pp. 257–288). Berlin: de Gruyter.

Bernstein, R. J. 1966. "Sellars' Vision of Man-in-the-Universe." *Review of Metaphysics*, 20: 113–143; 290–316.

Bradley, J. 1955. "On the Operational Interpretation of Classical Chemistry." *British Journal for the Philosophy of Science*, 6: 32–42.

Broad, C. D. 1925. *The Mind and Its Place in Nature*. London: Kegan Paul.

Bunge, M. 1985. *Treatise on Basic Philosophy*. (vol. 7, pp. 219–230). Dordrecht: Reidel.

Carnap, R. 1956. *Meaning and Necessity*. (pp. 206–221). Chicago: University of Chicago Press.

Carnap, R. 1963. "Reply to Beth." In P.A. Schilpp, ed. *The Philosophy of Rudolf Carnap*. La Salle, IL: Open Court.

Davidson, D. 1980. Mental Events. In *Essays on Actions and Events,* (pp. 207–224). Oxford: Clarendon.

Davidson, D. 1993. Thinking Causes. In J. Heil & A. Mele, eds. *Mental Causation* (pp. 3–18). Oxford: Clarendon.

Del Re, G. 1996. "The Specificity of Chemistry and the Philosophy of Science." In V Mosini, ed. *Philosophers in the Laboratory* (pp. 11–20). Rome: Euroma.

De Vidi, D., & Solomon, G. 1994. "Geometric Conventionalism and Carnap's Principle of Tolerance." *Studies in History and Philosophy of Science,* 25: 773–783.

Dirac, P. A. M. 1929. "Quantum Mechanics of Many-Electron Systems." *Proceedings of the Royal Society of London,* A123: 714–733.

Ehrlich, P. 1982. "Negative, Infinite and Hotter Than Infinite Temperatures." *Synthese,* 50: 233–277.

Endicott, R. P. 1998. "Collapse of the New Wave." *Journal of Philosophy,* 95: 53–72.

Forbes, G. 1985. *The Metaphysics of Modality.* Oxford: Clarendon.

García-Sucre, M., & Bunge, M. 1981. "Geometry of a Quantal System." *International Journal of Quantum Chemistry,* 19: 83–93.

Gell-Mann, M. 1994. *The Quark and the Jaguar: Adventures in the Simple and the Complex.* New York: W.E. Freeman.

Gutmann, M., & Hanekamp, G. 1996. "Abstraktion und Ideation—Zur Semantik chemischer und biologischer Grundbegriffe." *Journal for General Philosophy of Science,* 27: 29–53.

Hanna, M. W. 1966. *Quantum Mechanics in Chemistry.* New York: Benjamin.

Hare, R. M. 1984. "The Inaugural Address: Supervenience." *Proceedings of The Aristotelian Society,* (Suppl.) 58: 1–16.

Heisenberg, W. 1972. *Physics and Beyond.* New York: Harper & Row.

Hoffman, J., & Rosenkrantz, G. S. 1994. *Substance among Other Categories.* Cambridge: Cambridge University Press.

Hooker, C. A. 1981. "Towards a General Theory of Reduction." *Dialogue,* 20: 38–60; 201–235; 496–529.

Humphreys, P. 1996. "Emergence, Not Supervenience." *Philosophy of Science,* 64 (Proc.), S337–S345.

Janich, P. 1994. "Protochemie: Programm einer konstruktiven Chemiebegründung." *Journal for General Philosophy of Science,* 22: 71–87. Also in: *Chimica Didactica,* 21 (1995) 111–128.

Jordan, P. 1957. *Das Bild der modernen Physik.* Frankfurt: Ullstein.

Kemeny, J. G., & Oppenheim, P. 1956. "On Reduction." *Philosophical Studies,* 7: 6–19.

Kim, J. 1990. "Supervenience as a Philosophical Concept." *Metaphilosophy,* 21: 1–27.

Kitcher, P. 1984. "1953 and All That: A Tale of Two Sciences." *Philosophical Review,* 93: 335–373.

Kripke, S. 1972. *Naming and Necessity.* Oxford: Basil Blackwell.

Lévy, M. 1979. "Les relations entre chimie et physique et le problème de la réduction." *Epistemologia,* 2: 337–370.

Lewis, D. 1983. "New Work for a Theory of Universals." *Australasian Journal of Philosophy,* 61: 343–377.

Lowe, E. J. 1989. *Kinds of Being: A Study of Individuation, Identity and the Logic of Sortal Terms.* Oxford: Basil Blackwell.

Medin, D. L. 1989. Concepts and Conceptual Structure." *American Psychologist,* 44: 1469–1481.

Menzies, P. 1988. "Against Casual Reductionism." *Mind,* 97: 551–574.

Nagel, E. 1961. *The Structure of Science.* London: Routledge and Kegan Paul.

Nagel, E. 1974. "Issues in the logic of reducting explanations." In *Teleology Revisited* (pp. 95–113). New York: Columbia University Press.

Needham, P. 1996. "Macroscopic Objects: An Exercise in Duhemian Ontology." *Philosophy of Science,* 63: 205–224.

Needham, P. 1998. "Duhem's Physicalism." *Studies in History and Philosophy of Science*, 29: 33–62.

Needham, P. 1999a. "Macroscopic Processes." *Philosophy of Science*, 66: 310–331.

Needham, P. 1999b. "Reduction and Abduction in Chemistry—A Response to Scerri." *International Studies in the Philosophy of Science*, 13: 169–185.

Nickles, T. 1973. "Two concepts of Inter-Theoretic Reduction." *Journal of Philosophy*, 70: 181–201.

Oppenheim, P., & Putnam, H. 1958. "The Unity of Science as a Working Hypothesis." In H. Feigl, G. Maxwell, & M. Scriven, eds. *Minnesota Studies in the Philosophy of Science* (vol. 2, pp. 3–35). Minneapolis: University of Minnesota Press.

Ostwald, W. 1902. *Vorlesungen uber Naturphilosophie; Natural Philosophy*. London (1911).

Ostwald, W. 1904. "Elements and Compounds." *Faraday Lectures*. London: Chemical Society.

Ostwald, W. 1909. *The Fundamental Principles of Chemistry*. London: Longmans, Green,

Pepper, S. C. 1926. "Emergence." *Journal of Philosophy*, 23: 241–245.

Pirie, N. W. 1952. "Concepts out of Context." *British Journal for the Philosophy of Science*, 2: 269–280.

Prélat, C. E. 1947. *Epistemología de la química*. Buenos Aires: Espasa and Calpe Argentina.

Primas, H. 1983. *Chemistry, Quantum Mechanics and Reductionism*. Berlin: Springer.

Primas, H. 1985a. "Kann Chemie auf Physik reduziert werden? I: Das molekulare Programm." *Chemie unser Zeit*, 19: 109–119.

Primas, H. 1985b. "Kann Chemie auf Physik reduziert werden? II: Die Chemie der Macrowelt." *Chemie unser Zeit*, 19: 160–166.

Primas, H. 1991. "Reductionism: Palaver Without Precedent." In E. Agazzi, ed. *The Problems of Reductionism in Science* (pp. 161–172). Dordrecht: Kluwer.

Psarros, N. 1994. "Die 'Gesetze' der konstanten und der multiplen Proportionen." In P. Janich, ed. *Philosophische Perspektiven der Chemie* (pp. 53–64).

Psarros, N. 1995a. "The Constructive Approach to the Philosophy of Chemistry." *Epistemologia*, 18: 27–38.

Psarros, N. 1995b. "Stoffe, Verbindungen und Elemente—Eine methodische Annäherung an die Gegenstände der Chemie." *Chimica Didactica*, 21: 129–148.

Putnam, H., ed. 1975. "The Meaning of Meaning." In *Mind, Language and Reality* (pp. 215–271). Cambridge: Cambridge University Press.

Quine, W. V. 1992. "Structure and Nature." *Journal of Philosophy*, 89: 5–9.

Quine, W. V. 1993. "In Praise of Observation Sentences." *Journal of Philosophy*, 90: 107–116.

Reichenbach, H. 1978. "The Aims and Methods of Physical Knowledge." In M. Reichenbach and R.S. Cohen, eds. *Hans Reichenbach: Selected Writings 1909–1953* (vol. 2, pp. 81–225). Dordrecht: Reidel.

Sarkar, S. 1992. "Models of Reduction and Categories of Reductionism." *Synthese*, 91: 167–194.

Scerri, E. R. 1991. "The Electronic Configuration Model, Quantum Mechanics and Reduction." *British Journal for the Philosophy of Science*, 42: 309–325.

Scerri, E. R. 1997. "Has the Periodic Table Been Successfully Axiomatised?" *Erkenntnis*, 47: 229–243.

Scerri, E. R. 1998. "Popper's Naturalised Approach to the Reduction of Chemistry." *International Studies in the Philosophy of Science*, 12: 33–44.

Schlesinger, G. 1961. "The Prejudice of Micro-Reduction." *British Journal for the Philosophy of Science*, 12: 215–224.

Schröder, J. 1998. "Emergence: Non-Reducibility or Downwards Causation?" *Philosophical Quarterly*, 48: 433–452.

Schummer, J. 1997. "Towards a Philosophy of Chemistry." *Journal for General Philosophy of Science*, 28: 307–335.

Schweber, S. S. 1997. "The Metaphysics of Science at the End of a Heroic Age." In (R.S. Cohen *et al.*, eds.) *Experimental Metaphysics*, (pp. 171–198) Dordrecht: Kluwer.

Sellars, W. 1963. *Science, Perception and Reality*. London: Routledge and Kegan Paul.

Sklar, L. 1993. *Physics and Chance: Philosophical Issues in the Foundations of Statistical Mechanics*. Cambridge: Cambridge University Press.

Spencer-Smith, R. 1995. "Reductionism and Emergent Properties." *Proceedings of the Aristotelian Society*, 95: 113–129.

Stöckler, M. 1991. "A Short History of Emergence and Reductionism." In E. Agazzi, cd. *The Problem of Reductionism in Science*, (pp. 71–90). Dordrecht: Kluwer.

Timmermans, J. 1963. *The Concept of Species in Chemistry*. New York: Chemical Publishing.

van Brakel, J. 1986. "The Chemistry of Substances and the Philosophy of Natural Kinds." *Synthese*, 69: 291–324.

van Brakel, J. 1991. "Meaning, Prototypes and the Future of Cognitive Science." *Minds and Machines*, 1: 233–257.

van Brakel, J. 1992. "Natural Kinds and Manifest Forms of Life." *Dialectica*, 46: 243–263.

van Brakel, J. 1994. "Emotions as the Fabric of Forms of Life: A Cross-Cultural Perspective." In W. M. Wentworth & J. Ryan, eds. *Social Perspectives on Emotion*. (vol. 2, pp. 179–237). Greenwich, CT: JAI Press.

van Brakel, J. 1996a. "Interdiscourse or Supervenience Relations: The Priority of the Manifest Image." *Synthese*, 106: 253–297.

van Brakel, J. 1996b. "Empiricism and the Manifest Image." In I. Douven & L. Horsten, eds *Realism in the Sciences* Leuven: Leuven University Press.

van Brakel, J. 1997. "Chemistry as the Science of the Transformation of Substances." *Synthese*, 111: 253–282.

van Brakel, J. 1998. *Interculturele Communicatie en Multiculturalisme: Enige Filosofische Voorbemerkingen*. Assen: van Gorcum.

van Brakel, J. 1999a. "Supervenience and Anomalous Monism." *Dialectica*, 53: 3–25.

van Brakel, J. 1999b. "On the Neglect of the Philosophy of Chemistry." *Foundations of Chemistry*, 1: 4.

van der Waals, J. D. 1927. *Lehrbuch der Themostatik: Das heisst des thermischen gleichgewichtes materieller Systeme*. Leipzig: Johann Ambrosius Barth.

van Fraassen, B. C. 1998. "The Manifest Image." In D. Aerts, ed. *Einstein Meets Magritte*. New York: Kluwer Academic.

Wald, F. 1897. "Die chemischen Proportionen." *Zeitschrift für physikalische Chemie*, 22: 253–267.

Wimsatt, W. C. 1996. "Aggregativity: Reductive Heuristics for Finding Emergence." *Philosophy of Science* 64 (Proceedings):S371–S384.

Witt, C. 1989. *Substance and Essence in Aristotle: An Interpretation of Metaphysics VII–IX*. Ithaca; NY: Cornell University Press.

Woolley, R. G. 1978. "Must a Molecule Have a Shape?" *Journal of the American Chemical Society*, 100: 1073–1078.

Woolley, R. G. 1991. "Quantum Chemistry beyond the Born-Oppenheimer Approximation." *Journal of Molecular Structure (Theochem)*, 230: 17–46.

Part IV

Synthesis

11

Chemical Synthesis

Complexity, Similarity, Natural Kinds, and the Evolution of a "Logic"

STUART ROSENFELD
NALINI BHUSHAN

The goal of this chapter is to extract some of the conceptual underpinnings of the idea of synthesis and of the different aspects that constitute its practice. In so doing, we show why chemical synthesis should be of interest to metaphysicians and philosophers of science. To this end we (1) provide a provisional characterization of synthesis; (2) describe what chemists have understood to be the "logical" structure that underlies the modern practice of multistep synthesis; (3) explore the notions of molecular and synthetic complexity and the relationship between them; (4) analyze the use of similarity judgments in the categorization of compounds; and, related to this, (5) undertake a scrutiny of the notion of a natural kind in the context of the possibility of chemical synthesis.

These last two, intertwined, issues having to do with categorization are of particular interest, given that some philosophers have taken chemistry to be the science that, in its theoretical workings, dispenses with such disreputable concepts as similarity and the associated idea of a natural kind which are of "dubious scientific standing." For instance, Quine (1969) argues that the freedom from such imprecise means of categorization in chemistry is a marker of its status as a more "mature" science, one to which other domains aspire. However, by the same token, and ironically, this feature of chemistry in effect removes the discipline from the purview of philosophers, for reasons that will become clearer later on in the chapter. We argue against Quine, concluding that chemistry fails the test of maturity but becomes philosophically interesting in the process.

A Prologue

The mid-nineteenth-century defeat of vitalism and the subsequent unification of organic and inorganic chemistry came in large measure as a result of chemical synthesis. This early indication of the powerful implications that arise for chemistry from this unique field of investigation might well suffice for its continuing philosophical scrutiny.[1] There are at least two other reasons for undertaking a philosophical investigation of synthesis. Synthesis is, and has long been, pervasive in the practice of chemistry and is a unique, and defining, feature of this field. In fact, the synthesis and the analysis of chemical compounds is what chemists do. These two, often intertwined areas constitute the science of chemistry. (One might respond that chemists aim more generally to develop and refine their view of the physical world; true, but this is furthered precisely by the synthesis and analysis of chemical compounds.) Crucially, "chemistry creates its subject" (Berthelot, 1860); the objects of our study are often created with our own hands.[2] As the science that creates new "material entities," (van Brakel & Vermeeren, 1981, p. 541) chemistry has changed, and continues to change our world in a fundamental way. Indeed, one might argue that chemistry is alone among the sciences in this regard.[3] Because it is the more specific practice of synthesis within chemistry that produces these new substances or material entities, it is surprising (as pointed out by van Brakel & Vermeeren, 1981), that synthesis has been largely neglected as a subject for philosophical investigation. Our primary goal in this essay is to break the silence.

Synthesis was an active and significant area of investigation in the nineteenth-century. However, the concept and practice of synthesis up to and including the nineteenth-century was missing many key elements that characterize synthesis in its modern form. After World War II a dramatic shift on several related fronts enabled the birth and development of this exciting subfield of modern chemistry. It was not simply a matter of discovering more chemical reactions, although it is true that the synthesis of complex molecules required knowledge of a greater number of reactions than was available then. Crucially, it was a matter of understanding why and how those reactions occur.

The increasingly complex syntheses that were accomplished during the twentieth-century, and the notably greater sophistication in the practice during the post–World War II period, were due primarily to five factors (Corey & Cheng, 1989). The elucidation of the mechanisms of organic reactions (1) and the development of conformational analysis, along with its application to ground state and transition state structures, (2) were related, necessary developments. Here the need was for a deep understanding of molecular structure and the detailed pathways of chemical mechanisms. There are no more fundamental issues in chemistry than these, and their resolution was required for the birth and subsequent evolution of modern synthesis. Equally important was the development of spectroscopic and other physical methods for structural analysis (3) and the further evolution and use of chromatographic methods of analysis and separation (4). Each of these required crucial conceptual and technological developments, both within and beyond chemistry. To give an example, nuclear magnetic resonance (NMR) spectroscopy (circa 1945) required both the conceptual framework that allowed the phenomenon to be understood, along with the technology to construct the necessary instrumentation for its observation. It was important to understand the NMR phenomenon in terms of nuclear spin, spin angular momentum, and the associated nuclear magnetic moment to bring NMR spectroscopy to the service of molecular structure analysis. The additional limitation was the achievement of stable and (spatially) homogeneous magnetic fields through advancement in magnet technology, a requirement for the high resolution that is so central to the power of this method. Finally (the fifth factor), it was necessary to discover new reagents for selective reactions, and this depended to some extent on the other four factors.

Although it is not our intention to review the history of synthesis, this short sketch is included to underscore that the modern practice of synthesis awaited certain developments that spanned much of the rest of chemistry, and that it was only in the latter part of the twentieth-century that any sort of truly rational, logical approach could be applied generally.

Woodward, one of the most eloquent voices on the importance of synthesis and its role in chemistry, argued that synthesis provides a measure of the condition and powers of the science (Woodward, 1956).[4] If this is so, it might be further argued that the structures of successfully synthesized molecules themselves provide an index of the knowledge of the discipline because a structure conveys to the chemist the specific features and combinations of features that comprise the "synthetic problem." For example, if a successfully synthesized compound has a structure that contains two contiguous stereogenic centers of opposite chirality, it follows that we have the chemical knowledge required to construct a structure that contains two contiguous stereogenic centers of the desired relative stereochemistry. One might argue that as a marker of the sophistication of our technology, this is no different from other creations (e.g., bridges, books, aircraft, and virtual worlds) that require certain knowledge, or at least a certain level of knowledge, of the civilization that is responsible for their construction. However, there is a difference here because other "creations" do not typically *also* occur in nature. And if they do, our engineered analogue would not seek to reproduce the natural world in *all* respects although it may reproduce the function, for example, the case of a natural bridge. In synthesis, a close analogue of this situation occurs only in the special case of the intent to synthesize a compound that has a particular property (e.g., activity toward a certain virus). Here one might seek to discover "design principles" that are important for the desired property, and there may be a variety of designs that are successful, just as in the case of bridges. Respects of similarity and cataloging become significant. More generally, though, a synthesis will be directed toward a compound that already exists, and it and the synthetic target, once synthesized, are identical in all respects. No discovery of design principles is needed because one is literally creating the very thing that already exists. Even where the synthesis affords a compound that does not exist (or is thought not to exist), one presumes that if it were to exist it would be identical to the one that has been synthesized.

In modern chemical synthesis, the creations and our knowledge of them are special in that they are arguably creations and knowledge of the very fabric/furniture/stuff of the universe: this epistemological perspective is unique and could benefit from further philosophical commentary on the different ways, kinds, and objects of human making. We take up one implication of "making" at this fundamental level in the section entitled Natural Kinds.

Synthesis: A Provisional Characterization

The comments of Corey and Woodward on the nature of synthesis that have formed the basis for this discussion refer specifically to the synthesis of organic compounds of sufficient complexity to require multistep syntheses. However, the term *synthesis* also includes simple or short syntheses done in industrial or academic settings to produce a useful or valuable organic or inorganic compound, syntheses in which the precise identity of the expected product is not known, syntheses of mixtures rather than

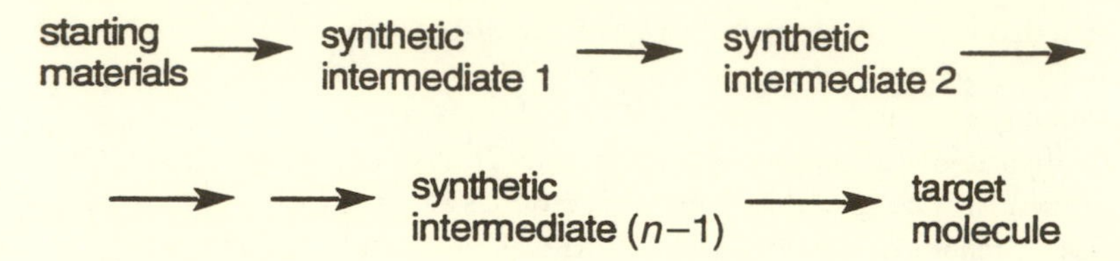

Figure 11.1 An *n* step (nonconvergent) synthesis.

single substances, and syntheses that produce something other than an intended or anticipated product. We focus here on the practice of multistep organic synthesis because it is the most developed, intricate, and philosophically interesting form of synthesis and because it largely encompasses the issues raised by a consideration of other varieties of synthesis.

A synthesis in the broadest sense is the creation of a chemical compound from other, often less complex, compounds. In outline (figure 11.1), the process involves selecting a starting compound or compounds ("starting materials"), converting it (them) through chemical reaction (often with the involvement of other reagents, solvents, and external agents such as heat or light) to a synthetic intermediate that in turn, may, be converted by chemical reaction to a second synthetic intermediate, and so on, until one reaches the desired final product (target molecule). The aspects of synthesis then that are candidates for discussion include the starting material, the intermediates, the target, the complexity of structure of the first three (molecular complexity), the similarity of structure of the first three, the attendant difficulty of synthesis (synthetic complexity), the nature and kinds of reactions, planning (i.e., the selection of reactions), and, finally, the role of intentionality. We will restrict ourselves to a discussion of only a subset of these topics.

We use the term *synthesis* to refer to (a) the methods and procedures applied in the preparation of a chemical compound, as well as to (b) the analysis and design that are employed in determining the specific procedures (reactions) to be used. This characterization has a philosophically interesting consequence (the ramifications of which we do not pursue in this essay), one that chemists have successfully negotiated in an automatic and intuitive fashion: although the procedures and the determination of the success of those procedures (i.e., the establishment of the identity of reaction products) rely on *macroscopic* description, it is depictions of molecular structure (*microscopic* description) that are central to design and some aspects of the establishment of identity (e.g., interpretation of spectra). What this means is that in our discussions we will, of necessity, move between the macroscopic and microscopic realms (and will do so without noting this shift each time).[5] This moving between realms is a practice that is deeply embedded in the structure of the science of chemistry itself. This is because chemists do and discuss manipulations on a macroscopic level but theorize largely on a molecular level. We adopt the terms *molar* (in place of the often used "macroscopic") and *molecular* (in place of "microscopic") following the suggestion of Jensen (1998). So, for instance, in speaking of "structure" we implicitly invoke a molecular level concept while a comment on a chemical reaction may include both molar and molecular meanings. Although we have focused on the synthesis of compounds, it is important to note that synthesis on the scale of the individual molecule is not precluded in principle and may even become practical through the use of developing tools for the imaging and manipulation of molecules (e.g., atomic force microscopy).[6]

Intentionality

Cornforth (1992) describes synthesis as "the intentional construction of molecules by means of chemical reactions."[7] This is certainly the case in modern multistep synthesis of organic compounds because the common practice is to select a target compound to be synthesized and then to decide the sequence of chemical reactions needed to complete the synthesis. Here, intention most often (though not always) means the intention to create a compound of particular structure rather than one having particular (molar) properties. It is possible that the inclusion of the newly developing area of combinatorial chemistry as a synthetic approach would change the precise meaning of "intent" still more because the specific structural features of products of those syntheses may not be of interest unless and until they are shown to have some desired molar property.

Further, combinatorial synthesis by definition results in multiple end products, "libraries" as it were of chemical compounds, in stark contrast to the usual goal of a single, pure, final product. We will discuss the implications of combinatorial chemistry in synthesis in some depth in the section entitled Similarity and Categorization.

Some would include "biosynthetic" transformations (i.e., the synthesis of compounds by living organisms) within the scope of synthesis and, therefore, would not require intent. (We exclude interpretations that would ascribe the intent to higher beings here.) We do not include biosynthesis within the scope of synthesis in this chapter. However, it is significant that synthetic plans have often been informed by a knowledge of biosynthetic pathways (so-called biomimetic synthesis) and that certain synthetic steps are commonly accomplished with assistance from naturally occurring enzymes or whole organisms. Further, it is increasingly the case that a catalysis that involves manipulation of biological systems to bring them into the service of a specific synthetic transformation (for example, the raising of catalytic antibodies) is employed.

Target Molecules

The selection of a particular target molecule, increasingly so in recent times, involves criteria that should be of interest to philosophers. Although most choices of target molecule are made in unsurprising ways—for example, compounds first produced by plants or animals or "nonnatural compounds" thought (or hoped) to have desirable molar properties—there have been numerous examples in recent times of targets that are modeled on features of macroscopic objects. More specifically, for instance, syntheses of so-called molecular devices use "technological objects" as the model (or at least inspiration) for the target molecule structure. So, one finds switches or brakes, for example, serving as models, somewhat ironically, because these relatively crude and often mechanical devices thereby motivate the plan for an otherwise highly refined and sophisticated practice.[8] We discuss this sort of synthesis in more detail in the section entitled Are Chemical Kinds Natural Kinds?

Regardless of the reasons for choosing a particular target molecule, whenever that target is likely to be one that is *not* previously represented in nature, the successful synthesis takes on a philosophically charged significance in that it appears that such synthetic activity adds entities to the basic furniture of the universe. One might respond skeptically to this claim, asserting that all one has here is simply a rearrangement of what was already there, and that in this way the activity is not fundamentally different from making new artifacts at the macroscopic level, cars or wrenches or what have you. But this is to miss the significance of the fact that the new entities are being added at the very fundamental level of material existence, at which level it becomes a very interesting and open *philosophical* question whether and in what sense the creation of these new entities (which then go on to be the objects of study for chemists) constitute an "adding" to the furniture of the universe (at a level of reality that ought to provoke metaphysical speculation) and not just to the furniture of our daily lives.

We offer the following as prima facie evidence in support of the stronger (and, of course, more provocative!) position. If one takes H_2O to be in the class of entities that constitute the essential features of our world, then here is its relative—another molecule, but newly created! One might respond that molecules like water and the air (oxygen/nitrogen) we breathe, the chlorophyll that makes leaves green, and so on, are different—that they have a special place in the way we view the fabric of our universe. But what justifies this treatment? At the very least, it is significant that these are tokens of the same broad "type," namely molecules. We will take up a somewhat different but related issue in the section entitled Natural Kinds.

Thus far we have characterized synthesis as an intentional process where the chemist actively conducts every step of the process. Newly developing fields may necessitate some modification. Indeed, some have suggested that in the rapidly evolving field of supramolecular chemistry, the concept of noncovalent synthesis may require that we expand, drastically revise even, our views on the nature of the end products of synthesis, given that these may now be supramolecular entities rather than simple molecules. However, as we argue next, this example by itself does not warrant such revision.

Consider the following characterization of this new kind of synthesis:

> *Supramolecular, non-covalent synthesis* consists in the generation of supramolecular architectures through the designed assembly of molecular components directed by the physicochemical features of intermolecular forces; like molecular, covalent, synthesis, it requires strategy, planning and control. . . . [But unlike covalent synthesis] It implies the conception of components containing already at the start the supramolecular project through a built-in programme. (Lehn, 1995, p. 185)

It is significant here that synthesis of a supramolecular structure may include some final step(s) that is(are) essentially designed into the later (traditionally synthesized) synthetic intermediates that immediately precede these final steps. Does this represent a significant departure from covalent synthesis? Indeed, one might be tempted in this direction by the language of the quote. However, we argue that noncovalent synthesis is in fact a very close analogue of a covalent synthesis in which one of the particular *properties* of the target molecule is *precisely* a propensity to associate with another particular molecule in a well-defined way. For example, a target that is designed to fit the active site of a specific enzyme becomes associated with that enzyme when the two are put together. Here one could quite reasonably claim that the synthetic work was *completed* with the synthesis of the covalent molecule; the subsequent association to form the enzyme–substrate complex, then, is simply a feature of the complementarity of structure that was a design feature of the covalently synthesized target molecule. At the very least, it remains an open question whether and to what degree we have here a very different kind of synthesis.[9]

We also note the recent first case of the method of combinatorial synthesis applied to end targets (the so-called library) that are supramolecular entities (Calama et al., 1998). Although this idea of there being multiple targets that are related in complex ways and categorized into "libraries" does not challenge the basic characterization of *synthesis* that we have provided here, it does raise an interesting challenge for the philosopher's characterization of *chemistry*. We take this up in our discussion of Quine in the section entitled Similarity and Categorization.

The "Logic" of Synthesis

We began this chapter by claiming that there is a "logic" to the modern practice of synthesis. That claim is based on the idea that optimal sequences of reactions may be employed in the construction of a particular molecule and the best sequences are dictated by the specific structural features of the molecule and are, therefore, to be discovered by an analysis of that structure. (Other conditions or concerns—for example, hazards, cost, and environmental impact—may also be considered in the determination of "best," but it could be argued that there are logical criteria such as shortest synthetic route that should supersede these.)

The "logic" (typically called retrosynthetic analysis) consists of the identification of the appropriate synthetic transformations (Corey, 1967, 1971) through a stepwise movement from more complex structures backward to less complex ones, with each step back corresponding to the reverse of a known chemical reaction (Ireland, 1969) (figure 11.2). It is not necessary that all retrosynthetic steps correspond to known reactions because reactions can be discovered or "invented" in some rare cases to fill a particular need.[10] In what follows, in describing the idea of a logic, we will discuss the notion of complexity, (which is integral to this logic), models of complexity that have been proposed by chemists, and the kinds of "strategies" they employ for "discovering" synthetic routes.

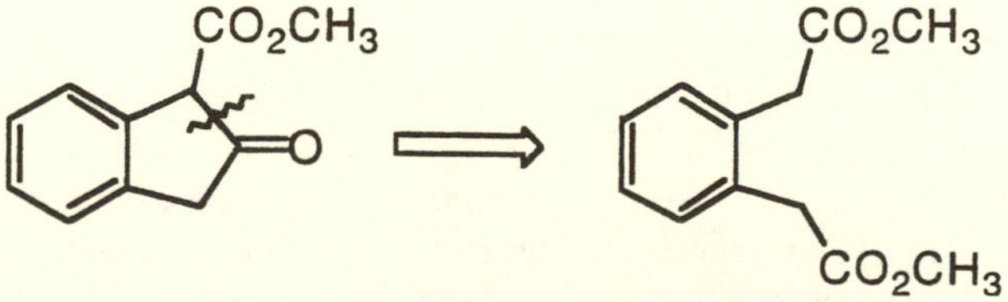

CO_2CH_3 $\Longrightarrow$ CO_2CH_3 CO_2CH_3

Figure 11.2 Example of a disconnection or transformation of the structure on the left, affording the synthon on the right. The reverse of this transform (induced by base) is the Dieckmann condensation reaction.

Synthesis has been likened to chess (e.g., Cornforth, 1992; Hoffmann, 1991; 1995, p. 103). The analogy seems apt in the sense that synthesis, like chess, has a logic; it involves "moves" of a sort intended to meet a certain goal; and the synthetic target (prior to successful synthesis) may be seen to be the analogue of the opponent (its complexity, the opponent's expertise, and so on) in chess. Nonetheless, the analogy is misleading when it is carried further, in that the end result of synthesis is a material entity, while chess yields only a history of the contest. The two are most comparable when the "process" aspect of each is brought to the fore.[11]

Complexity

In retrosynthetic analysis, the complexity of structure typically decreases with each retrosynthetic step,[12] the reverse being true for synthesis. Accordingly, the complexity of a particular structure that represents a synthetic target is of central concern to synthetic chemists because the difficulty of synthesis is considered to depend directly on structural complexity. Chemists recognize complexity and can describe the elements of molecular structure that contribute to complexity. These include the following features enumerated by Corey & Cheng (1989, p. 2) and others: molecular size; the identity and number of elements and functional groups; the specific connectivity pattern of the atoms, including the presence of cyclic structural features; the presence, nature, and number of stereogenic centers; chemical reactivity; and thermodynamic stability. Various topological indices (employing graph theory) have been used to describe molecular complexity (e.g., Merrifield & Simmons, 1980) but these focus on skeletal complexity.[13] The advantage of this sort of index is that it ignores all of the messy details of structure that arise out of functionality, placement of functionality, lability of functionality, and so on. In the process, one can work with a well-defined and simple representation of structure. Of course, graph theory was not specifically developed for the representation of molecular structure. It happens that the theory encompasses one very central feature of structure and thereby allows one to theorize about molecular complexity. Unfortunately for synthesis, however, those details ignored by topological indices of molecular complexity are often crucial and may direct the chemist to approach the problem in particular ways; not only this, but those details may create special impediments to particular synthetic routes. In a further step, Bertz (1981, 1982) described what was apparently the first "general" index of molecular complexity by taking additional qualitative descriptors into account using a combination of graph theory and information theory.[14]

It may be tempting to assimilate molecular complexity (complexity of structure) with synthetic complexity (complexity of the synthetic process). However, indices of molecular complexity, even ones such as that described by Bertz, are not equally good as indices of synthetic difficulty.[15] For one thing, there are features such as thermodynamic stability that cannot easily be incorporated into such an index. Also, because each individual feature of structure places constraints on the specific chemical reactions and conditions that may be applied during the time it is present in the synthetic sequence, and because synthetic difficulties may arise through the synergy of certain structural elements, complexity increases nonlinearly with the accumulation of structural elements. In other words, specific combinations of the same structural features can lead to synthetic targets with different degrees of difficulty. The comparison between molecular complexity and synthetic difficulty is further confounded by the fact that the transforms discovered in a retrosynthetic analysis will differ in "power"—that is, a particular transform will be more or less powerful, depending on the degree of simplification of structure that it affords. (In fact, as Bertz points out, his index allows assessment of the change in complexity caused by a particular chemical reaction and is, therefore, a measure of the power of a particular synthetic reaction.) So, the existence of a specific, "powerful" chemical reaction that provides a great jump in complexity during a synthesis may make a particular synthesis of a complex molecule relatively easy.

Strategies for Synthetic Analysis

In principle, one (or more likely a computer and appropriate algorithm) could generate all possible syntheses and then apply criteria to select the best one. However, for multistep synthesis this problem becomes unwieldy, for the following reason. In this "brute force" approach, one might view the initial problem as the creation of a so-called synthesis tree (figure 11.3) where each "branch" corresponds to a disconnection or functional group transformation, and each node (dot on diagram) is a synthetic intermediate. It is then evident that the tree grows rapidly in complexity with an increasing number of required synthetic steps. For a tree constructed of x levels (synthetic steps) with n disconnections or functional group transformations at each level, the number of possible syntheses is n^x. So, for example, a tree of five levels with only 10 disconnections or functional group transformations at each level would yield 100,000 possible syntheses. Because actual syntheses of 20 to 30 steps are not uncommon, and because the number of possible disconnections and functional group transformations for any synthetic intermediate is typically greater than 10, it is apparent that the generation of the synthesis tree is likely to be labor intensive and the sorting and selection of optimal syntheses even more so.[16]

We will discuss two general approaches to synthetic analysis that avoid the difficulties inherent in the brute force approach. The first seeks to discover "best" synthetic routes by employing various strategies, while the second extends the logic of syn-

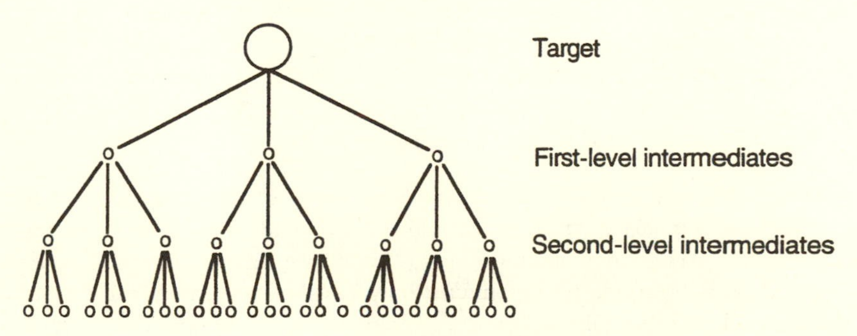

Figure 11.3 A simple synthesis tree in which three transforms are represented for the target molecule and each intermediate.

thetic analysis. Because the optimal synthesis is in a sense embedded in the structure of the target molecule, and because the "logic" of retrosynthetic analysis provides a means of determining potential syntheses, and because there are so many of these, the first general type of approach employs strategies for discovery of that synthesis.[17] These include the following categories of higher level strategies described by Corey (1988; Corey & Cheng, 1989, p. 16):

1. *Transform-based strategies.* These involve a search for a key transform or combination of transforms that afford great simplification.
2. *Structure-goal strategies.* Here a potential intermediate or starting compound aids in guiding the analysis.
3. *Topological strategies.* Here an attempt is made to locate a particular disconnection that plays a strategic role.
4. *Stereochemical strategies.* These focus on the removal of stereochemical features under stereocontrol.
5. *Functional group-based strategies.* A functional group or pair of functional groups may, for example, suggest a particular transform.

Strategies (1), (3), and (4) respond directly to elements of complexity present in the target. Retrosynthetic analysis and the companion strategies described here (and others) are incorporated in various computerized approaches that generally rely on the use of a database of known chemical reactions.[18] Guidance in the selection of retrosynthetic choices has also been gathered from empirical work such as chemical degradation studies and mass spectral fragmentation patterns (Kametani & Fukumoto, 1976).

There is by no means full agreement on the best strategies to apply to synthetic problems. As an example of one of several "logic-oriented" approaches (Ott & Noordik, 1992), Hendrickson's approach (1986 and references therein, 1989) describes synthesis as a skeletal concept: the efficient construction of the carbon skeleton (carbon–carbon sigma bond framework) of complex organic molecules should drive the discovery of optimal synthetic routes. Here, the goal is shortest route or smallest number of "construction reactions," an approach that favors fully convergent routes.[19] Uniquely, Hendrickson collapses the synthesis tree into a so-called construction tree that includes only skeletal changes, functional group identity being removed in this initial phase of analysis. Only construction reactions are obligatory in a synthesis because the target is put together from smaller pieces. An advantage of this approach is that the full synthesis tree need not be generated and, because one seeks shortest routes, a subset of the full construction tree consisting of the most highly convergent routes will suffice. So, rather than pruning a very much larger tree, one initially generates a simpler tree.

Similarity

It is desirable to view synthetic intermediates in terms of their similarity. In a purely synthetic sense, similarity can be taken to measure the amount of "synthetic work" required to convert one compound to another. Therefore, one might evaluate the pathways that potentially connect compounds through evaluation of the similarity of those compounds. This ability falls out naturally from certain aspects of the Hendrickson approach. Specifically, in that approach a numerical description of structure in which each carbon is identified and characterized by the nature of its attachments (e.g., to one or more other carbons or to a heteroatom) is used. The numerical description of structure then allows calculation of the "reaction distance"—that is, the minimum number of reactions needed to interconvert the two compounds. As pointed out by Hendrickson, this also gives rise to a system of categorization of reactions that has a mathematical basis. Whether or not this sort of approach succeeds more often than others, it represents a logical framework that offers insight for the problems of synthesis and also more generally in understanding structure and reactions. The use of similarity judgments by chemists has philosophical implications that we go on to discuss explicitly in the section that follows.

We return, finally, to the claim with which we began this section: that there is a "logic" to synthesis. This is a substantive claim. One might object that we are here conflating logic with methodology, because, strictly speaking, all we have here are different strategies for synthesis (and retrosynthetic analysis) based on the specific details of individual molecules or kinds thereof. There are no underlying principles *of logic* that guide this activity. This is a reasonable objection, and it raises the very interesting issue of what one takes to constitute a logic. Then the right question to ask is: Does synthesis have some of the elements of a logic? And: Does it have the potential to have them all? (See notes to this chapter.)

Similarity and Categorization

The emerging discipline of combinatorial chemistry employs the concept of similarity as a key notion. Combinatorial chemistry has already been adopted rather widely in synthesis for the purpose of drug discovery, and it may even represent a true paradigm shift in synthesis. Briefly, there are two general ways of employing combinatorial chemistry, referred to as parallel synthesis and split synthesis, but both result in the creation of an array of structurally related yet distinct compounds, the so-called combinatorial library. In parallel synthesis, simultaneously each end product is synthesized in its own container through a sequence of coupling reactions where the sequence is unique for each container. A knowledge of the sequence for a given container, the synthetic history of the end product in that container, specifies the structure of the end product. In split synthesis, a series of starting compounds of similar but unique structure are attached (chemically, via a "linker") to polymer beads. "Similar" here typically means having the same functional group; other aspects of structure may vary. Each bead is attached to a single compound and must somehow be coded so as to be identifiable later. The beads are then split into groups, with each group containing equal numbers of each type of bead. Each of these groups is then treated with a new reagent, and these reagents are again similar, but unique, compared to each other. All beads are then combined and redivided into equal groups. This procedure is repeated through the desired number of iterations, and in the end there exists a mixture of compounds bound to a collection of beads with each bead containing molecules of a single compound. Because this method does not provide a complete history for each bead automatically, some additional scheme for tagging along the way must be incorporated. In both parallel and split synthesis there is created a library of compounds that might, for example, be tested for a particular type of biological activity such as antiviral behavior. It then becomes important, and the library metaphor holds up rather well here, to understand where a specific compound is located in the library and to organize those individual compounds of the library for convenient use. Here the problem becomes interesting because one would like an organization that placed "similar" compounds near each other, and this requires some definition and an index of molecular similarity.

The concept of similarity in organizing combinatorial libraries of compounds has largely pragmatic roots (Wilson, 1998). In fact, the sense in which two compounds are considered to be similar depends on context. For example, one may wish to find compounds that are similar to a given compound in terms of the ability to bind to a particular receptor (because this might be an indicator of certain pharmacological properties). Here, the positioning of various functional groups, the binding sites, becomes the crucial characteristic for comparison and, therefore, this is a purely structural similarity where the important aspects of structure are things like the distance between two functional groups.[20] A different sort of descriptor might be a set of properties like whether the structure includes an aromatic ring, a chlorine atom, and so on. The selected properties could even include molar properties such as solubility. In a sense, and to extend the library metaphor a bit further, one might think of the set of all descriptors for a particular compound as being akin to keywords used in a (computer) search of a library of books.

The contextuality of a concept like similarity is, in a sense, built in: as Davidson has said quite aptly, without such appeal to context, "everything is like everything, and in endless ways" (Davidson, 1984, p. 254). This observation about the concept of similarity takes us back to a comment we made at the start of this chapter about the disreputability of concepts like similarity that we attribute to Quine, in an influential article he wrote in the late 1960s. We think it is worth mentioning because, in the course of discussing and critically evaluating this concept, Quine articulates a view about chemistry that we believe is widely shared among philosophers of science, although Quine puts it explicitly. It is a view that we think is not borne out in the methods and concepts of chemists currently working in the areas we have already discussed, a view that has in effect removed chemistry from the realm of philosophers.

At the start of his article entitled "Natural Kinds," Quine states, "I . . . suggest that it is a mark of maturity of a branch of science that the notion of similarity or kind finally dissolves, so far as it is relevant to that branch of science" (reprinted in Boyd et al., 1991, p. 162). Toward the end of his article he reveals the name of such a mature science: "chemistry . . . is one branch that has reached this stage [of maturity]" (p. 168). He explains what he means by this: "Comparative similarity of the sort that matters for chemistry can be stated outright in chemical terms, that is, in terms of chemical composition. Molecules will be said to *match* if they contain atoms of the same elements in the same topological combinations. . . . At any rate, a lusty chemical similarity concept is assured" (p. 168). Finally, "in general we can take it as a very special mark of the maturity of a branch of science that it no longer needs an irreducible notion of similarity" (p. 170), and, in "finally disappearing altogether [of the concept of similarity, that is], we have a paradigm of the evolution of unreason into science" (p. 170; brackets added). What is interesting here is that Quine is in one sense paying chemistry a big compliment: it is one of the "mature" sciences, perhaps indeed the only one that has made it this far. The evidence for this is that chemistry does not make use of "intuitive" concepts like similarity in their attempts to categorize things. So, if two molecules share chemical composition and structural features like the position of the atoms, they are automatically (or uncontroversially) similar. The criterion for similarity is very "real"—that is, one does not have to mention the respects in which molecules are similar, for instance. Indeed, and this is Quine's main point, in chemistry one does not even need to talk about similarity as a basis for categorization: one goes straight to the chemical composition and other structural features! It is in this sense that we have the "disappearing altogether [of the concept of similarity, and in this branch of human activity] we have a paradigm of the evolution of unreason into science."

In the kinds of work by synthetic chemists that we have been describing, however, we find that this mark of "maturity" is largely absent. *Similarity* is a robust concept that is very much in use by chemists, but in many cases it is intuitive and not the "lusty chemical similarity concept" that Quine talks about. It is true that similarity could be very well defined in a narrow context and satisfy the criteria for rigor required by Quine. The notion of reaction distance is a case in point. Here, talk of reaction distance in mathematical terms replaces talk of similarity, exactly as Quine wants. However, this quantitative expression of similarity of structure uses as a vehicle processes (called unit reactions) that interconvert structures, and, clearly, this is not the only way to frame structural similarity. To see how the means of comparison used here is embedded in the specific concerns of chemists, compare the following macroscopic analogue to this situation. One might compare a pair of pants to a bag and see the two as similar in virtue of the fact that simply cutting off the legs and sewing up the pant holes would do the interconversion trick. However this seems an arbitrary basis for similarity (between pants and bags) even though the "interconversion" is straightforward. The fact that chemists do use interconversion as a mark of similarity for its utility in synthesis, therefore, does not mean that there aren't other, perhaps better, measures of chemical similarity.[21] Thus, there is a real need for discussion and evaluation of the "respects" in which two molecules can be said to be similar—the very mark of its "disreputable" character! On the negative side, this means that one may have to rethink Quine's assessment of chemistry as a "mature" science; on the positive side, this means that chemistry, like the other sciences such as physics and biology, does, in fact, work with concepts that admit of some measure of slack in their relationship to the chemical world; this is precisely the space that philosophers occupy. So chemistry gives up maturity and gains a philosophy!

Are Chemical Kinds "Natural" Kinds?

The concept of a "natural kind" is often invoked to settle debates in the philosophy of science and in metaphysics, debates that raise skepticism about a variety of our practices: the justifiability of inductive reasoning, the projectibility of predicates, the correct meaning of some of our terms. Thus, for example, the predicates that are projectible are just the ones that are true of natural kind terms.[22]

But what makes a kind *natural*? And what distinguishes it from other kinds? The philosophical literature has traditionally recognized two sorts of distinction: (a) natural versus arbitrary groups of objects; and (b) natural versus artefactual groups of objects. What justifies these distinctions? In the first case, the idea is that there is some theoretically important property that members of the "natural" group share; in the second, natural is understood as "naturally occurring" (found in nature). "Natural" understood in the sense of (a), however, will not suffice to give us natural kinds because one could conceive of an arbitrary group of objects sharing a theoretically important property under some concocted theory. Thus, all things could potentially be part of a group, as long as one could come up with "respects" in which they are related (this is connected to our preceding discussion on categorization based on similarity). So it is the distinction of type (b) that has de facto been relied on to answer the skeptic, "natural" understood as the occurrence of the grouping (that shares a theoretically important property) *in nature*. In other words, (b) entails (a) in that if the grouping occurs in nature, it follows that there must be some theoretically important property that its members share. But the reverse does not hold: it does not follow that the sharing of a theoretically important property implies a natural grouping. Notice how Guttenplan explains the distinction: "to understand the concept of a natural kind, one must focus on the difference between kinds as represented by the set of typewriters and those as represented by the set of tigers" (*Companion to the Philosophy of Mind*, 1994, p. 449). His choice of examples is telling of the kind of distinction that is really doing the work in articulating the concept of a natural kind. Presumably, typewriters (or, for that matter, computers) share some theoretically important property that explains their operation and can be used to predict their behavior, but they do not occur in nature.

Not surprisingly, a belief in natural kinds tends to go hand in hand with a belief in a robust form of Realism. Here again is Guttenplan: "what is crucial to the notion [of a natural kind] is that the shared properties have an independence from any particular human way of conceiving of the members of the kind" (1994, p. 450). This tight connection between natural kinds and realism is relatively uncontroversial. For if the grouping is "natural," then it is reasonable to infer that there are underlying properties that are shared, even if we do not yet know what they are.

The natural kinds literature in philosophy has typically identified natural kinds with chemical kinds. For instance, Putnam's famous twin-earth thought experiment (Putnam, 1975) uses water (or the H_2O-ness of water) as a critical component in his argument for factoring "deep structure" properties of a natural object into the meaning of the term that serves to pick it out. Thus, the meaning of the term *water* is not exhausted by the description "the stuff that is colorless, odorless, flows in rivers, quencher of thirst," and so on; in addition, and, crucially, the H_2O-ness of water, in being essential to its makeup, is essential to its meaning as well. Although philosophers have used zoological kinds (tigers, *Homo sapiens*), botanical kinds (elms, beeches), and psychological kinds (colors like red and yellow) as examples of natural kinds, the chemical examples predominate. This is not surprising. For if there are any properties of a natural object that one may regard as essential (or necessary)[23] to its makeup, chemical properties would seem to be the most likely, in being the least controversial candidate.[24]

Might a closer look at the implications of the possibility of chemical synthesis bring fresh insights into the discussion regarding natural kinds? We think so. Natural kinds research has historically proceeded independently of a consideration of the implications of the possibility of synthesis. Ruminating on what it means to create entities at the microscopic level as chemists do, entities that behave in every way as do naturally occurring counterparts, forces us to reevaluate the weight traditionally given by philosophers to natural (or naturally occurring) kinds. Our goal in this section is to raise questions about the philosophical practice of automatically identifying natural kinds with chemical kinds. If our reasoning here is plausible, it should incline the reader to rethink the naturalness of being a realist in the field of chemistry as well.

Consider the following three cases of chemical synthesis.

1. *Synthesized compounds that have an analogue in nature.* For example, take naturally synthesized proteins versus artificially synthesized proteins that are in every other way identical to the former.
2. *Synthesized compounds that have no natural analogue.* This group might include molecules for which the analogue is a complex macroscopic object (brakes, switches, gears),[25] molecules for which the analogue is a conceptual or mathematical object (Platonic solids), molecules for which the analogue is a macroscopic object with aesthetic appeal, and molecules for which there is no analogue and where the molecular structure itself has aesthetic appeal. For example, pentacyclo[4.2.0.0^{2,5}.0^{3,8}.0^{4,7}]octane, also known as cubane, has eight carbons located with respect to each other as the vertices of a cube. First synthesized in 1964 (Eaton & Cole, 1964), cubane does not occur in nature nor was any use known or anticipated for it at the time of its synthesis.
3. *A compound that is synthesized first and later discovered to occur naturally.* Here there are two possibilities: (a) the compound was already present in the environment; or, (b) it was newly introduced into the environment, but by "natural" means. For example, around 1985, the compound now called buckminsterfullerene was prepared synthetically. This is a 60-carbon compound in which the carbons are at the vertices of five- and six-membered rings positioned like those inscribed on the surface of a soccer ball. Subsequently, buckminsterfullerene was found to be a component of soot, including ancient samples of soot.

In all of these cases, what is the relationship between the synthesized molecules and their naturally occurring counterparts? This is a discussion yet to take place in the philosophical literature, presumably because the answer has been taken to be obvious: they are all of a kind, synthesized or not, because the synthesized ones are indistinguishable from the "natural" ones. But it seems far from obvious that this is the case, once one thinks through, on the one hand, the notion of natural kind that is at work and, on the other, the extent of the "making" that is involved in synthesis and the level at which this occurs. So what is the basis for categorization?

It seems that one can adopt one of the following positions:

a. Bite the bullet and continue to stipulate natural kind to mean naturally occurring, but this now seems ad hoc, in light of the indistinguishability of the natural and synthetic compounds.
b. Broaden the notion of natural kind so that it now comes to mean naturally occuring things and those identical to them. This move has some possibilities. But, how about a molecule *first* synthesized by humans (so nonnaturally occurring) and then created by a plant or animal (3b) above)? Does its category change at that moment and become a natural kind where it wasn't before? One might respond by conceding that some chemical compounds are not natural kinds—those synthesized ones that don't occur in nature perhaps—but hold to the basic intuition that the paradigmatic examples of natural kinds are still chemical kinds. But this strategy is not persuasive. For there is

an additional feature of chemical kinds that distinguishes them from their zoological and botanical counterparts: any chemical compound is potentially makeable by us. This is not the case with the other examples of natural kinds. So there is a sense in which, far from being paradigmatic, chemical kinds are really more suspect as natural kinds than their zoological, botanical, and psychological cousins.

c. Conclude that the category of natural kind is suspect to begin with.

In cases (1) through (3), every attempt at categorization seems problematic. We are tempted to go the route of (c)—that is, to give up the category of natural kind as explanatorily, epistemologically, or ontologically basic. Indeed, philosophers have adopted critiques of type (c)—that one cannot get at a suitable criterion for separating off the natural from nonnatural kinds—but on grounds independent of the implications of chemical synthesis. We have shown, by going a different route—by exploring the nature of synthesis and its products—what is wrong with a metaphysic of categorization that takes as central the concept of a natural kind.

We have argued here the possibility of "making" at the chemical level serves to blur the natural/artefact divide. This has three consequences: (1) chemical kinds are not the unproblematic, paradigmatic instances of natural kinds they have been taken to be in the philosophical literature; (2) we are left without a clear and uncomplicated example of a "natural" kind; (3) brands of realism, which are tightly tied to natural kinds, are no longer automatically the status quo position to take vis-à-vis chemistry. Of course, chemists may come to realism for reasons that have nothing to do with natural kinds. Our point here is that philosophers who have tended to use natural kinds as a route to a defense of realism may have to do some rethinking.

Notice that the philosophical payoff here is consistent with that of the previous section: for if realism is not the automatic position to adopt toward chemistry, this implies that there is at least a potential gap between what there is and what chemists study, and so it is that chemistry becomes a domain with philosophical potential.

Why Engage in Synthesis?

Chemists often identify one or more reasons for attempting a particular synthesis. The most common reasons are as follows:

1. Creation of a useful (or potentially useful) product that does not occur naturally, or one that does, but in quantities that are too small for the proposed use or require more effort or expense to isolate from the natural source than to synthesize.
2. Proof of structure, or synthesis of a compound for comparison to naturally occurring material for which a structure has been proposed with the intention to either prove or disprove the proposed structure.
3. Creation of a compound with structural features that may serve to examine, test, or extend theories of structure.
4. As a challenge that may lead to the discovery of new chemistry.
5. As an exercise in creativity.
6. Creation of a compound with molecular structure analogous to the shape of a macroscopic object or entity, for example, a cubelike molecule.
7. Creation of a product for its potential utility based on molecular properties intended to mimic the functional properties of an entire macroscopic domain such as mechan-

ical objects, for example, the case of molecular devices like brakes, gears, bearings, ratchets, and so on.

This list of motivations illustrates that syntheses are conducted with both molar (e.g., polymers) and molecular (e.g., drugs, molecular devices) properties of the desired products in mind. These reasons for engaging in synthesis also serve in elucidating the role that synthesis plays in explanation. That role may be as simple as providing a reference material (proof of structure), but it may be much more profound as in the case of (3). Here we see a testing of the limits of theoretical knowledge through both successful and unsuccessful syntheses: we learn what is possible and, possibly, what is not. It is interesting that the aesthetic serves as a motivation, for both the appreciation of the structure of the resulting new compound (the hydrocarbon "analogues" of the platonic solids, cubane, dodecahedrane, and tetrahedrane are good examples) and the artistry (5) with which the goal was achieved.

With respect to (3), (6), and (7), in particular, the advent of precise and fast computational methods has given rise to the "design" of molecules and materials in a way that is akin to engineering. The field called computer-aided molecular design (CAMD)[26] is closely analogous to the general area of computer-aided design as it might be applied in engineering or architecture. In this regard, it is notable that we often think about "shape" in the design of new molecules because it is not clear to what degree shape, which is ubiquitous and primary for macroscopic objects, transfers over to molecular entities.[27] Finally, the urge to mimic the functional properties of an entire domain such as those implied in (7), friction, for example, which in turn depend on the more primary properties like shape and solidity for their operation, raises a whole new set of provocative issues for future discussion.

Conclusion

Synthesis is a practice that alters the world by introducing new materials. It is an activity that has no analogue in the other sciences, reason enough for attention from philosophers of science. It is a field where there is a developing logic. This is a scientific happening, and it is worth the while of philosophers to watch that logic unfold, to discuss the ways in which it constitutes a logic, and to explore the notions of complexity and similarity that are integral to its exercise. Although we do not claim to have touched on all philosophically interesting aspects of synthesis nor to have fully probed the ones that we do discuss, we do hope that this initial analysis demonstrates the richness of synthesis as a source of philosophical inquiry.

Notes

This chapter was in the penultimate stage of completion that is reproduced here when Stuart Rosenfeld died suddenly. I have decided not to make any changes, allowing the piece to stand as the final product of our joint collaboration. We were still talking about the issue of whether and to what extent synthesis involves a logic. Although the term clearly applies in a broad sense, and chemists use it as such, we were curious about whether one could make the case for synthesis involving the development of a logic in a narrower, more technical, sense. In this chapter we have left this an open question. In the end the really interesting, and pertinent, issue concerns the methods or heuristics (described here) that chemists use to explore the search space as one performs retrosynthetic analyses on chemical compounds: articulation of their nature, evaluation of their "goodness" (how often they lead one to the right target and/or how quickly) and rank ordering them accordingly, and, of course, the extent to which the use of such methods might be unique to synthesis (and to the chemical sciences).

We thank Ernie Alleva, Merrie Bergmann, and Jay Garfield of Smith College, and Jim Hendrickson of Brandeis University for helpful comments on an earlier draft of this chapter. We also thank Smith student Sorina Chircu for translating an earlier article that we had unearthed on the philosophical implications of chemical synthesis from the original Romanian.

1. This often-expressed view has been criticized by Brooke (1971) as overstating the role of synthesis and failing to acknowledge an important place for analogical argument in the unification of chemistry. Even accepting these reservations, there remains room for the clearly significant role that synthesis played in these transitional events.

2. Others have reminded us of Berthelot's thoughts Woodward, 1956; Seone, 1989; (Hoffmann, 1991; Lehn, 1995, p. 206.)

3. On one side of chemistry, we have physics where it might be argued that elements have been synthesized. Here the practice is quite limited both by the number of elements for which this is possible and by the generally transient nature of their existence following synthesis. On another border, polymer science and material science encompass the practice of synthesis, but here it is clear that these areas are properly in the domain of chemistry or are at least outgrowths of chemistry with respect to synthetic practice.

4. Here it is also interesting that Woodward recognized and deemed it important that the science at that time had achieved a predictive power that would unify the previously specialized domains of organic synthesis such as alkaloid chemistry and carbohydrate chemistry.

5. Hoffmann and Grosholz (Chapter 13) provide an analysis of this sort of shifting between domains.

6. Although we might in this case "do and discuss" only on the molecular level, this would presumably represent a special case rather than a complete shift in practice.

7. If Cornforth's meaning is taken to be "particular molecules," i.e., the way current practice works, then the meaning of "intentional" has undergone a shift, because the nineteenth-century mode of synthesis involved less specific targets. Also, with regard to intent, we agree with Cornforth that chemical transformations that do not involve intentionality are best described as reactions and not syntheses.

8. (Chapter 12) See Carpenter for a discussion of some perils of this approach to designing target molecules.

9. Some relatively recent developments, such as the discovery of ribonucleic acids with catalytic powers and the synthesis of peptides with the ability to direct and catalyze their own formation, suggest that chemists are advancing toward an understanding of self-replicating systems. Because this sort of progress moves us in the direction of understanding the chemical origins of life, and insofar as synthesis is a central player, we may have an example that will truly require some revision of our characterization of synthesis.

10. "Invention" is the more apt description, because the specific reaction invented did not "exist" prior to its invention, although it may be an example of a class or type of reaction that did exist. Perhaps an exception can be imagined for a reaction that occurs in a biological system but is not known to chemists (yet), so that the knowledge of that reaction when gained by a chemist is "discovered."

11. Architecture is another domain to which the activity of synthesis has been compared.

12. This is not strictly correct, because a temporary increase in complexity may be required by a given step even though the general trend is to lower complexity.

13. "Skeletal" here refers to the sigma bonded carbon "framework" of a molecule, ignoring functional groups.

14. The Bertz index incorporates skeletal complexity by representing the molecular structure as a graph (a skeletal molecular graph that represents all nonhydrogen atoms) that will have properties that are expressed as so-called graph theoretical invariants. Bertz uses information theory to incorporate the effects of symmetry, and also accounts for molecular size and the presence of heteroatoms in his general index of complexity.

15. A recent contribution to this problem (Whitlock, 1998) represents an interesting attempt to assess and compare the complexity of *synthetic routes*, although it does not address the issue of synthetic difficulty of a particular target.

16. The end of synthesis? The mechanical generation of all possible pathways, followed by a heuristic selection of the best synthetic route, possible in principle if not yet in practice, might be thought to signal the "end" of synthesis. In other words, if synthesis by definition involves an exercise in creativity and imagination, then the field might be judged to have reached maturity at this point. The argument against this is that the ideal synthesis is probably unachievable because it would entail strict conditions like the completion of the entire procedure in one vessel, the creation of no by-products (atom economy), etc. Also, the openness of the synthetic frontier results in part because new methods of promoting reactions, for example, using ultrasound, microwaves, improved photochemical methods, magnetic fields, and new catalysts, are continually discovered. (For further comment on the maturity of the field see Trost, 1985).

17. Chemists also use the term strategy to describe the synthetic plan itself or some key element that is central to the synthetic design. See, for example, Deslongchamps (1984).

18. For a review of the history of synthetic planning and especially computerized approaches, see Ott and Noordik (1992), and for an in-depth discussion of synthetic strategies see Smith (1994).

19. A nonconvergent route (see Fig. 11.1) is one in which starting materials are converted to a synthetic intermediate, which in turn, is converted to a next synthetic intermediate and so on until the target structure is reached. In contrast, in a convergent route two or more synthetic intermediates are derived from their own starting materials and these "pieces" (synthetic intermediates) are later joined.

20. For an in-depth description of the concept of molecular similarity and the mathematical methods that have been applied in assessing similarity see, Doucet Weber (1996, ch. 11).

21. See Johnson and Maggiora, 1990, for a range of approaches to the problem of molecular similarity.

22. Thus, "green" rather than "grue," for example, where "grue" means "observed before 2000 and green, or blue" (Goodman).

23. An examination of the relationships between the concepts "natural," "essential," "necessary," and "real" would take us too far afield, given the goals of this chapter. Here, we will simply accept these connections.

24. In general, this is perhaps why there has been so little philosophy of chemistry: because chemistry deals with "the real," all the fundamental questions in chemistry seem settled or at least settle able without deep controversy.

25. For the very interesting and provocative example of a "molecular ratchet," see the chapter by Carpenter (Chapter 12).

26. See Doucet and Weber (1996) for an authoritative treatment of CAMD.

27. See Ramsey (Chapter 7) for a discussion of the sense in which molecules can be said to have a shape.

References

Bellu, Elena. 1965. "Chemical Synthesis and Some Philosophical Implications." In *Revista de filozofie*. Trans. from Romanian by Sorina Chircu. 12(5): 615–625.

Berthelot, Marcelin. 1860. *Chimie organique fondée sur la synthèse* (vol. 2). Paris: Mallet Bachelier.

Bertz, Steven H. 1981. "The First General Index of Molecular Complexity." *Journal of the American Chemical Society*, 103: 3599–3601.

Bertz, Steven H. 1982. "Convergence, Molecular Complexity, and Synthetic Analysis." *Journal of the American Chemical Society*, 104: 5801–3.

Boyd, R., Gasper, P., & Trout, J. D. Eds. 1991. *The Philosophy of Science*. Cambridge; Mass.: MIT Press

Brooke, John Hedley. 1971. "Organic Synthesis and the Unification of Chemistry—A Reappraisal." *British Journal for the History of Science*, 3(20): 363–392.

Calama, Mercedes Crego, Hulst, Ron, Fokkens, Roel, Nibbering, Nico M. M., Timmerman, Peter, & Reinhoudt, David N., 1998. "Libraries of Non-Covalent Hydrogen-Bonded Assemblies; Combinatorial Synthesis of Supramolecular Systems." *Chemical Communications* 1021–1022.

Corey, E. J. 1967. "General Methods for the Construction of Complex Molecules." *Pure Applied Chemistry*, 14: 19–37.

Corey, E. J. 1971. "Computer-Assisted Analysis of Complex Synthetic Problems." *Quarterly Reviews*, 25: 455–482.

Corey, E. J. 1988. "Retrosynthetic Thinking—Essentials and Examples." *Chemical Society Reviews*, 17: 111–133.

Corey, E. J., & Cheng, X.-M. 1989. *The Logic of Chemical Synthesis*. New York: Wiley.

Cornforth, Sir John Warcup. 1992. "The Trouble with Synthesis." *Australian Journal of Chemistry*, 46: 157–170.

Davidson, Donald. 1984. "What Metaphors Mean." In *Inquiries into Truth and Interpretation*. Oxford: Clarendon.

Deslongchamps, Pierre. 1984. "The Concept of Strategy in Organic Synthesis." *Aldrichimica Acta*, 17(3): 59–71.

Doucett, Jean-Pierre & Weber, Jacques 1996. *Computer-Aided Molecular Design: Theory and Applications*. San Diego: Academic Press.

Eaton, P. E., & Cole, T. W., 1964. "Cubane." *Journal of the American Chemical Society*, 83: 3157.

Goodman, N. 1955. *Fact, Fiction & Forecast*. Cambridge, Mass.: Harvard University Press.

Gutenplan, S. 1994. "Natural Kind." Entry in *A Companion to the Philosophy of Mind*. Eds. Gutenplan. Oxford: Blackwell. (pp. 449–450).

Hendrickson, James B. 1986. "Approaching the Logic of Synthesis Design." *Accounts of Chemical Research*, 19: 274–281.

Hendrickson, James B. 1989. "The SYNGEN Approach to Synthesis Design." *Analytica Chimica Acta*, 235: 103–113.

Hoffmann, Roald. 1991. "In Praise of Synthesis." *American Scientist*, 79: 11–14.

Hoffmann, R. 1995. *The Same and Not the Same*. New York: Columbia University Press.

Ireland, Robert E. 1969. *Organic Synthesis*. Englewood Cliffs, NJ: Prentice Hall.

Jensen, William B. 1998. "Logic, History, and the Chemistry Textbook. 1: Does Chemistry Have a Logical Structure?" *Journal of Chemical Education*, 75(6): 679–687.

Johnson, M., & Maggiora, G., eds. 1990. *Concepts and Applications of Molecular Similarity*. New York: Wiley.

Kametani, T., & Fukumoto, K. 1976. "Total Synthesis of Natural Products by Retro Mass Spectral Synthesis." *Accounts of Chemical Research*, 9: 319–325.

Lehn, Jean-Marie. 1995. *Supramolecular Chemistry: Concepts and Perspectives*. Weinheim: VCH.

Merrifield, R. E., & Simmons, H. E. 1980. "The Structures of Molecular Topological Spaces." *Theoretica Chimica Acta*, 55: 55–75.

Ott, Martin A., & Noordik, Jan H. 1992. "Computer Tools for Reaction Retrieval and Synthesis Planning in Organic Chemistry: A Brief Review of Their History, Methods, and Programs." *Recueil des travaux chimiques des Pays-Bas*, 111(June): 239–246.

Putnam, Hilary. 1975. "The Meaning of 'Meaning.' " In K. Gunderson, ed. *Language, Mind and Knowledge* (Minnesota Studies in the Philosophy of Science, VII). Minneapolis: University of Minnesota Press.

Quine, W. V. O. 1969. "Natural Kinds." In *Ontological Relativity and Other Essays* (reprinted in Boyd, Gasper, and Trout, eds, 1991, *The Philosophy of Science*). Massachusetts: MIT Press.

Seoane, Carlos. 1989. "Teaching Organic Synthesis: Why, How, What?" *Aldrichimica Acta*, 22(2): 41–46.

Smith, Michael B. 1994. *Organic Synthesis*. New York: McGraw-Hill.

Trost, B. M. 1985. "Sculpting Horizons in Organic Chemistry." *Science*, 227(Feb. 22): 908–916.

van Brakel, J, & Vermeeven, H. 1981. "On the Philosophy of Chemistry." *Philosophy Research Archives*, 7: 00–00.

Whitlock, H. W. 1998. "On the Structure of Total Synthesis of Complex Natural Products." *Journal of Organic Chemistry*, 63: 7982–7989.

Wilson, E. K. 1998. "Computers Customize Combinatorial Libraries." *Chemistry and Engineering News* April 27, 31–37.

Woodward, Robert B. 1956. "Synthesis." In A.R. Todd, ed. *Perspectives in Organic Chemistry* (pp. 155–184). New York: Interscience Publishers.

Part V

Models and Metaphors

12

Models and Explanations

Understanding Chemical Reaction Mechanisms

BARRY K. CARPENTER

An Illustration of Some of the Issues

In 1997, Ross Kelly and his coworkers at Boston College reported their results from an experiment with an intriguing premise (Kelly et al., 1997; see also Kelly et al., 1998). They had synthesized the molecule shown in figure 12.1.

It was designed to be a "molecular ratchet," so named because it appeared that it should undergo internal rotation about the A—B bond more readily in one direction than the other. The reason for thinking this might occur was that the benzophenanthrene moiety—the "pawl" of the ratchet—was anticipated to be helical. Thus, in some sense, this might be an inverse ratchet where the asymmetry dictating the sense of rotation would reside in the pawl rather than in the "teeth" on the "wheel" (the triptycene unit) as it does in a normal mechanical ratchet. Kelly and coworkers designed an elegant experiment to determine whether their molecular ratchet was functioning as anticipated, and they were (presumably) disappointed to find that it was not—internal rotation about the A—B bond occurred at equal rates in each direction.

In 1998 Davis pointed out that occurrence of the desired behavior of the molecular ratchet would have constituted a violation of the second law of thermodynamics (Davis, 1998). With hindsight, I think most chemists would agree that Davis's critique is unassailable, although the appeal of the mechanical analogy was so strong that I imagine those same chemists would also understand if Kelly et al. had overlooked the thermodynamic consequences of their proposal in the original design of the experiment.

But now comes the interesting question: Suppose Kelly et al. had been fully aware that their experiment, if successful, would undermine the second law of thermodynamics, should they have conducted it anyway? Davis, in his critique writes:

> Some would argue that this experiment was misconceived. To challenge the Second Law may be seen as scientific heresy (a nice irony, considering the Jesuit origins of Boston College), and the theoretical arguments against molecular ratchets and trapdoors are well-developed. . . . However, as scientists we should take the view that nothing is sacred, that experimental results outweigh all theoretical considerations, and that it is quite appropriate to revisit old questions as new techniques become available. The Boston College

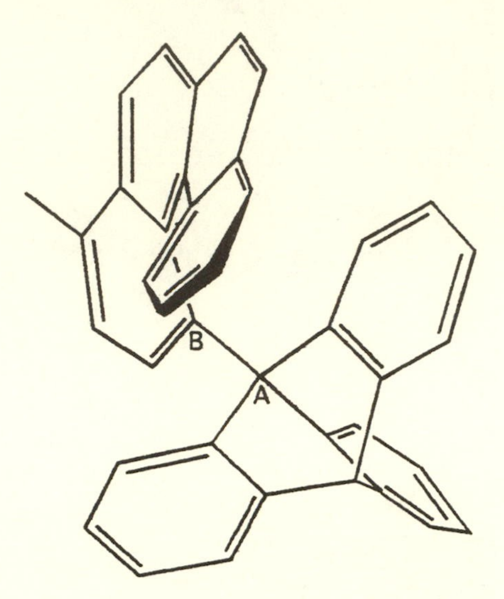

Figure 12.1 "Molecular ratchet," a molecule synthesized by Ross Kelly and his coworkers at Boston College.

> group should be congratulated on their courageous, if perhaps Quixotic, attempt at the 'impossible.' Wishful thinking will not make the Second Law disappear, but if there is any chance we can escape its deathly grip, surely we should take it. (p. 910)

For the purposes of the present discussion, let us take Davis's exhortation to be serious rather than ironic. The implication would seem to be that we should be spending our time trying to overthrow the most fundamental laws of our science, because each success would represent a major intellectual advance. I think that there can be little argument that the second clause of the preceding sentence is correct; however, it is much less obvious that the implied strategy is sound. Experiments are not without cost—real cash and human effort. If the probability of success of any experiment we choose to undertake is, say, one in a million, then we may well spend our entire scientific careers encountering nothing but failure. Repeated and unremitting failure is likely to have both psychological and professional consequences for most people!

The opposite end of the conceptual scale from the fundamental laws may initially seem to provide more fruitful arenas for our experiments. If our investigations sought to test the validity of some extremely tentative hypothesis, then either consistency or falsification might be deemed a successful outcome (depending, in part, on whether the hypothesis was one's own or somebody else's) and so the probability of success would apparently rise from one in a million to unity. However, if the hypothesis under scrutiny were of extremely limited scope, the consequences for the science of its falsification or support would probably be very small. Thus, if the measure of success includes some component of overall significance of the results, this, too, is likely to be a losing strategy. I say "likely to be" because one must allow for the possibility that an investigation whose aim may be to address a question of limited scope could serendipitously turn up a phenomenon of much wider significance.

The analysis here implies two things that bear further scrutiny. First, natural laws and tentative hypotheses reside at opposite ends of a single spectrum. This is actually not a universally held belief, as I will discuss in the next section. Second, natural laws are of broad scope in their applicability, whereas tentative hypotheses are of limited scope. This is obviously not rigorously correct, although I think it does have some operational validity in a mature science like chemistry. Breadth of applicability is part of what makes the natural laws so important. New hypotheses rarely have comparable coverage.

In the end, I would like to ponder the rather complex interactions between the design of experiments, particularly in the field of reaction mechanisms, and the various levels of conceptual construct that science employs. However, to begin, it is necessary to define the meanings that I will imply by the use of various terms. Those meanings may not always be identical with the ones that others assign to them.

Hypotheses, Models, Theories, Laws, and Explanations

It may be useful, I think, to start this discussion not by considering what professional philosophers (to which community, the reader will by now have recognized, the author obviously does not belong) mean by the terms, *hypotheses, models, theories, laws,* and *explanations* but, rather, by reviewing what we teach beginning chemistry students. These are the foundations on which most practicing chemists have built their understanding of their science. I excerpt below the relevant material from some popular freshman chemistry books.

> A **theory**, which is often called a **model**, is a set of tested hypotheses that gives an overall explanation of some natural phenomenon. . . .
>
> As scientists observe nature, they often see that the same observation applies to many different systems. For example, studies of innumerable chemical changes have shown that the total observed mass of the materials involved is the same before and after the change. Such generally observed behavior is formulated into a statement called a **natural law**. . . .
>
> Note the difference between a natural law and a theory. A natural law is a summary of observed (measurable) behavior, whereas a theory is an explanation of behavior. A *law summarizes what happens; a theory (model) is an attempt to explain why it happens.* (Zumdahl, 1993, p. 5)

> The key to the [scientific] method is to make no initial assumptions, but rather to make careful observations of natural phenomena. When enough observations have been made that a pattern begins to emerge, one then formulates a generalization or **natural law** describing the phenomenon. This process, observations leading to a general statement or natural law, is called **induction**. . . .
>
> A **hypothesis** is a tentative explanation of a natural law. If a hypothesis survives by experiments, it is often referred to as a theory. We can use this term in a broader sense, though. A **theory** is a model or way of looking at nature that can be used to explain and to make further predictions about natural phenomena. (Petrucci & Harwood, 1993, p. 3)

> A **scientific law** is *a concise verbal statement or a mathematical equation that summarizes a broad variety of observations and experiences.* . . .

> A **theory** is *an explanation of the general principles of certain phenomena with considerable evidence or facts to support it.* (Brown et al., 1994, p. 7)

> When enough information has been gathered, a **hypothesis**—*a tentative explanation for a set of observations*—can be formulated. . . .
>
> After a large amount of data has been collected, it is often desirable to summarize the information in a concise way. A **law** is *a concise verbal or mathematical statement of a relationship between phenomena that are always the same under the same conditions.* . . .
>
> A **theory** is *a unifying principle that explains a body of facts and those laws that are based on them.* (Chang, 1994, p. 6)

As is often the case in freshman chemistry textbooks, there is much that is very similar from one author to another. Thus, all of the cited authors seem to agree that *laws* are based entirely on observations and contain no theoretical component. Zumdahl is probably the most explicit on this point. All of the cited authors who address the term also seem to agree that a *hypothesis* provides some tentative explanation of a set of observations, although Petrucci and Harwood, if taken at face value, would appear to be implying that one has to wait until there is some law on the books before one can begin to hypothesize.

When we contemplate the relationship among *theories, models,* and *explanations,* the situation in the freshman texts is less clear cut. For Zumdahl and Petrucci and Harwood, the terms *theory* and *model* are synonymous. For Brown, LeMay, and Bursten, *theory* is synonymous with *explanation,* whereas for Chang, a *theory* serves to provide an *explanation* but, presumably, need not be synonymous with an explanation.

I would like to consider these widely promulgated views in more detail. Let us first consider the suggested theory independence of laws. Although the question of whether any observations can be treated as truly free from interpretation has been much discussed by philosophers, the specific issue of whether the natural laws are also theory dependent has apparently received less attention. Of relevance, though, is the work of Cartwright, who, in her 1980 discussion of laws and explanations, recognizes the existence of "phenomenological laws" and "theoretical laws." The latter would apparently be an oxymoron for our freshman chemistry authors.

My own view is that many, but not all, of the laws of science may well have arisen from the kind of empirical roots that are suggested in the freshman texts. Boyle's Law and Charles's Law on the behavior of gases probably did, but Newton's Laws of Motion clearly did not, because the uniform motion posited in his First Law could not be demonstrated on Earth in the eighteenth-century.

I believe that empirical observations only attain the status of laws when they are accompanied by theories that support their generality or, sometimes, show their limitations. The second law of thermodynamics, with which I began this chapter, is surely a case in point. The claim that spontaneous processes will always increase the entropy of the universe is something that one can accept as a law *primarily* because of the statistical argument that there are more ways for things to be disordered than ordered. Certainly, one wants experimental studies to bolster the claim, but such studies by themselves would not be convincing because of the general inductive problem of empirical science—one could never rule out the possibility that some particular set(s) of conditions not studied in the experimental tests would lead to an exception to the general claim. In the case of Boyle's and Charles's Laws, the kinetic theory of gases showed both why the laws ought to exist and why they should be inexact (as they are).

The definition of a *theory* as a set of hypotheses that has passed a test of experimental verification is uncontroversial. However, the equation of *theories* and *models* proposed by Zumdahl and Petrucci and Harwood is less straightforward, I think. It surely is true that all but the most grandiose of scientists would admit that their theories were approximations to reality, and so, to the extent that a model requires a specified list of approximations, all theories are models. However, not all models are theories. If I make the approximation of treating molecules as perfect spheres or springs as massless, I am creating a model that will make subsequent calculation easier or comprehension of the results easier, but I presumably do not believe these approximations to be true in my theory of what is occurring in reality. Chemists will talk of the harmonic-oscillator *model* as a mathematically convenient approximation for the interpretation of vibrational spectra, but I do not think many people would consider this to be a *theory* of vibrational spectroscopy.

Most troublesome to the professional philosophers is the role that *explanation* has to play in all of this. Are theory and explanation synonymous, and if not what is a scientific explanation? There is a nice collection of essays on these questions to be found in the 1991 book edited by Boyd, Gasper, and Trout. I will here only very briefly summarize the disagreements, which appear to be profound, but I may begin by noting that the one thing on which all but one of these philosophers appear to agree is that *theory* and *explanation* are *not* synonymous.

The holdout for identity of theory and explanation is Hempel, one of the most prominent proponents of the *covering-law* or *deductive-nomological* model of scientific explanation (Hempel, 1966). In this view, explanations are nothing more nor less than deductive arguments that link the thing to be explained (the *explanandum phenomenon*) to one or more of the natural laws. A theory that allowed successful computation of an observed quantity on the basis of existing laws would thus be equivalent to an explanation in this view. The covering-law model is not favored by many contemporary philosophers because in some ways it seems overly restrictive. One can think of examples of statements that seem like acceptable explanations without being linked to any existing natural laws (Gasper gives some instances—see Boyd et al., 1991). In other ways it appears too permissive. There exist logically rigorous deductive arguments that link an observed phenomenon to natural laws, but the result does not feel like an explanation.

Perhaps most startling for practicing scientists is the view of van Fraassen (1977), who is opposed to the notion that explanation is even the principal purpose of scientific inquiry. He argues that what we choose to explain and how we try to explain it are pragmatic issues determined by our particular interests, and that there exists no objective basis for selection of either. As I will suggest later, it seems right to me that there is indeed a pragmatic component to scientific explanation, but I do not think that explanation is quite as personal and devoid of value for the institution of science as van Fraassen would seem to be suggesting. Most scientists could probably agree on the phenomena that merit explanation—they are typically ones for which there seems to be some general importance (as judged by the impact that the answers might have on other aspects of science or life in general) and for which intuitively obvious explanations are not readily perceived: What causes so-called gravitational attraction between objects? What caused the disappearance of the dinosaurs? What gives some ceramics but not others the property of high-T_c superconductivity? I suspect that there would also be some agreement about what constitutes a satisfactory explanation, but I will defer that discussion for the moment.

Kitcher (1981) has argued the purpose of scientific explanation is *unification*—the ability to show that apparently disparate phenomena are linked to some common cause. There is, I think, no doubt that the power or importance of an explanation is related to its ability to unify. However, it is not as clear that this is a necessary condition for an argument to be considered an explanation. Nevertheless, there appear to be scientific arguments of quite limited scope that are considered to be explanations. If I propose a new mechanism for a particular chemical reaction, that mechanism could be considered an explanation even if it applied only to the one example I have studied experimentally. It may gain plausibility and stature if I can show that it applies to other examples as well, but that doesn't seem to be a necessary part of its being considered an explanation.

Boyd makes a proposal that, on first encounter, seems fully as startling as van Fraassen's does. It can be summarized in the following quote:

> If the methods of actual scientific practice for resolving questions about sampling in experimental design rely upon prior (approximate) theoretical knowledge of unobservable factors, then, in particular, knowledge of such factors is actual and therefore possible. Thus, the empiricist conception that experimental knowledge cannot extend to unobservable causal powers and mechanisms must be mistaken and the philosophical justification of the Humean definition of causation rests upon a false epistemological premise. (Boyd, 1985, p. 73)

Boyd goes on to develop a non-Humean description of explanatory power. His replacement to Hume's analysis of causation (which, I think it is fair to say, most practicing scientists have adopted) is difficult for me to grasp, and it is therefore entirely possible that I do it injustice here. Nevertheless, it seems to me that Boyd's principal reason for wanting to overthrow Hume does not take full account of the *conditional* acceptance that scientists give to their theories and even their laws. The confidence that they have in them is dependent in large measure on the extent to which the "theoretical knowledge of unobservable factors" has played a role in limiting the sample size of the experiments conducted to test the model. The greater the contribution from such factors, the lower the confidence that the scientific community has in the theory. Thus, although it is true that nonexperimental factors play a necessary role in formulation of a theory, that role, as I see it, is not always one that is considered to bolster the theory. Quite to the contrary, I think that most scientists see some of the extra-evidentiary components of their theories as little more than necessary evils and as the sources of potential fatal flaws in them.

The principal nonexperimental factor that does appear to be considered to *increase* the prestige of a theory, and would seem, therefore, to be a good example of the issue that concerns Boyd, is simplicity (or elegance). The significance of this criterion of explanatory power has been discussed elsewhere, in the context of chemical reaction mechanisms (Hoffmann et al., 1996). However, even in the case of this apparently clear example of Boyd's claim, the situation is not so clear-cut. The acceptance of the simplest explanation as the best (among contenders that all fit the observed facts equally well) must be recognized as having some pragmatic and some empirical components. If I make a series of measurements of some property y that changes in response to a control factor, x, and I find that a plot of y versus x reveals all of the points to fall on a straight line, then I will probably choose to draw a straight line through those points and thereby interpolate (perhaps even extrapolate) to predict values of y at values of x where experiments were not actually conducted. There exist an infinity of possible jagged lines that would pass through the experimental data points, but I (*conditionally*) reject them for one pragmatic and one empirical reason. The pragmatic reason is that the straight-line interpolation is unique, whereas there is, on the basis of the available data, no way to choose among the infinity of jagged-line fits. If I am going to set down a relationship that will form the basis for further experiment, I must choose one, and it seems natural to choose the unique one. The empirical reason is that one has learned from the history of scientific investigation that most physical phenomena seem to be relatively smooth—a change in control parameter x rarely causes wild and erratic variations in the response, y. I assume a similar smooth relationship when I choose the straight-line rather than one of the jagged-line interpolations.

There is, I think, also a psychological preference for simplicity that is related to comprehensibility, but which also may have an esthetic component. I will return to the psychology of explanation after finishing this brief review of the principal proposals for scientific explanation from the community of professional philosophers. In completing the summary of Boyd's position, though, I would say that the evidence for the important role of nonexperimental factors that Boyd cites as the reason to overthrow Humean causation is not clear-cut. I guess I remain a Humean at heart.

Last of the points of view that I will summarize here is that of Cartwright (1980). It is the one with which I find greatest resonance. She argues, and I agree, that the empiricist philosophers have had greater reverence for laws than practicing scientists typically do. Furthermore, she claims, it is the causal relationships, which the empiricists have attacked, that form the actual "truths" of scientific explanation. This seems right to me. There may be many models, differing in their particular sets of approximations, that seek to explain the same phenomenon. Often, the underlying causation that they try to describe is the same. Under these circumstances, science will generally permit a parallel existence of the competing models without any feeling of dissonance—the various approximations may be seen to have complementary strengths and weaknesses. However, explanations that differ in their underlying causal structure are not generally allowed to exist long in the scientific books without attempts to devise experiments to distinguish between them. If this description of the way science operates is accurate, and I think it is, then it argues that scientists assign greater importance to the causal relationships of their explanations than they do to the specific technical models that seek to approximate them. If, as I suggest in this essay, the physical laws are only models with an unusually large amount of supporting experimental evidence, then they are not inviolable. They are not the ultimate "truths" of science—that is, the place of the causal relationships. I will finish this section on definition of terms, by expanding on Cartwright's thesis.

My own belief is that there is an essential subjective component to the evaluation of explanations. There is a phenomenon—the "Eureka" phenomenon—that I imagine most people involved in problem-solving enterprises have experienced. It occurs when one has discovered (or created!) an explanation for some long-standing puzzle, and it is especially powerful if that explanation has certain qualities. For me, those qualities are simplicity, generality, and subtlety. The appearance of both simplicity and subtlety in this list is intended to convey the idea that the explanation should have ready comprehensibility (simplicity) but not be so obvious that it elicits a disappointed response—the feeling that this is just another example of some well-known phenomenon. The nature of the "Eureka" response is, I believe, physiological. I would not be surprised if the release of one or more specific neuromodulators could be associated with it. The feeling can be very powerful and pleasurable, and it is certainly a component of what drives me to do research. There also seems (based, again, just on how it feels) to be a similarity between the "Eureka" phenomenon and the esthetic response to music or other works of art with which one finds a particular resonance. It is not, I think, coincidental that scientists use esthetic language, "elegant" or even "beautiful," to describe explanations that they find to be particularly appealing.

Despite this claimed connection between higher—order behaviors such as scientific discovery and esthetic response to art, I believe that the pleasure in explanation may have deep evolutionary roots. An unexplained phenomenon is potentially dangerous for a creature that largely depends on its intelligence for survival because "unexplained" translates easily into "unpredictable," and an unpredictable environment is one that is insecure. I suspect that our striving to fit things into predictable patterns through the act of explanation has its origin in this history. Curiously, though, when faced with things that are too complex to be explained (or perhaps are inherently incapable of being explained), humankind has apparently preferred to codify its uncertainty into religion rather than to accept it at face value. It is apparently more satisfying to ascribe unexplained events to the will of an inscrutable deity than to accept a role for randomness or chaos in the universe.

Having then admitted to my belief in the subjectivity of explanation, I would appear to have joined the camp of van Fraassen or even the antirealists who deny that there is any external validity at all to scientific explanation. I have not. Realism is a profound, perhaps even the most profound, philosophical question—one that I am ill equipped to discuss, and which anyway strays too far from the topic of this essay. Nevertheless, I cannot resist offering a counter to an argument raised by Fine, who has suggested that, just as Gödel showed the impossibility of proving the consistency of mathematics using the methods of mathematics, so it is inherently impossible to prove the objective reality of scientific explanations using the methods of science (Fine, 1984). Imagine, then, a future in which by some means we have been able to contact several other civilizations throughout the universe and have learned to communicate with them. We discover that the phenomena they have tried to explain and the explanations they have provided for them are largely the same (given some translation algorithm) as our own. How would the antirealists deal with that? Would they say that these parallel, independent convergences are coincidence, or that we have fudged the translation to make things fit? I wouldn't find either of those responses very satisfactory. Even if one wanted to claim that the other civilizations were all "wired" in the same way as us and, hence, predisposed to interpret things in the same way (a thoughtful suggestion from one of the editors to this book), wouldn't this very conformity of biology itself be a strong pointer to the existence of universal laws that were external to the observer?

In contrast, suppose the sciences of these other civilizations were profoundly different from each other and from our own, while being equally able to rationalize and predict observable phenomena, then what would the realists say? Whatever the answers from the realist and antirealist camps, it seems to me that one can conceive of this very scientific (empirical even!) experiment whose results would allow a strong preference for one view or the other, and so to that extent I disagree with Fine.

In the absence of extraterrestrial data points, I would characterize my position on the question of realism as agnosticism. It seems to me that the business of constructing scientific explanations has some similarity to trying to fit a multidimensional function to a set of data points using, say, a nonlinear least-squares algorithm. One often finds that there is more than one minimum in the plot of variance versus adjustable parameters, and it is quite difficult to determine whether the particular solution on which one has converged is a local minimum or the global one, or indeed whether a single global minimum even exists. The criterion for choosing between solutions in a nonlinear least-squares fit—the magnitude of the variance—is just the sort of pragmatic one that van Fraassen would apparently like, but the criteria for choosing among scientific explanations are, I think, more complex and include the esthetic components that I have discussed. Whether the scientific enterprise in which I participate is converging toward a single universal truth or just some local minimum is not of enormous concern to me. Not even the most extreme antirealist or postmodern deconstructionist could defend the view that all explanations of physical phenomena are equally valid, and so to the extent that I am part of an enterprise that seems to give the best currently available explanations, I am content.

A relatively modern phenomenon that has not, to my knowledge, been much discussed in the philosophical literature is the ability to carry out computational *simulation*. Where does this fit into the scheme of scientific understanding? In a typical simulation, the physical principles are generally embodied in quite simple and comprehensible equations. However, the application of simple equations to complex systems, often including a random component in the sampling of initial conditions, can result in a massive quantity of information that is beyond human capacity to comprehend in its full detail. Assuming that the simulation has been constructed so that its general results *are*, comprehensible, should one treat them as constituting a theory? Presumably not, because the random component means that the next simulation on the same system will give somewhat different results, and so simulations do not have quite the ability to make testable predictions that one expects of a theory. Although it is true that, given proper sampling statistics for the initial conditions, the general results of repeated simulations of the same system should come out close to each other, the extrapolation from these specific simulations to the general case is inductive, not deductive, as one would expect for a theory. I like to think of simulations as *numerical experiments*. They tell one approximately how things might occur in a model universe (defined by the underlying equations of the simulation), and one can decide by comparison with the observed facts how close the model universe is to the real one.

Let me end this section, then, by summarizing where I stand on the relationship among the conceptual constructs in the heading. I view *laws*, *theories*, and *hypotheses* as forming a continuum in which the ratio of empirical data to inductive speculation is, in general, largest for the laws and smallest for the hypotheses. Because the susceptibility to overthrow is generally inversely related to the amount of factual information to be encompassed (it's easy to come up with ideas when there are few hard facts to be explained), it follows that the laws are likely to be the most robust, although not necessarily impregnable, features of science. All of the constructs in this continuum are models, but there is a more specific use of the term *model* that allows it to be included in my claimed spectrum. If I have a set of phenomena that I want to explain, I may begin by coming up with some hypothesis. Conversion of the hypothesis to a theory involves testing its predictions by experiment, and that often requires putting it into some sort of mathematical framework where the predictions can be made quantitative. The conversion of general ideas into mathematical equations frequently necessitates making some well-specified, and, one hopes, well-chosen set of approximations. This set of approximations defines the *model* in the more specific meaning of the word. Within this limited definition, the model is just a device with which one hopes to turn a hypothesis into a theory.

Least well defined and (necessarily, I would argue) most subjective is the *explanation*. Unlike van Fraassen, I think that explanation *is* the end goal of science, and that it is a great deal more than a mere pragmatic fitting of equations to observations. Like Cartwright, I think that it is the causal relationships of our explanations that form the core of their "truths."

That the end result of supposedly objective effort should itself be subjective is perhaps a little strange, but I think inevitable for this human enterprise. Whether the subjective explanations of science have external validity is something that is, I believe, knowable with some less-than-complete level of certainty, but for which we currently have insufficient data to make any claim.

Do Models Guide or Misguide Scientific Investigation?

The answer to whether models guide or misguide scientific investigation, I think, is both.

Obviously, building and testing of models is a large part of what science is about. Furthermore, if we adopt here the broader meaning of the word *model*, to encompass hypotheses, theories, and laws, then there is a way in which the conceptual spectrum that I have discussed suggests a strategy for conducting experimental investigation. It brings us back to the molecular ratchet and the second law of thermodynamics.

At this stage in human history, there are very few important phenomena for which we have no model. There are, however, plenty of technical details, each one of which is perhaps of rather limited significance to science as a whole, for which explanations are still lacking. Many researchers are engaged in studying these details, and their efforts help clarify the picture whose gross features are framed by the laws and major theories of science. Sometimes these detailed investigations reveal that a significant piece of theory is wrong, and the experiments then take on much greater importance than was originally expected, even by their designers. But one could justifiably question whether this rather haphazard approach is really the most efficient way to advance scientific understanding. Wouldn't we be better off attacking the major theories directly rather than relying on luck to show us where they need to be modified or replaced? Shouldn't Kelly and coworkers have proceeded with their molecular ratchet experiment precisely *because* success would indicate overthrow of the second law of thermodynamics? Probably most practicing chemists would answer "no" to the second question, for the reason that I cited earlier: the probability of success is too low to justify the effort involved in conducting the experiment. However, that is the case because of the rather secure place that the second law seems to have in our conceptual structure—it has both a history of resistance to experimental assault (all those failed perpetual-motion machines, for example) and a theoretical rationale that seems so convincing as to be almost self-evident. Still, there are very respectable and widely adopted theories in chemistry—transition-state theory, the Woodward-Hoffmann rules for pericyclic reactions, the Pauling model of enzyme catalysis, and the Marcus theory of electron transfer come to mind—that are less securely rooted and might be better candidates for experimental inquisition. Indeed, all of the examples listed have received considerable attention from experimentalists and have been modified as a result (although none has yet been shown to be entirely incorrect).

Recognition of the place of our various conceptual models in the continuum of ideas is thus of some value in guiding the design of scientific experiments. But can those models *mis*guide experimentation?

There is a way in which this happens, I think. It occurs through a phenomenon that one might call *false rigor*, in which a model masquerades as a more respectable theory than it really is. This is caused not by duplicity on the part of the model's proponents, but most frequently by the fuzzy thinking of those of us who use it. A somewhat related idea has been presented by Bhushan and Rosenfeld (1995). I will cite here an example that will be developed in more detail in the next section.

Most organic chemistry professors have taught (and some of us have written in books) (Carpenter, 1984) that an optically active reactant undergoing a reaction occurring by way of an achiral intermediate must give only achiral or racemic products. Let us specify that the reaction be unimolecular and occurring in the gas phase, so there can be no chiral auxiliaries brought into the picture to confuse things. Under those circumstances, the conventional wisdom seems irrefutable; an achiral molecule possesses, by definition, an improper rotation axis of symmetry. Unless our whole concept of symmetry is wrong, this structural element would appear to demand that our molecule give any chiral products as an equal mixture of enantiomers. This is false rigor.

Misguided Reaction Mechanisms?

To see where false rigor can get us in trouble, let's first examine another use of symmetry. Near the end of the book *The Conservation of Orbital Symmetry*, in which Woodward and Hoffmann described their rules for understanding pericyclic reactions, the authors included a section entitled "Violations" (Woodward & Hoffman, 1970). The first, single-sentence paragraph proclaims: "There are none!"

This bold assertion might have served to inhibit experimentalists from looking for exceptions to the rules, but it did not. Indeed, it apparently stimulated the search, and several investigators were delighted to quote the claim in articles where they reported that they had found examples of violations (see, for e.g., Berson, 1972, and Dewar & Ramsden, 1974). Why did the community react in this way rather than meekly accept the proclamation from Woodward and Hoffmann? Part of the answer is that most people recognized a certain tongue-in-cheek quality in the claim of infallibility. Indeed, the authors themselves went on to list ways in which there could be *apparent* violations to their rules by reactions that appeared to be pericyclic but, in fact, were not. However, it was not merely a search for *apparent* violations that was stimulated, several researchers went on to look for *real* violations—reactions that were truly pericyclic and yet did not follow the rules. Obviously, this kind of research can sensibly be undertaken only if one has some reason to doubt the universal applicability of the rules. If the rules were based *solely* on symmetry, there would be little room for skepticism, but of course, they are not. They *use* symmetry to help classify arguments based on molecular orbital theory. In fact, several authors, including Woodward and Hoffmann, recognized that molecules of C_1symmetry (i.e., possessing no symmetry element beyond the identity) could not be miraculously freed from following rules that very similar but merely more symmetrical molecules had to obey. It was really orbital *topology* that formed the basis for the rules, and the orbitals whose topology were deemed to be important were, in the 1960s and 1970s, calculated using quite approximate quantum mechanical methods. If the wavefunctions were wrong, then the rules based on them could be also (Buenker et al., 1971; George & Ross, 1971; Baldwin, 1972; Berson, 1972; Waschewsky et al., 1994).

The recognition that there were ancillary approximations in the rules for conservation of orbital symmetry was straightforward because the experimental search for violations began right after the rules had first been published, and so all participants were aware of the details of their formulation. When the ancillary approximations come from models of older vintage, they can be better camouflaged. This is the situation in the chirality problem that I raised at the end of the preceding section. As I will describe in more detail later, the chemical behavior of an achiral intermediate need reflect the presence of its defining symmetry element only if the so-called statistical approximation, embodied in kinetic models such as transition state theory, is valid for that case. Organic chemists have become so used to the presumption of the *general* validity of the statistical approximation that most no longer question whether it applies in any specific case. It is the covert nature of additional assumptions such as this that can lead to the phenomenon of false rigor.

Even when the approximations that we employ in our models are right before our eyes, we can fail to appreciate their consequences, or even their existence, simply because of their familiarity. A case in point can be found in the potential-energy (PE) (or sometimes standard free-energy) profiles that mechanistically oriented organic chemists draw. We are all aware that the label "Reaction Coordinate" or sometimes just "RC" on the abscissa is an ill-defined metric, but we may not be fully aware of the consequences of its use. Take the question of whether a reaction occurs in a "stepwise" fashion, involving an intermediate, or is "concerted," lacking an intermediate. As depicted in typical PE profiles (figure 12.2), the question seems clear cut and worthy of investigation—the profile either contains a little dip in the top (stepwise) or it doesn't (concerted). One might imagine a certain ambiguity about how *deep* the local minimum

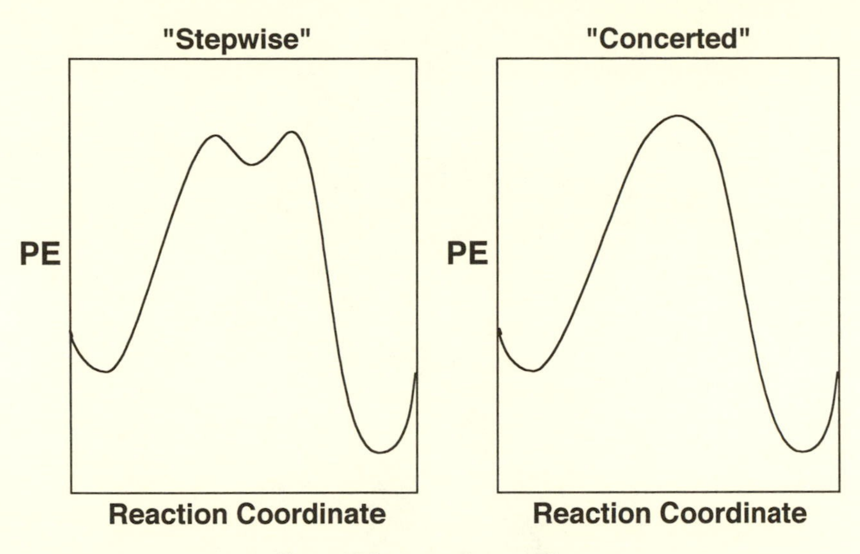

Figure 12.2 Typical PE profiles.

needs to be in order to constitute a "real" intermediate, but that seems to be the limit to which the distinction between the mechanistic alternatives can be blurred. However, consider what happens if one adds just one more dimension to the picture. For reactions of large organic molecules the result is still woefully compressed from the full hypersurface, which for a unimolecular reaction of a nonlinear, *N*-atom molecule requires 3N – 6 geometrical coordinates plus the energy coordinate. Nevertheless, the addition of just a third dimension to our graph is sufficient to let one recognize that for at least some reactions the distinction that seemed so clear between stepwise and concerted mechanisms is in fact moot (Carpenter, 1992). Take, for example, the surface depicted in figure 12.3: the potential energy is now plotted as a function of two geometrical coordinates, Q_1 and Q_2. There are four local minima on the surface, labeled **A**, **B**, **C**, and **I**. One could plausibly argue that **I** is an intermediate in the interconversion of **B** and **C**. In fact, if one took a section of the surface parallel to the Q_2 coordinate and included **B** and **C**, the result would be quite similar to the classic "stepwise" PE profile. However, the conversion of **A** to **B** or **A** to **C** is more problematic. Is **I** an intermediate in either of these reactions? There certainly exist paths across the surface from **A** to **B** and from **A** to **C** that include passage through **I**, but there are also paths that do not. The activation barriers associated with the two kinds of paths are identical, because **I**, **B**, and **C** are are all lower in PE than **A**. There is no way to specify in any general sense whether **I** is an intermediate in these reactions or not (and, hence, whether the conversion of **A** to **B** and **C** is stepwise or concerted) because the answer would be different for different molecules.

This example illustrates the problems associated with models of too low a dimensionality. One always knew that the single-dimensional "reaction coordinate" was an approximation, but perhaps did not recognize the difficulties that it could create. The

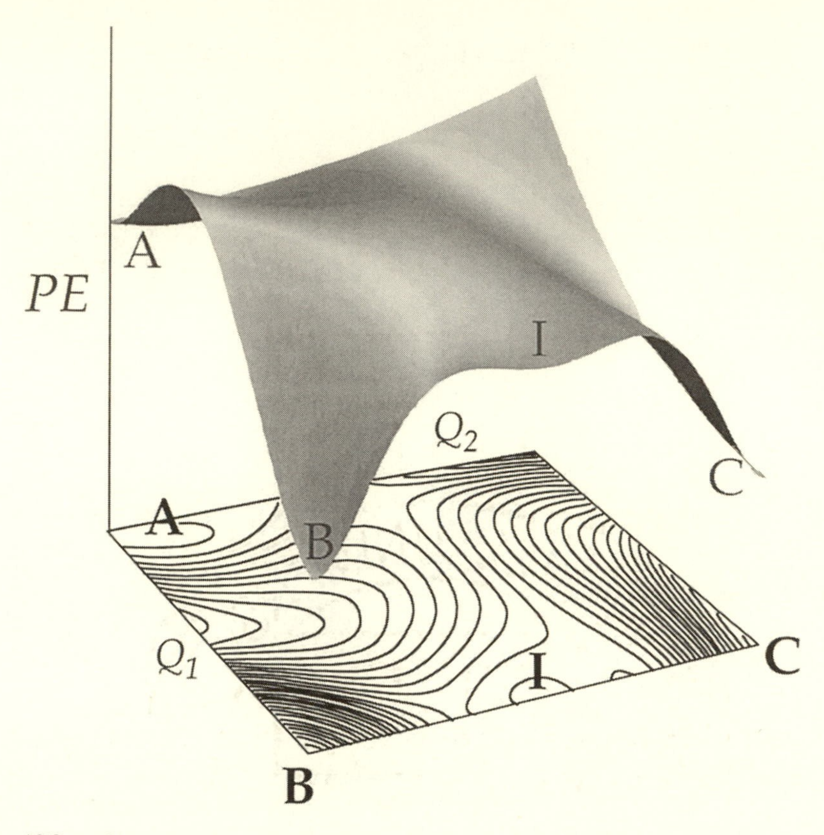

Figure 12.3 A three-dimensional potential energy surface that illustrates the difficulty of distinguishing between a "stepwise" reaction, which involves an intermediate, and "concerted" reaction, which does not. The local minimum labeled **I** on the potential energy surface could plausibly be called an intermediate for the **B** → **C** reaction since the minimum energy path must traverse the region of **I**. However, it is not clear whether it should be called an intermediate for the **A** → **B** or **A** → **C** reactions since some paths include **I** but some do not. The concepts of stepwise and concerted reactions are thus not well defined for reactions of the **A** → **B** or **A** → **C** type.

chirality problem, to which we will now return for the final time, carries better-hidden approximations. These are conveniently illustrated by using a contour plot (similar to the one under the surface in Figure 12.3) of a PE surface (Carpenter, 1998). It is shown in Figure 12.4. This surface has obvious symmetry—there is a vertical mirror plane that reflects the PE minimum of reactant **A** into that of reactant **B**, and of product **C** into that of product **D**. the mirror plane bisects the intermediate **I**, making it necessarily achiral. The two stars indicate the locations of the symmetry-related transition structures for conversion of the enantiomeric reactants to the achiral intermediate, and the two dots show where the transition structures for conversion of the intermediate

Figure 12.4 A contour plot of a general two-fold symmetric potential energy surface. The chart should be viewed like a topographical map, except that the contours reflect potential energy instead of altitude. The symmetry can be seen by recognizing that the left-hand and right-hand halves of the diagram are mirror images of each other. The intermediate **I** sits in a shallow local minimum that is bisected by the vertical mirror plane.

to the products occur. It seems obvious that the intermediate "sees" two symmetry-related transition structures for product formation and that it must consequently choose one or the other with exactly equal probability. This would seem to be the case whether **I** was generated from pure **A**, pure **B**, or any mixture of the two. However, let us now depict the reaction coordinates for the overall transformation, starting from pure **A**, by arrows (vectors). The result is shown in figure 12.5.

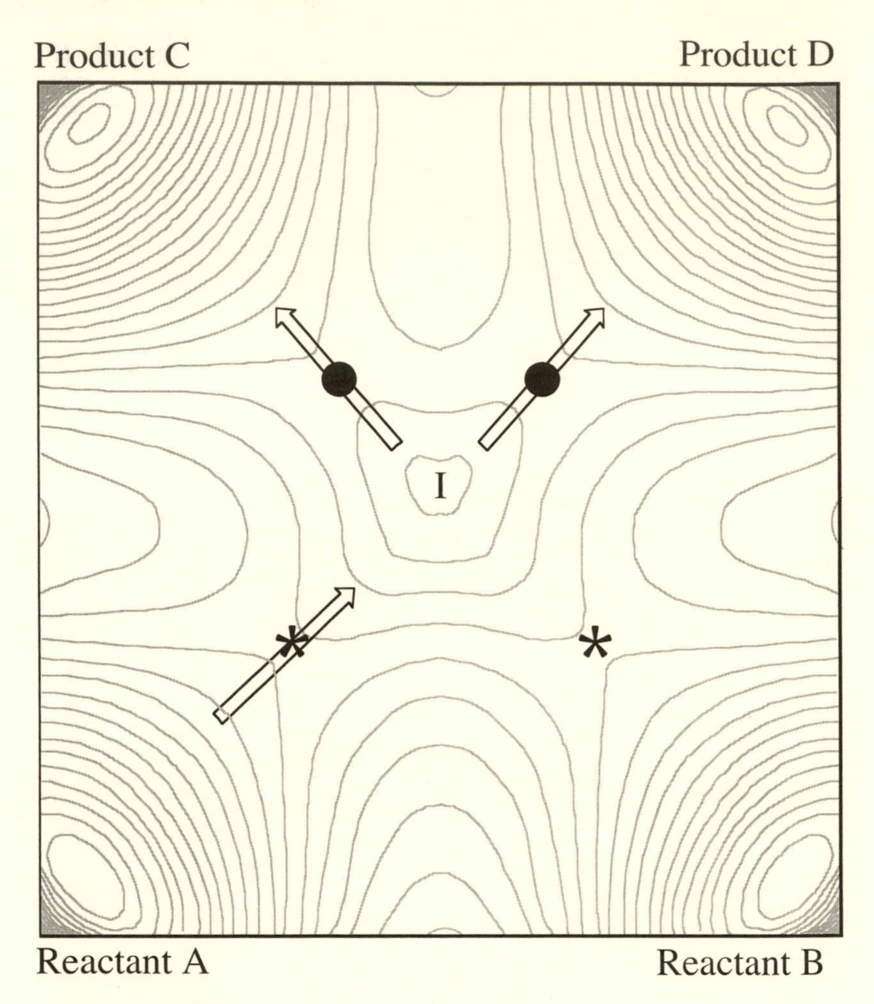

Figure 12.5 The same potential energy surface depicted in figure 12.4 with now the addition of arrows that symbolize the reaction coordinates for the conversion of one reactant to the intermediate and for conversion of the intermediate to both possible products.

The reaction—coordinate vectors serve to remind us that there is some sort of atomic *motion* that is obviously associated with any chemical transformation. The motion of the atoms (projected onto our two geometrical coordinates) for conversion of **A** to **I** is summarized by the arrow that connects these two local minima on the surface. It is apparent that the direction of this motion is quite similar to that converting **I** to **D**, but very different from that converting **I** to **C**. In other words, the symmetry of the PE surface is not apparent to a molecule on the **A**-to-**I** trajectory, because of its unsymmetrical approach to the intermediate region. Indeed, one might imagine that if there were any kind of Newtonian mechanics associated with the moving atoms, the product **D** would be formed preferentially when **I** was prepared from **A**, whereas **C** would be preferred if **B** were the reactant. Molecular dynamics simulations and variety of experimental studies suggest that this is indeed the case.

So what happened to the expectation of a 1 : 1 **C**:**D** product ratio that was so obvious a few moments ago? We can now recognize that the expression of the PE-surface symmetry in the product ratio can occur only if the intermediate "forgets" its origins. This would correspond to the molecules following more or less chaotic trajectories once they reached the intermediate region of the PE surface. Such scrambling of atomic motion (or, more precisely, rapid redistribution of internal vibrational energy) is exactly what kinetic theories using the so-called statistical approximation say should happen. Unfortunately for these theories, it now seems that the statistical approximation is not universally valid. It was the hidden attachment of the statistical approximation to our apparent symmetry analysis of the chirality problem that gave it its false rigor.

The consequences of accepting the symmetry argument for the behavior of achiral intermediates can be quite severe. The failure to see the expected 1 : 1 ratio of products **C** and **D** can then only be explained by changing the PE surface—adding channels or barriers that can give one the observed preference for formation of one product or the other. These additions to the PE surface may be entirely spurious if the model that mandated their creation is wrong. This is how the construction of mechanistic explanations can be misguided by models.

Conclusions

To my mind there is little doubt that most mechanistic chemists, and indeed most scientists, are in the business of finding explanations. In this belief I am clearly at odds with van Fraassen. Whether these explanations represent approximations to a single universal truth or have validity only within the context of the local scientific structure is, I think, an open question, but one for which I believe—disagreeing with Fine—the answer is knowable in principle. Whichever may be the case, it seems clear there are some explanations that are viewed by most knowledgeable participants to be "better" than others. What characterizes the quality of an explanation is, again disagreeing with van Fraassen, more than just its pragmatic value for fitting data. Nor do I believe, as Hempel does, that explanations are just dry prescriptions linking observations to laws. There is an esthetic response that one has when an explanation feels "right." Curiously, the explanation need not to be too obvious for one to react in this way. Why that should be I don't understand, but it seems similar to the qualities that characterize great artistic works—one's enjoyment of them depends in some measure on their message being somewhat indirect or even initially hidden.

The methods by which the mechanistic chemist goes about finding satisfactory explanations are much like those of other scientists. The construction and testing of models is an integral part of the enterprise. However, as I have tried to show in this essay, those models can become so well integrated that one forgets their existence and, hence, their underlying approximations. When that happens, the result can be that one searches for answers in too small a conceptual space. Of course, it is not practicable to consider all of science open to reinterpretation with every new problem—in the real world, we cannot afford to retest the validity of the second law of thermodynamics over and over again—but the theories residing lower down in the conceptual spectrum may well be worthy of reinvestigation. The trick is to know when they are being invoked. When an argument *seems* to be based only on the most robust laws of science, it is probably worth spending a few moments to question whether it really is, or whether there may not be some less respectable theories that are quietly holding up the scenery.

References

Baldwin, John E. 1972. "Orbital-Symmetry-Disallowed Energetically Concerted Reactions." *Accounts of Chemical Research* 5: 402–406.

Berson, Jerome A. 1972. "Orbital-Symmetry-Forbidden Reactions." *Accounts of Chemical Research* 5: 406–414.

Bhushan, Nalini & Rosenfeld, Stuart. 1995. "Metaphorical Models in Chemistry." *Journal of Chemical Education* 72: 578–582.

Boyd, Richard. 1985. "Observations, Explanatory Power, and Simplicity: Toward a Non-Humean Account." In P. Achinstein and O. Hannaway, eds. *Observation, Experiment, and Hypothesis in Modern Physical Science* (pp. 47–94). Cambridge, MA: MIT Press.

Boyd, Richard, Gasper, Philip, & Trout, J. D. 1991. *Philosophy of Science*. Cambridge, MA: MIT Press.

Brown, Theodore L., Eugene LeMay, Jr., H., & Bursten, Bruce E. 1994. Chemistry (6th ed.). Englewood Cliffs, NJ: Prentice Hall.

Buenker, Robert J., Peyerimhoff, Sigrid D., & Hsu, Kang. 1971. "Analysis of Qualitative Theories for Electrocyclic Transformations Based on the Results of Ab Initio Self-Consistent-Field and Configuration-Interaction Calculations." *Journal of the American Chemical Society* 93: 5005–5013.

Carpenter, Barry K. 1984. *Determination of Organic Reaction Mechanisms*. New York: Wiley-Interscience.

Carpenter, Barry K. 1992. "Intramolecular Dynamics for the Organic Chemist." *Accounts of Chemical Research* 25: 520–528.

Carpenter, Barry K. 1998. "Dynamic Behavior of Organic Reactive Intermediates." *Angewandte Chemie, International Edition in English* 37: 3340.

Cartwright, Nancy. 1980. "The Reality of Causes in a World of Instrumental Laws." In P. Asquith & R. Giere, eds. *Philosophy of Science Association* (vol. 2, pp. 38–48).

Chang, Raymond. 1994. Chemistry (5th ed.). New York: McGraw-Hill.

Davis, Anthony P. 1998. "Tilting at Windmills? The Second Law Survives." *Angewandte Chemie, International Edition in English* 37: 909–910.

Dewar, Michael J. S., & Ramsden, Christopher A. 1974. "Stevens Rearrangement. Antiaromatic Pericyclic Reaction." *Journal of the Chemical Society, Perkin Transactions* 1: 1839–1844.

Fine, Arthur. 1984. "The Natural Ontological Attitude." In J. Leplin, ed. *Scientific Realism* (pp. 83–107). Berkeley: University of California Press.

George, Thomas, F., & Ross, John. 1971. "Analysis of Symmetry in Chemical Reactions." *Journal of Chemical Physics* 55: 3851–3866.

Hempel, Carl. 1966. *Philosophy of Natural Science*. Englewood Cliffs, NJ: Prentice Hall.

Hoffmann, Roald, Minkin, Vladimir I., & Carpenter, Barry K. 1996. "Ockham's Razor and Chemistry." *Bulletin de la société chimique de France* 133: 117–130.

Kelly, T. Ross, Tellitu, Imanol, & Pérez Sestelo, José. 1997. "In Search of Molecular Ratchets." *Angewandte Chemie, International Edition in English* 36: 1866–1868.

Kelly, T. Ross, Pérez Sestelo, José, & Tellitu, Imanol. 1998. "New Molecular Devices: In Search of a Molecular Ratchet." *Journal of Organic Chemistry* 63: 3655–3665.

Kitcher, Philip. 1981. "Explanatory Unification." *Philosophy of Science* 48: 507–531.

Petrucci, Ralph H., & Harwood, William S. 1993. *General Chemistry* (6th ed.) New York: Macmillan.

van Fraassen, Bas. 1977. "The Pragmatics of Explanation." *American Philosophical Quarterly* 14: 143–150.

Waschewsky, Gabriela C. G., Kash, Philip W., Myers, Tanya L., Kitchen, David C., & Butler, Laurie J. 1994. "What Woodward and Hoffmann Didn't Tell Us: The Failure of the Born-Oppenheimer Approximation in Competing Reaction Pathways." *Journal of the Chemical Society, Faraday Transactions* 90: 1581–1598.

Woodward, Robert B., & Hoffman, Ronald 1970. *The Conservation of Orbital Symmetry.* Weinheim: Verlag Chemie.

Zumdahl, Steven S. 1993. *Chemistry* (3rd ed.) Lexington, MA: D.C. Heath.

13

How Symbolic and Iconic Languages Bridge the Two Worlds of the Chemist

A Case Study from Contemporary Bioorganic Chemistry

EMILY R. GROSHOLZ
ROALD HOFFMANN

Chemists move habitually and with credible success—if sometimes unreflectively—between two worlds. One is the laboratory, with its macroscopic powders, crystals, solutions, and intractable sludge, as well as the things that are smelly or odorless, toxic or beneficial, pure or impure, colored, or white. The other is the invisible world of molecules, each with its characteristic composition and structure, its internal dynamics and its ways of reacting with the other molecules around it. Perhaps because they are so used to it, chemists rarely explain how they are able to hold two seemingly disparate worlds together in thought and practice. And contemporary philosophy of science has had little to say about how chemists are able to pose and solve problems, and, in particular, to posit and construct molecules, while simultaneously entertaining two apparently incompatible strata of reality. Yet chemistry continues to generate highly reliable knowledge, and indeed to add to the furniture of the universe, with a registry of over ten million well-characterized new compounds.

The philosophy of science has long been dominated by logical positivism, and the assumptions attendant on its use of predicate logic to examine science, as well as its choice of physics as the archetype of a science. Positivism thus tends to think of science in terms of an axiomatized theory describing an already given reality and cast in a uniform symbolic language, the language of predicate logic. (See especially the locus classicus of this position, Carnap, 1937.)

We here wish to question certain positivist assumptions about scientific rationality, based on an alternative view brought into focus by the reflective examination of a case study drawn from contemporary chemistry. Our reflections owe something to Leibniz (1686, 1695, 1714), Husserl (1922), Kuhn (1970), and Polanyi (1960, 1966), and draw on the earlier writings of both of us—Hoffmann (1995; Hoffmann & Laszlo, 1991) and Grosholz (1991; Grosholz & Yakira, 1998). We will offer a nonreductionist account of methods of analysis and synthesis in chemistry. In our view, reality is allowed to include different kinds of things existing in different kinds of ways, levels held in intelligible relation by both theory and experiment, and couched in a multiplicity of languages, both symbolic and iconic.

We argue that there is no single correct analysis of the complex entities of chemistry expressed in a single adequate language, as various reductionist scripts require; and yet the multiplicity and multivocality of the sciences, and their complex "horizontal" interrelations, do not preclude but in many ways enhance their reasonableness and success. Nor is this view at odds with our realism; we want to distinguish ourselves quite strongly from philosophers engaged in the social construction of reality (see, e.g., Pickering, 1992; Shapin, 1992; Fuller, 1994; for a balanced analysis of the problem, see Labinger, 1995). We understand the reality whose independence we honor as requiring scientific methods that are not univocal and reductionist precisely because reality is multifarious, surprising, and infinitely rich.

Formulating the Problem

The article drawn from the current literature in chemistry that we shall consider is "A Calixarene with Four Peptide Loops: An Antibody Mimic for Recognition of Protein Surfaces," authored by Yoshitomo Hamuro, with Andrew Hamilton, Mercedes Crego Calama, Hyung Soon Park, and published in December 1997 in the international journal *Angewandte Chemie* (Hamuro et al., 1997. We will refer to this article as Hamuro et al.) The subfield of the article could be called bioorganic chemistry. One way to look at biology is to examine its underlying chemistry, in a well-developed program that is both one of the most successful intellectual achievements of the twentieth-century, and a locus of dispute for biologists. For many years, organic chemists had let molecular and biochemistry "get away" from chemistry; recently, there has been a definite movement to break down the imagined fences and reintegrate modern organic chemistry and biology. The article we examine is part of such an enterprise.

We have learned something about the structure of the large, enigmatic, selectively potent molecules of biology. But describing their structure and measuring their functions do not really answer the question of how or why these molecules act as they do. Here, organic chemistry can play an important role by constructing and studying molecules smaller than the biological ones, but which model or mimic the activities of the speedy molecular behemoths of the biological world.

The article opens by stating one such problem of mimicry, important to medical science and any person who has ever caught a cold. The human immune system has flexible molecules called antibodies, proteins of some complexity that recognize a wide variety of molecules including other proteins.

> The design of synthetic hosts that can recognize protein surfaces and disrupt biologically important protein–protein interactions remains a major unsolved problem in bioorganic chemistry. In contrast, the immune system offers numerous antibodies that show high sequence and structural selectivity in binding to a wide range of protein surfaces. (Hamuro et al., 1997, p. 2680)

The problem is thus to mimic the structure and action of an antibody; but antibodies in general are very large and complicated. Hamuro et al. ask the question, Can we assemble a molecule with some of the structural features of an antibody, simplified and scaled down, and, if so, will it act like an antibody? But what are the essential structural features in this case?

Prior investigation has revealed that an antibody at the microscopic level is a protein molecule that typically has a common central region with six "hypervariable" loops that exploit the flexibility and versatility of the amino acids that make up the loops to recognize (on the molecular level) the near infinity of molecules that wander about a human body. The article remarks that "this diversity of recognition is even more remarkable, because all antibody fragment antigen binding (FAB) regions share a common structural motif of six hypervariable loops held in place by the closely packed constant and variable regions of the light and heavy chains" (p. 2680).

What is recognition at the microscopic level? It is generally not the strong covalent bonding that makes molecules so persistent, but is rather a congeries of weak interactions between molecules that may include bonding types that chemists call hydrogen bonding, van der Waals or dispersion forces, electrostatic interactions (concatenations of regions of opposite charge attracting or like charge repelling), and hydrophobic interactions (concatenations of like regions attracting, as oil with oil, water with water). These bonding types are the subject of much dispute, for they are not as distinct as scientists would like them to be. (For an introduction to chemistry and molecular interactions, see Joesten et al., 1991). In any case, the interactions between molecules are weak and manifold. Recognition occurs as binding, but it is essentially more dynamic than static. At body temperature, recognition is the outcome of many thermodynamically reversible interactions: the antibody can pick up a molecule, assess it, and then perhaps let it go. In the dance of holding on and letting go, some things are held on to more dearly.

Whatever happens has sufficient cause, in the geometry of the molecule, and in the physics of the microscopic attractions and repulsions between atoms or regions of a molecule. The article remarks,

> Four of these loops . . . generally take up a hairpin conformation and the remaining two form more extended loops. X-ray analyses of protein–antibody complexes show that strong binding is achieved by the formation of a large and open interfacial surface (>600 Å) composed primarily of residues that are capable of mutual hydrophobic, electrostatic, and hydrogen bonding interactions.* The majority of antibody complementary determining regions (CDRs) contact the antigen with four to six of the hypervariable loops.* (pp. 2680–81)[1]

The foregoing passage is a theory about the structure and function of antibodies, but it is asserted with confidence and in precise detail. Standing in the background, linking the world of the laboratory—where small (but still tangible) samples of antibodies and proteins are purified, analyzed, combined, and measured—and the world of molecules are theories, instrumentation, and languages. There is no shortage of theories here; indeed, we are faced with an overlapping, interpenetrating network of theories backed up by instrumentation. These include the quantum mechanics of the atom along with a multitude of quantum mechanically defined spectroscopies, chemistry's highly refined means for destructively or nondestructively plucking the strings of molecules and letting the "sounds" tell us about their features (Hoffmann & Torrence, 1993, pp. 144–147). Three are equally ingenious techniques for separating and purifying molecules, that we will loosely term chromatographies. They proceed at a larger scale and, when traced, are also the outcome of a sequence of holding on and letting go, like antibody recognition.

Further, statistical mechanics and thermodynamics serve to relate the microscopic to the macroscopic. These theories are probabilistic, but they have no exceptions because of the immensity of the number of molecules—10^{23} in a sip of water—and the rapidity of molecular motion at ambient temperatures. Thus, the average speed of molecules "scales up" to temperature, their puny interactions with light waves into color, the resistance of their crystals to being squeezed to hardness, their multitudinous and frequent collisions into a reaction that is over in a second or a millennium (Atkins, 1984, 1987, 1991; Joesten et al., 1991; Hoffmann, 1995).

These theories are silent partners in the experiments described in the article, taken for granted and embodied, one might say, in the instruments. But a further dimension of the linkage between the two worlds is the languages employed by the chemists, and that is what we now propose to examine at length.

Solving the Problem

The construction of "a calixarene with four peptide loops" serves two functions in this article. It serves as a simplified substitute for an antibody, though we doubt that the intent of the authors is the design of potential therapeutic agents. More important, the calixarene serves to test the theory of antibody function sketched in the preceding discussion: Is this really the way that antibodies work? The authors note that earlier attempts to mimic antibodies have been unsuccessful, and they propose the alternative strategy, which is the heart of the article: The search for antibody mimics has not yet yielded compact and robust frameworks that reproduce the essential features of the CDRs.* Our strategy is to use a macrocyclic scaffold to which multiple peptide loops in stable hairpin-turn conformations can be attached (p. 2681).

Stage 1a: The Core Scaffold

The experiment has two stages. The first is to build the antibody mimic, by adding peptide loops to the scaffolding of a calix[4]arene—a cone-shaped concatenation of four benzene rings, strengthened and locked into one orientation by the addition of small-length chains of carbon and hydrogen (an alkylation), with COOH groups on top to serve as "handles" for subsequent reaction. (The benzene ring of six carbons is a molecule with a venerable history, whose structure has proved especially problematic for the languages of chemistry, as we point out later in this chapter.) The authors write,

> In this paper we report the synthesis of an antibody mimic based on calix[4]arene linked to four constrained peptide loops. . . . Calix[4]arene was chosen as the core scaffold, as it is readily available* and can be locked into the semirigid cone conformation by alkylation of the phenol groups. This results in a projection of the para-substituents onto the same side of the ring to form a potential binding domain.* (p. 2681)

Diagram 1 is given to illustrate this description, as well as the following "recipe:" "The required tetracarboxylic acid 1 was prepared by alkylation of calix[4]arene* (*n*-butyl bromide, NaH) followed by formylation (Cl_2CHOCH_3, $TiCl_4$) and oxidation ($NaClO_2$, H_2NSO_3H)*" (p. 2681).

The iconic representation offered is of a microscopic molecule, but the language is all about macroscopic matter, and it is symbolic. The symbolic language of chemistry is the language of formulas employed in the laboratory recipe. It lends itself to the chemist's bridging of the macroscopic and the microscopic because it is thoroughly equivocal, at once a precise description of the ingredients of the experiment (e.g., *n*-butyl bromide is a colorless liquid, with a boiling point of 101.6°C, and is immiscible with water), and a description of the composition of the relevant molecules. For example, *n*-butyl bromide is construed by the chemist as $CH_3CH_2CH_2CH_2Br$; it has the formula C_4H_9Br, a determi-

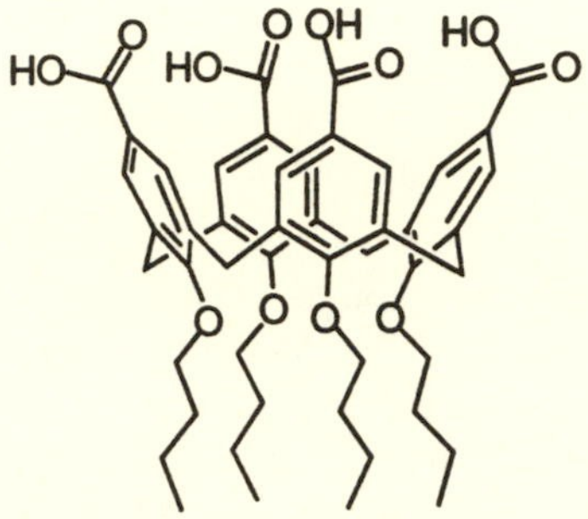

1

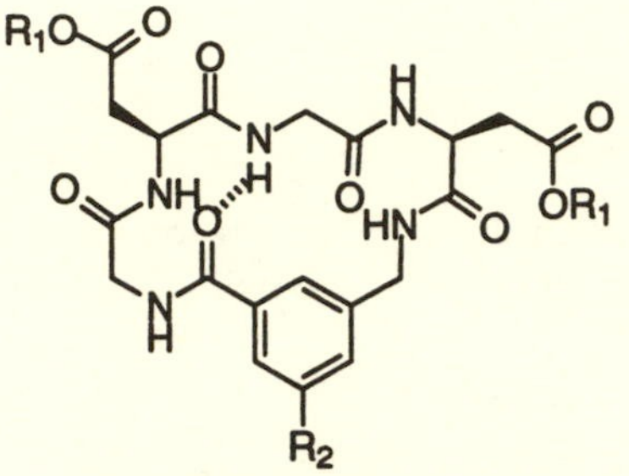

2. $R_1 = Bu^t$, $R_2 = NH_2$,

4. $R_1 = H$, $R_2 = NO_2$,

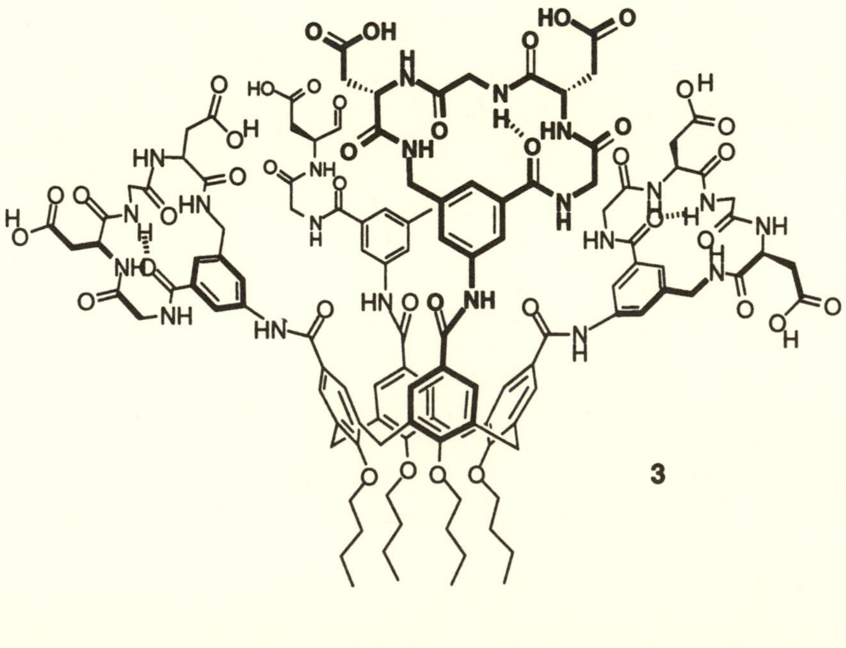

3

nate mass relationship among the three atomic constituents, a preferred geometry, certain barriers to rotation around the carbon–carbon bonds it contains, certain angles at the carbons, and so forth (Atkins 1987; Joesten et al., 1991; Hoffmann, 1995).

The laboratory recipe is thus both the description of a process carried out by a scientist, and the description of a molecule under construction: a molecule generic in its significance, because the description is intended to apply to all similar molecules, but particular in its unity and reality. There are parallels in other fields of knowledge. Thus, in mathematics, the algebraic formula of a function applies equally to an infinite set of number pairs and to a geometric curve; its controlled and precise equivocity is the instrument that allows resources of number theory and of geometry to be combined in the service of problem solving (Grosholz, 1991, chaps. 1 and 2). Likewise, here the algebra of chemistry allows the wisdom of experience gained in the laboratory to be combined with the (classical and quantum) theory of the molecule, knowledge of its fine structure, energetics, and spectra.

But the symbolic language of chemistry is not complete, for there are many aspects of the chemical substance/molecule that it leaves unexpressed: (1) We cannot deduce from it how the molecule will react with the enormous variety of other molecules with which it may come in contact; (2) We cannot even deduce from it the internal statics, kinematics, and dynamics of the molecule in space.[2] Molecules identical in composition can differ from each other because they differ in constitution, the manner and sequence of bonding of atoms (tautomers), in spatial configuration (optical or geometrical isomers), and in conformation (conformers). (See Joesten et al., 1991; Zeidler & Hoffmann, 1995; and Sobczynska, 1995/6.) An adequate description of the molecule must invoke the background of an explanatory theory, but to do so it must also employ iconic languages. Thus, the very definition of the calixarene core scaffold involves a diagram. (It was also necessary for the authors to identify C_4H_9Br as *n*-butyl bromide, a nomenclature that implies a specific connectivity of atoms.)

This diagram of calixarene is worth careful inspection, as well as careful comparison with its counterparts in the more complex molecules (for which it serves as core scaffold) furnished to us, the readers, by means of computer-generated images. First, it leaves out most of the component hydrogens and carbons in the molecule; they are understood, a kind of tacit knowledge shared even by undergraduate chemistry majors. The hexagons are benzene rings, and the chemist knows that the valence of (the number of bonds formed by) carbon is typically four and so automatically supplies the missing hydrogens. But this omission points to an important feature of iconic languages: they must always leave something out, because they are only pictures, not the thing itself, and because the furnishing of too much information is actually an impoverishment. In a poor diagram, one cannot see the forest for the trees. Not only must some things remain tacit in diagrams, but also the wisdom of experience that lets the scientist know how much to put in and how much to leave out—wisdom gleaned by years of translating experimental results into diagrams for various kinds of audience—is itself often tacit. It can be articulated now and then but cannot be translated into a complete set of fixed rules.

Second, the diagram uses certain conventions for representing configurations in three-dimensional space on the two-dimensional page, like breaking the outlines of molecules which are supposed to be behind other molecules whose delineation is unbroken. (In other diagrams, wedges are used to represent projection outward from the plane of the page, and heavy lines are used to represent molecules that stand in front of other molecules depicted by ordinary lines.) Sometimes, although not in the context of a journal article, chemists show three-dimensional representation, such as "ball-and-stick" models. But then, in addition, one may want to see more precise angles and interatomic distances in correct proportion and so resort to the images produced by X-ray crystallography. Or one may want some indication of the motion of the molecules, because all atoms vibrate and rotate. Arrows and other iconographies of dynamic motion are used in such diagrams. The cloudy, false-color, yet informative photographs of scanning tunneling microscopy come in here, as well as assorted computer images of the distribution of electrons in the molecule.[3]

Finally, the convention of a hexagon with a perimeter composed of three single lines alternating with three double lines to represent a benzene ring deserves a chapter in itself. This molecule has played a central role in the development of organic chemistry. No single classical valence structure was consistent with the stability of the molecule. Kekulé solved the problem by postulating the coexistence of two valence structures in one molecule. In time, practitioners of quantum mechanics took up the benzene problem, and to this day it has served them as an equally fecund source of inspiration and disagreement. The electrons in benzene are delocalized, that much people agree on; but the description of its electronic structure continues to be a problem for the languages of chemistry (Brush, 1998).

An Interlude on Symbolic and Iconic Languages

Philosophers of science working in the logical positivist tradition have had little to say about iconic languages. Symbolic languages typically lend themselves to logical regimentation, but pictures tend to be multiform and hard to codify; thus, if they proved to be indispensable to human knowledge, the logical positivist would be quite vexed. As any student of chemistry will tell you, conventions for producing "well-formed icons" of molecules exist and must be learned, or else your audience will misread them. But no single iconic language is the correct one or enjoys anything as precisely determined as a "wff" in logic. The symbolic language of chemistry is, to be sure, a precisely defined international nomenclature that specifies in impressive detail a written sequence of symbols so as to allow the unique specification of a molecule. But, significantly, the iconic representations of a molecule are governed only by widely accepted conventions, and a good bit of latitude is allowed in practice, especially a propos what may be omitted from such representations. Symbolic languages lend themselves to codification in a way that iconic languages don't.[4]

Symbolic languages, precisely because they are symbolic, lend themselves best to displaying relational structure. Like algebra, they are tolerant or relativistic in their ontological import: it doesn't so much matter what they pertain to, as long as their objects stand in the appropriate relations to each other. But iconic languages point, more or less directly, to objects; they are not ontologically neutral but, on the contrary, are ontologically insistent. They display the unity of objects, a unity that might metaphysically be called the unity of existence. But there is no way to give an exhaustive summary of the ways of portraying the unity of existence; it is too infinitely rich, and thought has too many ways of engaging it. We should, therefore, not jump to the conclusion that knowledge via an iconic language is impossible or incoherent: iconic languages despite being multiform employ intelligible conventions, they are constrained by the object itself, and they are made orderly by their association with symbolic language. An inference cannot be constructed from icons alone, but icons may play an essential role in inference.[5]

How does the iconic form of the chemical structure expressed as a diagram that displays atom connectivities and suggests the three-dimensionality of the molecule, bridge the two worlds of the chemist? The most obvious answer is that it makes the invisible visible, and does so, within limits, reliably. But there is a deeper answer. It seems at first as if the chemical structure diagram refers only to the level of the microscopic since, after all, it depicts a molecule. But in conjunction with symbolic formulas, the diagram takes on an inherent ambiguity that gives it an important bridging function. In its display of unified existence, it stands for a single particular molecule. Yet we understand molecules of the same composition and structure to be equivalent to each other, internally indistinguishable. (In this, the objects of physics and chemistry are like the objects of mathematics.)

Thus, the icon (hexagonal benzene ring) also stands for all possible benzene rings, or for all the benzene rings (moles or millimoles of them!) in the experiment, depending on the way in which it is associated with the symbolic formula for benzene. The logical positivist in search of univocality might call this obfuscating ambiguity, a degeneracy in what ought to be a precise scientific language that carries with it undesirable ontological baggage. And yet, the iconic language is powerfully efficient and fertile in the hands of the chemist.

Now we can better understand why the kind of world-bridging involved in posing a problem/construction in chemistry requires both symbolic and iconic languages for its formulation. On the one hand, the symbolic language of chemistry captures the composition of molecules, but not their structure (constitution, configuration, and conformation), aspects that are dealt with better, though fragmentarily, by the many iconic idioms available to chemists. Moreover, the symbolic language of chemistry fails to convey the ontological import, the realism, intended by practitioners in the field. Hamuro et al. are not reporting on a social construction or a mere computation, but a useful reality: the diagram confidently posits its existence. On the other hand, icons are too manifold and singular to be the sole vehicle of scientific discourse. Their use along with symbolic language embeds them in demonstrations and gives to their particularity a representative and well-defined generality, sometimes even a universality.

We return to our reading of the Hamuro et al. article.

Stage 1b: The Peptide Loops

Hamuro et al. chose cyclic hexapeptides to mimic the "arms" of the antibody because they can be modified so as to link up easily with the core scaffold, and because they form hairpin loops: "The peptide loop was based on a cyclic hexapeptide in which two residues were replaced by a 3-aminomethylbenzoyl (3amb) dipeptide analogue* containing a 5-amino substituent for facile linkage to the scaffold" (p. 2681). The recipe for constructing the peptide loops is then given; the way in which it couples the macroscopic and the microscopic is striking, for it describes a laboratory procedure and then announces that the outcome of the procedure is a molecule, pictured in diagram **2**.

> The 5-nitro substituted dipeptide analogue was formed by selective reduction (BH_3) of methyl 3-amidocarbonyl1-5-nitrobenzoate, followed by deesterification (LiOH in THF) and reaction sequentially with Fmoc-Asp-(tBu)-OH and H-Gly-Asp(tBu)-Gly-OH (dicyclohexyl carbodiimide (DCC), N-hydroxysuccinimide) to yield Fmoc-Asp(tBu)-5$NO_2$3 amb-Gly-Asp (tBu)-GlyOH. Cyclization with 4-dimethylaminopyridine (DMAP) and 2-1H-benzotriazole-1-yl-1,1,3,3,-tetramethyluronium tetrafluoroborate (TBTU) was achieved in 70% yield, followed by reduction (H_2, Pd/C) to give the amino-substituted peptide loop **2**. (p. 2681)

Working in the lab, the chemist has constructed a molecule, at least a dizzying 20 orders of magnitude "below" or "inward." To be sure, what was made was a visible, tangible material—likely less than a gram of it—but the interest of what was made lies in the geometry and reactivity of the molecule, not the properties of the macroscopic substance. So it is not by accident that the leap to the level of the molecule is accompanied by iconic language. Such language also accompanies the final step in the assembly of the antibody mimic.

Stage 1c: The Antibody Mimic

Four of the peptide loops are attached to the core scaffold; the laboratory procedure begins and ends with a pictured molecule. But this time the resultant new molecule is pictured twice in complementary iconic idioms.

> Amine **2** was coupled to the tetraacid chloride derivative of **1** ($(COCl)_2$, DMF) and deprotected with trifluoroacetic acid (TFA) to give the tetraloop structure **3**. The molecular structure of this host (Fig [13.]1) resembles that of the antigen binding region of an antibody but is based on four loops rather than six.* (p. 2681)

To someone who understands chemical semiotics, the iconic conventions in diagram **3** (the tetraloop molecule called structure **3** in the quote above) do allow a mental reconstruction of the molecule. But the shape of the molecule is so important that the authors decide to give it again, in another view, in figure 13.1. The figure is even printed in color in the original!

Why should the reader be offered another iconic representation? In part, it is part of a rhetorical strategy to persuade the audience of the cogency of a research program that involves mimicry. The computer-generated image of figure 13.1 is actually the result of a theoretical calculation in which the various molecular components are allowed to wiggle around any bonds that allow rotation and to reach a geometry that is presumably the most stable. In that image, the general shape of the molecule (in particular, the loopiness of the loops) is beautifully exhibited, emphasizing its resemblance to an antibody. Note that the experimentalist trusts the ability of a theoretical computer program to yield the shape of a molecule sufficiently to insert it—in color—in an article; that would not have been the case 20 years ago.

Diagram **3** and figure 13.1 are meant to be seen in tandem; they complement each other. Both representations are iconic, though perhaps figure 13.1 is more so. Diagram **3** has a symbolic dimension due to the labels, and thus serves to link figure 13.1 to the

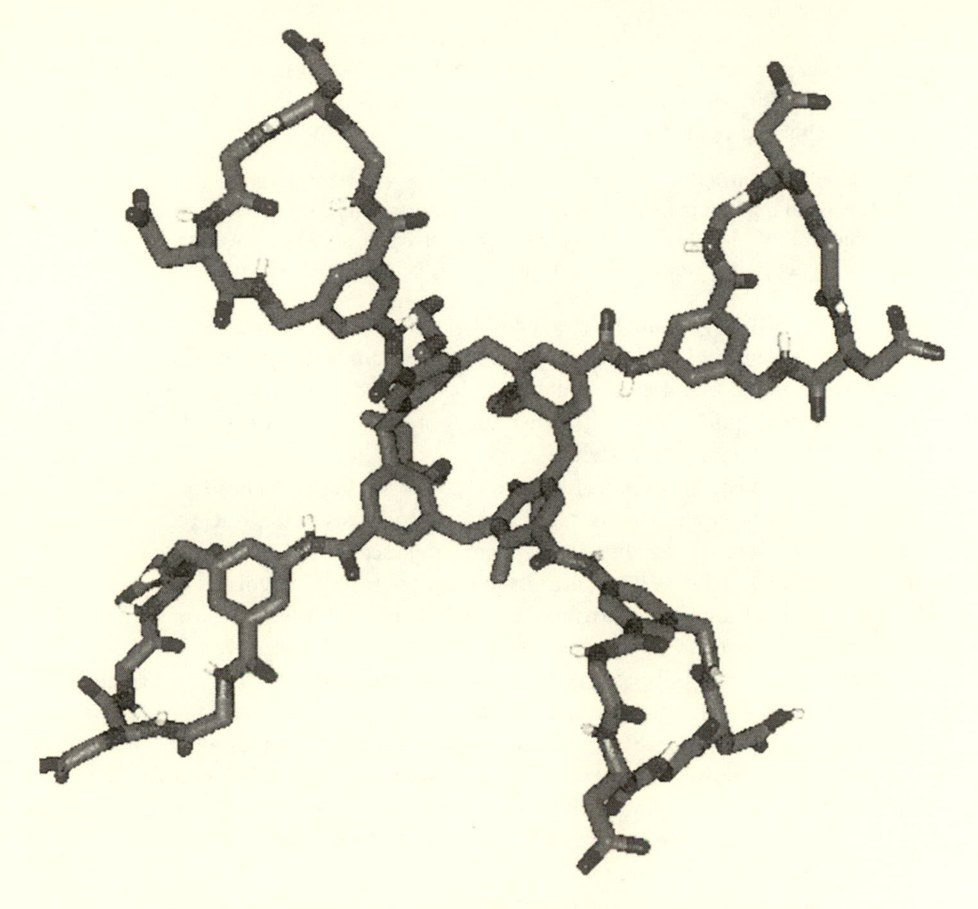

Figure 13.1 Figure 1 of Hamuro et al. article. See explanation in text.

symbolic discourse of the prose argument. Together with the reproducible laboratory procedure—given in more detail at the end of the article—Hamuro et al. give a convincing picture of this new addition to the furniture of the universe. There it stands: Ecce.

Stage 2

Once the antibody mimic has been assembled, it can be tested to see whether it in fact behaves like an antibody, a test which, if successful, in turn provides evidence supporting the theory of the action of antibodies invoked by Hamuro et al. Note the usefully—as opposed to viciously—circular reasoning here (Hoffmann, 1988): the antibody mimic correctly mimics an antibody if it behaves like an antibody; but how an antibody behaves is still a postulate, which stipulates what counts as the correctness of the antibody mimic's mimicry. To see if the antibody mimic—the base scaffold of calixarene with four peptide loops—will bind with and impair the function of a protein (the essence of what an antibody does), Hamuro et al. chose the protein cytochrome, an important molecule that plays a critical role in energy production and electron transport in every cell, and has thus been thoroughly investigated. Moreover, it has a positively charged surface region that would likely bond well with the negatively charged peptide loops.

> We chose cytochrome *c* as the initial protein target, since it is structurally well-characterized and contains a positively charged surface made up of several lysine and arginine residues.* In this study the negatively charged GlyAspGlyAsp sequence was used in the loops of **3*** to complement the charge distribution on the protein. (p. 2681)

Note that the antibody mimic is referred to by means of the diagram **3**. In a sense this is because the diagram is a shorthand, but its perspicuity is not trivial or accidental: as a picture that can be taken in at a glance, it offers schematically the whole configuration of the molecule in space. Its visual unity stands for, and does not misrepresent, the unity of the molecule's existence.

Does the antibody mimic, in fact, bind with the cytochrome? The affinity of the two is tested by an experiment that is neither analytic nor synthetic, but, rather, a matter of careful physical measurement—an aspect of chemical practice central to the science since the time of Lavoisier. The "affinity chromatography" involves a column filled with some inert cellulose-like particles and cytochrome *c* linked to those particles (for a description of chromatography, see Laszlo, 1997). The concentration of NaCl, simple salt, controls the degree of binding of various other molecules to the cytochrome *c* that is in that column. If the binding is substantially through ionic forces (as one thinks it is for the antibody mimic), then only a substantial concentration of ionic salt solution will disrupt that binding. At the top of the column one first adds a control molecule (diagram **4**). It is eluted easily, with no salt. But the antibody mimic **3** turns out to be bound much more tightly—it takes a lot of salt to flush it out.

A second kind of chromatography, "gel permeation chromatography," gives more graphic evidence for the binding of cytochrome *c* to **3**. In this ingenious chromatography, the column is packed with another cellulose-like and porous fiber, called Sephadex G-50. The "G-50" is not just a trade name; it indicates that molecules of a certain size will be trapped in the column material, but molecules both larger and smaller will flow through the column quickly.

The results of this experiment are shown in figure 13.2, replete with labeled axes. The vertical axis measures the absorption of light at a certain wavelength; this is related to the concentration of a species, the bound cytochrome *c*–**3** complex. The horizontal axis is a "fraction number" that is related to the length of time that a given molecule (or compound? the equivocity here pervades chemical discourse) resides on the column. The pores in the Sephadex retard cytochrome *c*; it stays on the column longer (has a higher fraction number). The molecular complex of the mimic and cytochrome *c* comes out in a different peak, at lower fraction number. This means it is too large to be caught in the pores of the Sephadex, which in turn, constitutes evidence for some sort of binding between the cytochrome *c* and the antibody mimic, creating a larger molecular entity.

So there is binding; but does it impair the function of the cytochrome *c*? Evidence for that is provided by reacting the cytochrome *c* with ascorbate (vitamin C), with which it normally reacts quite efficiently; here, on the contrary, it doesn't.

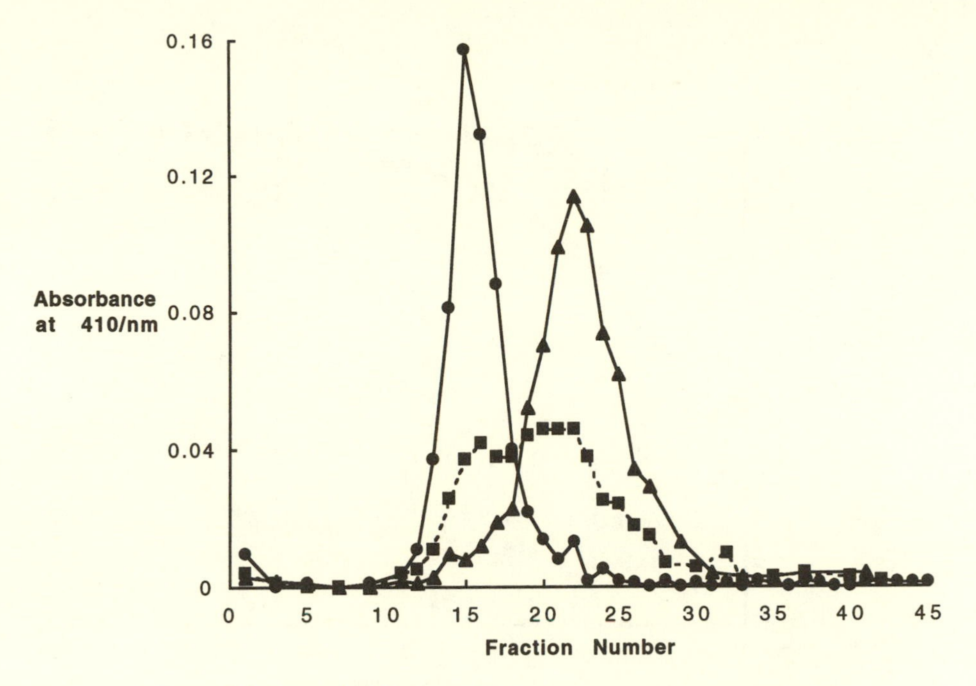

Figure 13.2 Figure 2 of Hamuro et al. article. See explanation in text.

> We have investigated the effect of complexation with **3** on the interaction of FeIII-cyt *c* with reducing agents.* In phosphate buffer Fe^{III}-cyt *c* (1.57×10^{-5}M) is rapidly reduced by excess ascorbate (2.0×10^{-3}M) with a pseudo-first-order rate constant 0.1090 ± 0.001 (Figure 4 [figure 13.3]). In the presence of **3** (1.9×10^{-5}M) the rate of cyt *c* reduction is diminished tenfold ($k_{obs} = 0.010 \pm 0.001$ s^{-1}), consistent with the calixarene derivative's binding to the protein surface and inhibiting approach of ascorbate to the heme edge (Figure 3 [figure 13.4]).

Figure 13.3 is another measurement, with the concentrations measured on the vertical axis, the time on the horizontal; it displays the outcome of an experiment on the kinetics of ascorbate reduction by cytochrome *c*, which supports the claim that the antibody mimic does impair the function of the protein, in this case its ability to react with ascorbate.

More interesting is another iconic representation, figure 13.4 which is a picture of the antibody mimic binding with cytochrome *c*. Since the authors admit, "The exact site on the surface of the cytochrome that binds with **3** has not yet been established," this image is a conjecture; and it is the outcome of the same computer program that generated figure 13.1. It "docks" "a calculated structure for **3**" at the most likely site on the cytochrome *c*, where the four peptide loops "cover a large area of the protein surface." Figure 13.4 is a remarkable overlay of several types of iconic representation. The antibody mimic (at top) is shown pretty much as it was in figure 13.1, but from the side. The atoms of cytochrome *c* are legion, and so are mostly not shown; instead, the essential helical loops of the protein are schematically indicated. But in the contact region, the atoms are again shown in great detail, not by ball-and-stick or rod representations

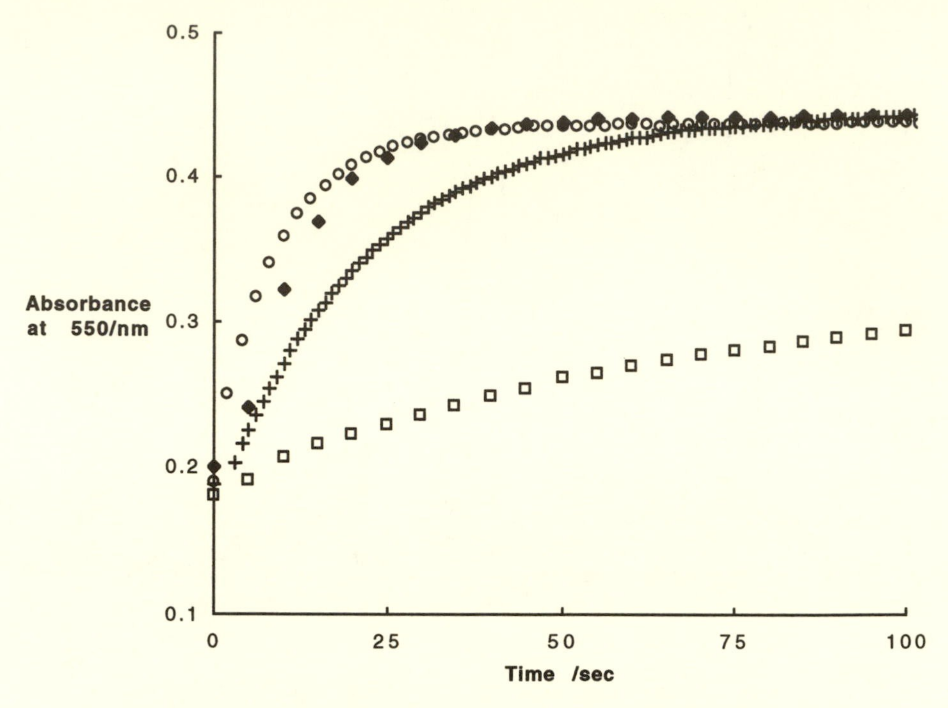

Figure 13.3 Figure 4 of Hamuro et al. article. See explanation in text.

but by tenuous spheres indicating roughly the atomic sizes or electron densities. The reader can make sense of these superimposed iconic idioms only by reference to a cognitive framework of words and symbols.

Iconic representations in chemical discourse must be related to a symbolic discourse; our access to the microscopic objects of chemistry, even our ability to picture them, is always mediated by that discourse rather than by our "natural" organs of perception. So the objects of chemistry may seem a bit ghostly, even to the practitioners for whom their existence is especially robust (see also, Laszlo, 1998). But, conversely, symbolic discourse in chemistry cannot dispense with iconic discourse as its complement, nor can it escape its own iconic dimension. The side-by-side distinction, iteration, and concatenation of letters in chemical formulas echo the spatial array of atoms in a molecule. Otherwise put, the iconic combination of symbols often articulates otherness, which is exhibited by the things of chemistry as their spatial externality and proximity. The iconic array in figure 13.3 also translates into spatial relations among symbols the temporal spread of stages of a chemical event, where otherness is priority or posteriority. Just as icons evoke existence, the unity of existence, so they evoke otherness as side-by-sideness, as externality. Identity and difference, *pace* the logicians, cannot be fully represented without the use of iconic as well as symbolic languages.

The icon in figure 13.4 stands for a molecular complex that may or may not exist. It is a possibility, a guide to future research. For the authors of the article, it is some-

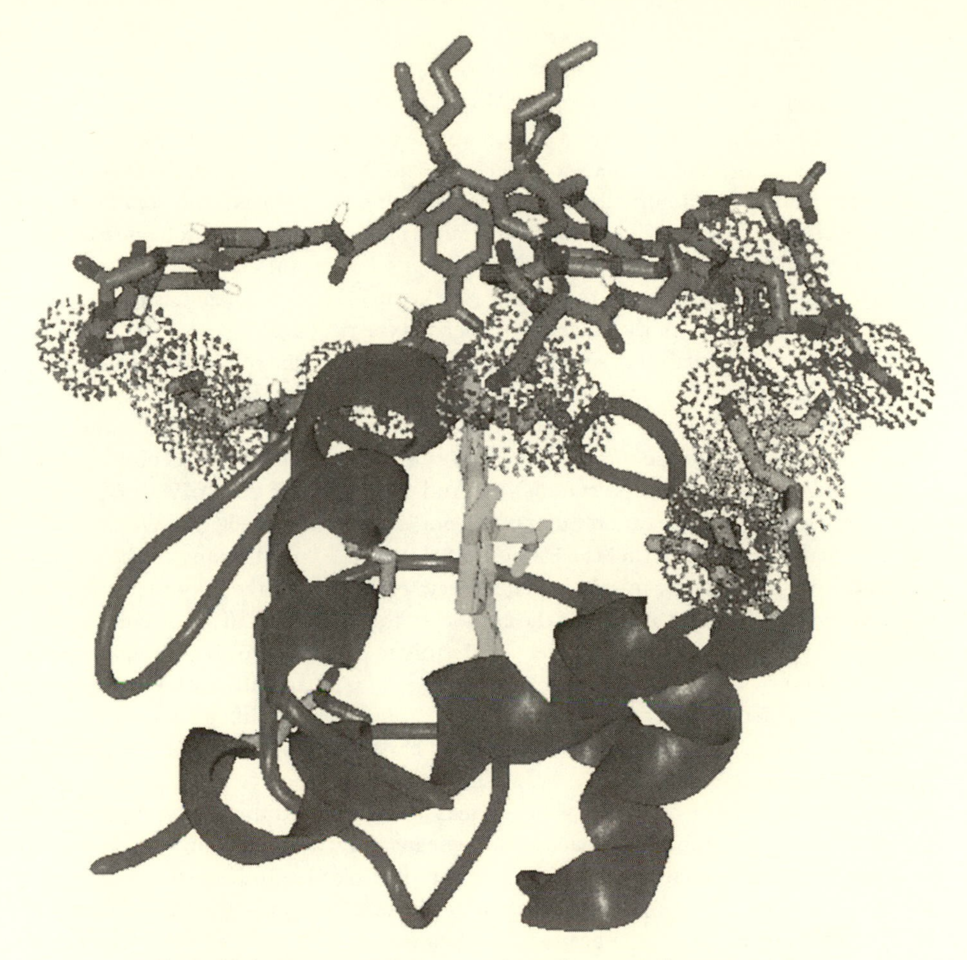

Figure 13.4 **Figure 3 of Hamuro et al. article. See explanation in text.**

thing they very much hope does exist, a wish that can perhaps be read in the bright, imaginary colors of the image. Chemical icons work their magic of asserting and displaying the unity of existence only when the symbolic discursive context and the experimental background allow them to do so.

Whatever remains still to be worked out, the authors of the article declare a positive result, and its generalization to a broader research program.

> The new type of synthetic host **3** thus mimics antibody combining sites in having several peptide loops arrayed around a central binding region. The large surface area in the molecule allows strong binding to a complementary surface on cytochrome *c* and disrupts, in a similar way to cytochrome *c* peroxidase, the approach of reducing agents to the active site of the protein. We are currently preparing libraries of antibody mimics from different peptide loop sequences and screening their binding to a range of protein targets. (p. 2682)

Conclusions

Angewandte Chemie, where Roald Hoffmann found the article closely read in this article, is no longer especially concerned with applied chemistry; indeed, it is arguably the world's leading "pure" chemistry journal. The December 15, 1997, issue of the journal in which the Hamuro article appears contains one review, two comments or highlights, several book reviews, and 38 "communications" articles, each one to three pages in length that, in principle, present novel and important chemistry.

Without question, the Hamuro, Calama, Park, and Hamilton article is a beautiful piece of work, deserving of the company it keeps in the pages of *Angewandte Chemie*; it caught Hoffmann's attention even though the subject is not one of his fields of specialization. But is this work typical of chemistry, and sufficiently so that any close reading of it might elicit generalities valid for the field? After all, it is not clear what counts as "typical" in a science whose topics range from cytochrome *c*, to reactions occurring in femtoseconds, to inorganic superconductors. And perhaps work that strives to redefine the boundaries of a science cannot fully represent what Kuhn called "normal science."

We nonetheless believe that the Hamuro et al. article exhibits many of the important features of most work in modern chemistry, especially in the way that it moves between levels of reality. On one line the authors of the article talk of a molecular structure, and on the next of a reaction; a certain linguistic item (symbol or icon) may stand for either or both. Theory and experiment, expressed in beautifully intertwined symbolic and iconic languages, relate the world of visible, tangible substances and that of the molecule. Is this sloppiness, an ambiguity that hard science must ultimately abolish? We think not.

The intuitive may be analyzed, the tacit may be articulated, but never completely and all at once: certain indeterminacies and logical gaps always remain, even as scientists achieve a consensual understanding of complex reality. Indeed, the indeterminacy and "gappiness" of knowledge may serve a useful purpose in that it allows the double vision where creative endeavor often takes place. If, as we have claimed in this chapter, chemists habitually think at both the level of macroscopic substances and their transformations in the laboratory, and the level of the statics and dynamics of microscopic molecules, the very equivocity of the field—the way it brings physics and mathematics into the service of chemistry—may be a source of its productivity. The logical gap between the two levels of description is never closed (by some kind of reduction) but, rather, is constantly and successfully negotiated by a set of theories embodied in instruments and expressed in symbolic and iconic languages.[6]

Precisely because these languages are abstract and incomplete (in the sense of being noncategorical, not capturing all there is to say and know about the entities they describe) they are *productively* ambiguous, and can be understood in reference to both the macroscopic and microscopic. This bridging function—carried out in different but complementary ways by symbolic and iconic idioms—is the special interest of this chapter. We have tried to explain how it allows chemists to articulate and to solve problems, a task that often takes the form of imagining and then trying to put together a certain kind of molecule. Thus, our account emphasizes what happens at the frontiers of knowledge rather than retrospective codification, and the investigation and creation of objects rather than the testing of theories.

Notes

A version of this chapter was delivered under the title "Comment les langages symboliques et iconiques servent de passerelle entre les deux mondes du chimiste: une étude de cas de chimie bioorganique contemporaine," by the authors on April 19, 1998, at the Maison Rhône-Alpes des Sciences de l'Homme, Université Stendhal, Grenoble, as part of a year-long seminar on the languages of science. We thank the organizers of that seminar, especially Françoise Létoublon, for the invitation and their efforts at realizing a complex and truly interdisciplinary project. We are also grateful to Andrew Hamilton for supplying the original illustrations for the article.

The nontrivial translation of this chapter into French was masterfully accomplished by Carole Allamand. We thank her.

Emily Grosholz thanks the American Council of Learned Societies and the Pennsylvania State University for their support of her research as a Visiting Fellow at Clare Hall and a Visiting Scholar in the department of history and philosophy of science at the University of Cambridge. She thanks François De Gandt (University of Lille) and Audrey Glauert (Clare Hall) for their useful suggestions on the essay, as well as Jeremy K. M. Sanders and Darren Hamilton of the department of chemistry at the University of Cambridge for their enlightening discussion of the practice of chemistry.

Roald Hoffmann is grateful to Cornell University for a grant in support of his research, scientific and otherwise, and to Bruce Ganem for a helpful explanation of chromatography.

1. The asterisks in the quoted passages are bibliographic end notes in the original Hamuro et al. article.

2. A reductionist might argue that given great computing power and perfected quantum mechanical calculations, one could start from a chemical formula and predict observations accurately. But in practice, the number of isomers for a given formula grows very rapidly with molecular complexity, so the goal is not realistic for a molecule the size of the calixarene. Moreover, complete computability may not be equivalent to understanding. Much of what a chemist means by understanding is couched in terms of fuzzy chemical concepts—the result of horizontal and quasi-circular reasoning—for which a precise equivalent in physics cannot be found (Hoffmann, 1988, 1995; Grosholz, 1994; Scerri, 1994).

3. For an excellent account of the language of chemistry, and its parallels to linguistics, see Laszlo (1995), as well as Hoffmann and Laszlo (1991) and Weininger (1998).

4. L. Kvasz (1998) helped us think about the distinction and the interactions between symbolic and iconic languages in mathematics and chemistry, but we disagree with Kvasz about the extent to which iconic languages may be codified.

5. G. G. Granger has an interesting discussion of the languages of chemistry in his book (Granger, 1983, chap. 3), where he focuses on the distinction between natural languages and formal languages. He makes the important observation that scientific language will always be partly vernacular and partly formal. Rejecting the claim that science might someday be carried out in a pure formalism, he writes. "The linguistic process of science seems to me essentially ambiguous: for if science is not at any moment of its history a completely formalized discourse, it is not to be confused with ordinary discourse either. Insofar as it is thought in action, it can only be represented as an attempt to formalize, commented on by the interpreter in a nonformal language. Total formalization never appears as anything more than at the horizon of scientific thought, and we can say that the collaboration of the two languages is a transcendental feature of science, that is, a feature dependent on the very conditions of the apprehension of an object." (p. 33). However, Granger does not go on to consider the further linguistic aspect of chemistry, that is, its iconic aspect.

6. Laszlo (1998) has cogently argued that in their practice of analysis, modern-day chemists "dematerialize" the substances they handle, so that the transactions of the contemporary laboratory mostly involve mental representations. He goes on to argue that our age of masterly

synthesis doesn't achieve the rematerialization one might desire. Although we think Laszlo verges perilously close to denying realism, his argument nevertheless is an intriguing one, and covers some of the same representational ground that we do.

References

Atkins, Peter W. 1984. *The Second Law*. New York: Scientific American.

Atkins, Peter W. 1987. *Molecules*. New York: Scientific American.

Atkins, Peter W. 1991. *Atoms, Electrons, and Change*. New York: Scientific American.

Brush, Stephen G. 1998. "Dynamics of Theory Change in Chemistry: The Benzene Problem." Forthcoming.

Carnap, Rudolf. 1937. *The Logical Syntax of Science*. London: Routledge and Kegan Paul.

Fuller, Steve. 1994. "Can Science Studies Be Spoken in a Civil Tongue? *Social Studies of Science*, 24: 143–168.

Granger, Gilles Gaston. 1983. *Formal Thought and the Sciences of Man*. Dordrecht: Reidel.

Grosholz, Emily. 1991. *Cartesian Method and the Problem of Reduction*. Oxford: Clarendon.

Grosholz, Emily. 1994. "Reduction in the Formal Sciences." *Proceedings of Conference on Physical Interpretations of Relativity Theory IV (Late Papers)*. London: British Society for the Philosophy of Science 28–51.

Grosholz, Emily, & Yakira, E. 1998. "Leibniz's Science of the Rational." Sonderheft 26. Studia Leibnitiana, Stuttgart: Franz Steiner Verlag.

Hamuro, Yoshimoto, Hamilton, Andrew D., Calama, Mercedes Crego, & Park, Hyung Soon. 1997., "A Calixarene with Four Peptide Loops: An Antibody Mimic for Recognition of Protein Surfaces." *Angewandte Chemie, International English Edition*, 36(23): 2680–2683.

Hoffman, Roald. 1988. "Nearly Circular Reasoning." *American Scientist*, 76: 182–185.

Hoffmann, Roald & Laszlo, Pierre 1991. "Representation in Chemistry." *Angewandte Chemie, International Edition English*, 30: 1–16.

Hoffmann, Roald & Torrence, Vivian 1993. *Chemistry Imagined*. Washington, DC: Smithsonian Institution Press.

Hoffmann, Roald. 1995. *The Same and Not the Same*. New York: Columbia University Press.

Husserl, Edmund. 1922. *Logische Untersuchungen*. Halle: Niemeyer.

Joesten, Melvin. D., Johnston, David O., Netterville John T., & Wood, James L., 1991. *World of Chemistry*. Philadelphia Saunders.

Kuhn, Thomas. S. 1970. *The Structure of Scientific Revolutions*. Chicago: Chicago University Press.

Kvasz, Ladislav. 1998. "History of Mathematics and the Development of Its Formal Language."

Labinger, Jay. A. 1995. "Science as Culture: A View from the Petri Dish." *Social Studies of Science*, 25(2): 285–306.

Laszlo, Pierre. 1993. *La parole des choses*. Paris: Hermann.

Laszlo, Pierre. 1997. "Chromatographie." In Michael Serres & Nayla Farouki:, eds. *Tresor. Dictionnaire des sciences*. Paris: Flammarion.

Laszlo, Pierre. 1998. "Chemical Analysis as Dematerialization." *Hyle*, 4(1): 29–38.

Leibniz, Gottfried Wilhelm. 1686. *Discourse on Metaphysics*. In Carl Immanuel Gerhardt ed., *Leibniz: Die philosophische Schriften* (7 vols.). Hildesheim: G. Olms (First published in Berlin, 1875–1990). vol. 4, pp. 427–463.

Leibniz, Gottfried Wilhelm. 1695. *A New System of Nature*. In Carl Immanuel Gerhardt ed., *Leibniz: Die philosophische Schriften* (7 vols.). Hildesheim: G. Olms (First published in Berlin, 1875–1990). vol. 4, pp. 477–487.

Leibniz, G. W. 1714. *Principles of Nature and Grace, based on Reason*. In Carl Immanuel Gerhardt ed., *Leibniz: Die philosophische Schriften* (7 vols.). Hildesheim: G. Olms (First published in Berlin, 1875–1990). vol. 6, pp. 598–606.

Pickering, Andrew., ed. 1992. *Science as Practice and Culture*. Chicago: University of Chicago Press.

Polanyi, Michael. 1960. *Knowing and Being*. Chicago: Chicago University Press.

Polanyi, Michael. 1966. *The Tacit Dimension*. New York: Doubleday.

Scerri, Eric R. 1994. "Has Chemistry Been at Least Approximately Reduced to Quantum Mechanics?" *PSA*, 1: 160–170.

Shapin, Steven. 1992. "Discipline and Bounding: The History and Sociology of Science as Seen through the Externalism/Internalism Debate." *History of Science*, 30: 333–369.

Weininger, Stephen J. 1998. "Contemplating the Finger: Visuality and the Semiotics of Chemistry." *Hyle*, 4(1): 3–25.

Zeidler, Pawel, & Sobczynska, Danuta. 1995–96. "The Idea of Realism in the New Experimentalism and the Problem of the Existence of Theoretical Entities in Chemistry." *Foundations of Science*, 4: 517–535.

Part VI

The Chemical Senses

14

Archiving Odors

THOMAS HELLMAN MORTON

nam unguentum dabo, quod meae puellae
donarunt Veneres Cupidinesque,
quod tu cum olfacies, deos rogabis
totum ut te faciant, Fabulle, nasum.

Gaius Valerius Catullus (*c*. 84–*c*. 54 BCE)

In an ode addressed to his friend Fabullus, the Roman poet Catullus speaks of a fragrance so pleasing that "when you smell it you will beg the gods to make you all nose." Would that the recipe for such a scent had been transmitted through the ages! Even today, however, it is not possible to document chemical composition with adequate fidelity to reconstruct an odor perfectly.

Catullus writes that the gods of love gave the perfume to his girlfriend. Suppose such gods existed and could list the ingredients of its aroma. The list would contain hundreds—perhaps thousands—of chemical structures and their relative proportions. Very likely, many of the structures would stand for compounds that are currently unknown, but they could be synthesized in the laboratory. Would that knowledge permit me to reproduce the odor? This chapter argues that the answer remains uncertain. The current state of chemical knowledge can neither account for why an odor smells the way it does nor what determines its intensity. The recipe for replicating a sensory experience—what is essential and what is superfluous—remains obscure.

The sense of smell challenges chemical understanding. On the one hand, given the structure of a new molecule a chemist can predict its spectroscopic properties over a wide domain of electromagnetic frequencies. A mixture ordinarily displays a spectrum that superimposes the spectra of its individual components, unless they physically interact with each other. In the chemical senses, on the other hand, perceptions of mixtures often cannot be inferred from their constituents, even though the components do not interact at the molecular level. Moreover, no one can reliably predict the organoleptic properties (taste or smell) of a new molecule from its structure. Even if that were possible, the English language does not offer a vocabulary with which to describe new smells, except by analogy to odors that are already familiar.

The poverty of descriptors means that, in talking about olfactory stimuli, many people allude to direct experiences. These allusions call on memories of characteristic odors of familiar objects, which represent "unitary percepts." A unitary percept stands for a specific whole that resists straightforward decomposition in terms of the perceptions of its individual components. A smell often represents the response to a chemical mixture of dozens of different compounds, no one of which corresponds to a dominant signal carrier. As an example, when I mention the aroma of coffee, the reader brings to mind a set of odors that unambiguously fall into a clearly recognizable category; but none of those odors corresponds to any pure organic compound yet discovered (Czerny et al., 1999). Available evidence does not support the thesis that unitary percepts form as simple superpositions of responses to individual chemical constituents (Livermore & Laing, 1998). Although the context of an odor often confuses people when they try to identify it (Engen, 1991), unitary percepts, nevertheless, *seem* to constitute fundamental entities in discussing olfaction.

Chemists traditionally classify organic molecules according to functional groups. A systematic study found that chemists do moderately well at correlating functional group with odor (Brower & Schafer, 1975). Their near-perfect performance in identifying divalent sulfur compounds must be discounted, however, because that functional group has a rich oxidation chemistry in the presence of air, which produces impurities that have a characteristic stench. The reader can confirm the effect of these impurities by contrasting commercial dimethyl sulfide with a sample that has been freshly washed with saturated aqueous mercuric chloride to remove di- and polysulfides and mercaptans! The correlation for other functional groups demonstrates an apparent but imperfect connection between the categories of molecules and categories of smells.

In speaking of a molecule, a chemist calls to mind a structure, which often stands for a collection of equilibrating tautomers or conformational isomers. I think of molecules in terms of the notion of pure compounds. The equation $A + B \rightarrow C$ refers to the reaction of pure A with pure B, regardless of the fact that C may stand for a mixture of products. The notion of a pure compound depends upon context: in one circumstance the molecule called "ethyl acetoacetate" designates a substance purified by distillation and available from a commercial vendor, which is, in fact, a mixture of interconverting tautomers. In another circumstance, the same name explicitly denotes the keto tautomer only. Olfaction recognizes unitary percepts instead of pure compounds (although some pure compounds turn out to correspond to unitary percepts). The names for unitary percepts have a degree of latitude comparable to the names of molecules.

Surely the science of chemistry will some day reconcile the sense of smell with a reductionist approach in terms of pure compounds. In the interim, this chapter seeks to raise two questions related to the one posed in the opening paragraph: what can another organism detect by means of olfaction? How might an odor be archived so as to convey it to posterity? These questions address the issue of reproducibility of sensation. Unless we take the position that all sensory experiences are unique, some criterion must be advanced to assess whether two stimuli are perceived as similar. Ideally, this criterion should apply to other air-breathing vertebrates as well as ourselves, particularly because we can envisage controlled experiments that cannot be ethically be performed on human subjects. Once the possibility of such a criterion has been established, the question of how to replicate a smell becomes meaningful. In the discussion below I shall differentiate between the nouns used to designate olfactory stimuli (odor, odorant, aroma, fragrance, etc.) and the one used to denote the perception (smell).

What Is Olfaction?

All living creatures respond to chemicals in their environment. There can be little doubt that other mammals use their noses in much the same way as humans do. Drawing parallels between our sense of smell and that of another terrestrial species (which may have sensory modalities that we lack) requires a set of guidelines for assessing features of chemoreception that correspond to human olfaction. These guidelines seek to segregate olfaction from other chemical senses to discuss how other animals respond to chemical stimuli. Future investigation may prove the impossibility of such a compartmentalization, so I shall attempt to circumscribe olfaction rather than to define it.

Any operational definition of olfaction must pertain to the behavioral responding of an organism to volatile compounds. Although olfaction may correlate with electrical responses of nerves, which can be measured independently of responding, behavior has to be considered as the primary standard for deciding whether an organism perceives an odor. This postulate permits one to differentiate olfaction from the enormous variety of other responses that living creatures exhibit to chemicals in their environment. The set of compounds that elicit behavior does not coincide perfectly with the set that produces electrical activity, even in comparatively primitive terrestrial vertebrates (Mason et al., 1980; Dorries et al., 1997). Given a discrepancy between two observables—behavior on the one hand and electrophysiology on the other—the former trumps the latter.

Consider the following propositions, which provide additional boundaries that separate olfaction from other types of chemoreception:

1. The stimuli have appreciable vapor pressures.
2. The association between stimulus and response is learned.
3. Severing the first cranial nerve on both sides eliminates the behavior. (Or, more precisely, olfaction designates those features of responding that bilateral nerve transection abolishes.)
4. The organism can discriminate different molecules and can discriminate different stimulus intensities without necessarily confusing intensity with the qualitative features of an odor.

This list embraces much of what I believe to be true of olfaction in human beings, and is put forth as a set of necessary conditions for the analogous chemical sense in other vertebrate species. Given the range of biological diversity, it probably does not represent a set of conditions sufficient to define olfaction. In discussing these four guidelines I shall make reference not only to my own perceptions of odors, but also to published reports about people who display peculiarities in their sense of smell. Some are anosmic (i.e., "odor blind"), while at least one may be said to be "odor deaf."

The Stimuli Have Appreciable Vapor Pressures

In the present context, a compound can be defined as having an appreciable vapor pressure if it has a normal boiling point or sublimation temperature below 350°C. Some compounds decompose before they boil at atmospheric pressure, but normal boiling points (or sublimation temperatures) can be extrapolated from vapor pressures at lower temperatures. This definition excludes all molecules with electric charges and also zwitterions such as amino acids. Limiting olfaction to the sensory detection of volatile compounds does not mean stimuli must necessarily be presented in the gas phase. Olfactory receptor neurons are naturally immersed in a fluid layer (mucus, in the case of mammalian noses), and do not come into direct contact with vapors. Plate 1 reproduces a scanning electron micrograph of the olfactory mucosa of an adult tiger salamander (*Ambystoma tigrinum*). The receptor neurons of this amphibian morphologically resemble those of virtually every other animal, from tiny nematodes to human beings. In tiger salamanders (unlike many mammals) the receptor neurons cluster in distinct patches, separate from other types of cells. This species has been widely studied, because the surface represented in the micrograph is readily accessible to the experimenter without the necessity of dissection.

Several cilia attach to the outer end of each receptor neuron, so that (as the photomicrograph shows) the surface of the olfactory mucosa looks like a tangle of spaghetti. The receptor surface is ordinarily covered with mucus (to a depth of 10–100 μm). (To prepare the tissue for microscopy, the mucus layer was removed by washing with a solution of EDTA.) Adult tiger salamanders are the only cold-blooded, terrestrial vertebrates that have yet been trained in the laboratory to respond behaviorally to vapors from reagent-grade chemicals. Extensive experimentation has been reported on the electrical responses of their receptor neurons and brains to chemical stimuli. The olfactory mucosa responds electrically to volatiles presented either in an airstream or in aqueous solution (Arzt et al., 1986). At this point, there is no reason to suspect that these animals could not be trained to respond behaviorally to aqueous presentations, though such experiments have yet to be reported.

The distinction between volatile and nonvolatile corresponds roughly to the separation between smell and taste in human beings, but this pair of distinctions may not be perfectly congruent. I cannot smell glycerine, sodium chloride, quinine, and dilute sulfuric acid, but they taste sweet, salty, bitter, and sour. By contrast, many volatile organic compounds produce a burning sensation when placed on the tongue without eliciting the sensation of taste. Most of the flavor of food comes from olfaction (as you can demonstrate by holding your nose while you eat), but there might be volatiles that can be detected by taste, as well as nonvolatile compounds that might elicit olfactory responses when their dust impinges on the olfactory mucosa.

If volatile molecules have to diffuse through a layer of mucus that is at least 10 μm thick before reaching the neuron, molecular diffusion implies a temporal resolution no less than 0.1 second. In other words, fluctuations of odorant concentrations will not be detectable if they occur on a time scale much less than 0.1 second. This conclusion derives from the value of the diffusion constant for small molecules in solution (about 0.001 mm^2 per second), which shows little dependence upon the identity of the solvent or of the solute. This time constant represents a intrinsic constraint that results from the physics of the stimulus and its receptor.

The vapor pressure criterion pertains especially to animals that possess sensory apparatus that humans lack. Many mammals have a morphologically distinct chemosensory organ called the vomeronasal organ (VNO) (Meredith, 1998), which is vestigial in human beings. The VNO appears to respond to nonvolatile chemical stimuli that arrive as airborne particles. The VNO may also have the capacity to detect volatile stimuli, and the remaining guidelines help differentiate its inputs from olfaction.

Association between Stimulus and Response Is Learned

Many natural languages contain words that describe the painful sensations associated with harsh chemical stimuli. Strong vinegar is pungent. So is ammonia. Most people recoil from them without having learned to do so. The pain can be sensed elsewhere than in the nose. Fumes from strong vinegar or ammonia sting the eyes. Therefore, pungency refers to sensations that have similar effects on other sensitive tissues. Pungency and other sensations that are not localized to the nose, such as the thermal "feel" of menthol, must be viewed as separate from olfaction, particularly because congenitally anosmic human subjects can detect and discriminate among odorants with such properties (Laska et al., 1997). The "common chemical sense" comprises this set of chemosensory inputs.

The English language contains few words to describe smells that do not allude to specific examples. One often hears "That smells like . . .", and even the vocabulary of perfumers is filled with similes such as "floral" or "woody." The world "putrid," however, has a more general meaning. But it denotes a hedonic judgment in addition to its descriptive function. Even a slight acquaintance with the variety of world cuisine suffices to refute any claim that humans agree about what is putrid and what is not.

There is no evidence that small children classify odors as smelling good or bad (apart from the irritation produced by pungent odors, which, I assert, does not constitute olfaction) (Engen, 1978; Mennella & Beauchamp, 1998). They certainly distinguish between familiar and unfamiliar and form hedonic categories on that basis. At some stages of development, for instance, nearly all children display neophobia, an aversion to unfamiliar stimuli. But I submit that the olfactory associations, which represent a prominent feature of the adult sense of smell, represent a set of entirely learned responses, and that there are no odors that are intrinsically pleasant or repellent (except insofar as they produce pain) (Engen, 1988). In any event, I wish to exclude from olfaction any components of smell that elicit responding that has not been learned.

Pheromones are examples of compounds that evoke behavior that has not been acquired by associative learning. For instance, during their estrus, domestic sows respond to the volatile steroid androstenone in a characteristic fashion. Androstenone constitutes a pheromone for swine, which appears to be detected via the olfactory system (Dorries et al., 1997). It is possible that humans may also respond to volatile steroids as olfactory stimuli that cannot be consciously perceived (Sobel et al., 1999). The above criteria of olfaction do not include this type of responding. In choosing to exclude pheromone detection from the definition of olfaction I am following Engen (1978, 1988, 1991), who views pheromones as airborne hormones, which represent a category of chemical messengers physiologically distinct from olfactory stimuli.

Bilateral Transection of the First Cranial Nerve Abolishes Olfaction

Olfaction represents a subset of the sense of smell, which this chapter seeks to circumscribe by means of four propositions put forth as guidelines. Of these guidelines, the first is chemical, and the second is behavioral. The third guideline addresses the relationship between anatomy and behavior. Applying it requires that the organism possess a well developed central nervous system (CNS). The CNS of terrestrial vertebrates includes a spinal cord and a brain, from which emanates a set of cranial nerves. The first cranial nerve is often called the olfactory bulb. If the connections between the nose and the olfactory bulb are completely severed, the ability to sense and to discriminate among volatile stimuli do not necessarily vanish utterly. Those capabilities that are completely lost, however, include olfaction.

Regeneration of the olfactory nerve after transection occurs in many species. Subsequent reacquisition of olfactory-mediated behavior does not take place without reconditioning. Hamsters, for example, lose their ability to discriminate odors when the olfactory nerve is cut, but they recover their ability to smell over a period of a few weeks. Even if trained to discriminate a pair of odors prior to surgery, they need to relearn the task postoperatively. Relearning the familiar pair of odors follows the same time course as learning a new pair of odors (Yee & Costanzo, 1998).

In tiger salamanders, bilateral olfactory nerve transection (abbreviated ONX) greatly reduces—but does not totally abolish—discriminative responding to airborne chemical stimuli (Mason et al., 1981, 1984, 1985). The brain of this amphibian has been well explored by anatomists (Herrick, 1948), and the animals' responding to the vapors of pure organic compounds fulfills the guidelines discussed so far (Mason & Stevens, 1981; Dorries et al., 1997): the discriminative capacity destroyed by ONX is a learned behavior. In many species, though, the first cranial nerve also transmits impulses from the VNO, so anatomy does not, by itself, necessarily segregate olfaction from other chemosensory modalities. Therefore, in addition to the other guidelines, an additional one must be included in order to account for what is known from studies of human subjects.

Olfaction Discriminates Different Molecules and Different Stimulus Intensities

A widely accepted dictum holds that humans enjoy three chemical senses—the common chemical sense (which records irritation, pungency, and other "feels"), taste, and smell. A recent study of human subjects who lack the last (anosmics) has shown that they can still distinguish among volatile chemical stimuli by means of the first (Laska, et al., 1997). For several odorants the anosmic subjects' verbal responses mirror those of people with a normal sense of smell (normosmics). However, the anosmics do not distinguish ethanol and propanol easily. I find that those alcohols smell quite different, but (to extrapolate from the published data) an experimeter could prepare concentrated solutions of the two such that an anosmic person could not differentiate them, while a normosmic person could.

Although this insight may not appear very profound, its implications warrant some reflection. Some molecules convey similar odors—citral (from lemongrass) and limonene (from lemon peel) smell the same to me—but I do not know of many examples where two compounds with different odors become confused with one another when their relative concentrations are adjusted. That sort of problem does often occur, though, in electronic instrumentation in chemical laboratories: a signal from an impurity can be misinterpreted as coming from the analyte. The comparative rarity of interferences of this kind in olfaction argues against any simple analogy between the nose and a spectrophotometer.

The ability to discriminate intensity differences raises a very complicated issue, because we do not know what physical parameter determines the strength of an olfactory stimulus. My subjective impression of an odor is that it fades with time, even when the concentration of stimulus does not change. This phenomenon (known as adaptation) has been studied for over a century (Stuiver, 1958). Olfaction must register something different from the ambient concentration of an odoriferous chemical; otherwise, the sensation would remain constant. Single-celled organisms (such as bacteria) swim towards attractive chemical stimuli by responding to the change in concentration with time (Koshland, 1977; Koshland et al., 1982). In other words, they are rate-sensitive detectors. Do humans respond in a similar fashion?

Some years ago we modeled the properties that olfaction should display if intensity is coded for humans in the same way as for a simple model of bacterial chemotaxis (Nachbar & Morton, 1981). The implications of this model include the expectation that more rapid transit of an odorant pulse through the nose should result in a greater perceived intensity, as has been demonstrated experimentally by examining the consequences of increasing airflow (Rehn, 1978) or decreasing nasal resistance (Hornung et al., 1997). Rate-sensitive detection has recently been documented in human subjects during eating. (Baek et al., 1999). For a fixed set of airflow characteristics, our model also predicts a dependence of perceived intensity upon concentration that fits neither of the commonly used two-parameter mathematical scaling functions—Stevens's Law and the Beidler Equation—perfectly, but gives approximately the same goodness of fit to either expression. Despite the substantial differences between those two mathematical functions, our expectation has also been recently confirmed (Chastrette et al., 1998), although a three-parameter expression based on the Hill Equation appears to fit the experimental relationship between concentration and perceived intensity better than either of the two-parameter fits. In any event, perceived intensity is not a simple function of concentration. Studies of insect chemoreception have been interpreted in terms of two different types of detectors—flux detectors and concentration detectors (Kaissling, 1998)—but our model accounts for the same results without demanding that sort of dichotomy.

If we cannot quantify stimulus intensity in terms of some well-understood physical parameter, still less can we discuss odor quality. From the organic chemist's standpoint, the sense of smell remains a mystery. Undoubtedly, the flavor and fragrance industry has amassed a huge body of pertinent data, which remains proprietary. But information that stays unavailable does not equate to scientific knowledge.

Patient H. M., an Odor-Deaf Subject

The ability to discriminate different molecules constitutes a criterion for olfaction. Because, as mentioned, anosmic persons can tell some pairs of odors apart based on nonolfactory cues, an experimenter must choose with care the compounds for study. β-Phenethyl alcohol has an odor that many people find reminiscent of roses, and vapors from dilute solutions are widely accepted as an olfactory stimulus that does not interact with other chemosensory modalities in humans (Betcher & Doty, 1998). Consider a human subject who can detect β-phenethyl alcohol with the same sensitivity as normosmics and can also detect *n*-butanol (another alcohol often used for testing olfactory sensitivity (Hummel et al., 1997), which has an odor very different from that of β-phenethyl alcohol) with normal acuity. Suppose this subject cannot distinguish the two odors. How can an experimenter assess whether the subject exhibits the sense called olfaction?

This question arises in the case of H. M., a patient who underwent experimental brain surgery in the 1950s to alleviate a severe epileptic disorder. Many experimenters have published postsurgical studies of case H. M., because he has lost the ability for declarative learning while retaining the capacity for procedural learning. Tests of H. M.'s sense of smell have documented that he has normal acuity but cannot tell one odor from another (Eichenbaum, et al., 1983). He can describe what he smells in some detail, but the descriptions do not correlate with the stimulus. Descriptions of the same odor vary widely from one presentation to another, and show no obvious trend when compared to his descriptions of different odors. When given verbal cues, H. M. can identify pieces of fruit with his hands without looking at them. When given olfactory cues, he selects haphazardly.

H. M. has demonstrated that he can consciously differentiate chemical vapors, but not by means of olfaction. In a randomized series of trials, I presented H. M. with the odor of a dilute solution of β-phenethyl alcohol or of *n*-butanol. I then asked him to choose which of two flasks—one a "target" and the other a "distractor"—contained the same odor. One of those flasks contained β-phenethyl alcohol solution and the other *n*-butanol solution. In half of the trials, the target odor solution had the same concentration as the initial presentation, and the distractor was more concentrated. In the other half, the distractor solutions had the same concentrations as the initial presentations, and the target odor solutions were more concentrated. Thus, we had four possibilities for the presented stimulus and the target: dilute β-phenethyl alcohol with a dilute target; dilute β-phenethyl alcohol with a more concentrated target; dilute *n*-butanol with a dilute target; and dilute *n*-butanol with a more concentrated target. Unscrambling the results of 40 trials I found that H. M. had scored a perfect 10/10 in matching dilute β-phenethyl alcohol with a dilute target, but that he had performed at random (half right answers and half wrong) for the other three matches. I rationalize his performance as based on the ability to match perceived intensities. Suppose dilute β-phenethyl alcohol represents a "weak" stimulus, but the other three—more concentrated β-phenethyl alcohol or *n*-butanol at either concentration—are perceived as equally "strong." Then H. M.'s successes would correspond to his ability to match "weak" with "weak," and his failures represent an inability to distinguish among the "strong."

H. M. clearly understands how to discriminate chemical stimuli. His impairment limits only his recognition of odors, but no other aspect of his sense of smell. He is "odor deaf," by analogy to stroke victims who can read and write and retain an intact sense of hearing but cannot recognize words aurally (and are said to be "word deaf") (Takahashi et al., 1992). Based on the evidence discussed so far, H. M.'s behavior does not satisfy the fourth of the guidelines for olfaction.

Could one devise a strategy for differentiating *n*-butanol from β-phenethyl alcohol using H. M. as a detector, without confounding odor quality with odor intensity? An affirmative answer comes from H. M.'s adaptation to odors. When H. M. sniffed *n*-butanol for a period of time, his experimentally measured sensitivity to both odors decreased, but adaptation attenuated his sensitivity to β-phenethyl alcohol much less than his sensitivity to *n*-butanol. Similarly, sniffing β-phenethyl alcohol for a period of time attenuated his sensitivity to that odorant (self adaptation) but did not affect his sensitivity to *n*-butanol significantly (no crossadaptation). The selectivity of his adaptation demonstrates that H. M. indeed uses olfaction to detect odors. Moreover, his adaptation (like that of normosmics) occurs in the CNS, as detailed in the following discussion.

The experiments with H. M. employed a battery of tests. One set of experiments measured his sensitivity by means of a technique derived from signal detection theory (Corbit & Engen, 1971), in which I asked H. M. to sniff 20 presentations of dilute odorant solution randomly interspersed with 20 presentations of odorless blank. The odor was so faint as to make it hard to tell it apart from blank. Figure 1 compares some of the data for H. M. with a male normosmic (P. D.) matched for age and race. After each presentation I asked H. M. whether he could smell an odor. His pattern of responding was the same as that of normosmics: sometimes he gave affirmative responses to blanks (false alarms, symbolized by open symbols in fig. 14.1), but he did not always respond affirmatively to the dilute sample (correct affirmatives are symbolized by solid symbols in fig. 14.1).

The circles in figure 14.1 symbolize responses to *n*-butanol without adaptation. Using both nostrils H. M. gave 28 correct answers out of 40 presentations. Using just one nostril he gave 26 correct answers. The important datum is not the total number of correct answers, but rather the distribution of false alarms (open symbols) and correct affirmatives (solid symbols), from which can be extracted a measure of his signal-to-noise ratio (d') for olfactory detection. H. M.'s one-nostril d' was virtually the same as his two-nostril (bilateral) d', in agreement with a recent report that normosmics' bilateral sensitivity is the same as the sensitivity of their more sensitive nostril (Betcher & Doty, 1998). H. M.'s sensitivity to *n*-butanol was lower than that of P. D., but the difference between them is well within the normal range of human variation.

When P. D. sniffed a more concentrated solution of *n*-butanol in the opposite nostril, his one-nostril sensitivity decreased (symbolized by squares in fig. 14.1). But when he subsequently sniffed concentrated β-phenethyl alcohol in the opposite nostril, his senstivity was slightly greater than in the unadapted condition (symbolized by triangles).

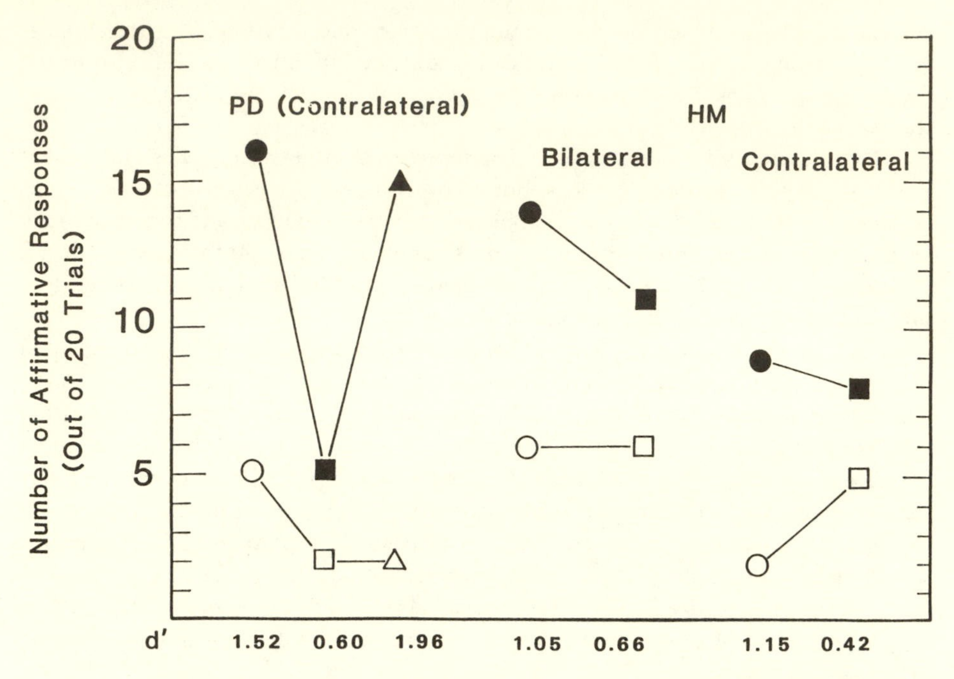

Figure 14.1 Comparison of an odor-deaf patient (H. M.) with an age-matched normosmic (P. D.) in responding to randomized presentations of weak samples of *n*-butanol (*solid symbols*) interspersed with blank (*open symbols*). Circles denote unadapted condition; squares denote adaptation to *n*-butanol; triangles represent adaptation to ∅-phenethyl alcohol (P. D. only).

Thus, P. D. exhibited contralateral self adaptation to *n*-butanol without any crossadaptation to β-phenethyl alcohol. It was not possible to test H. M. as thoroughly as P. D., but, as figure 14.1 summarizes, his contralateral self adaptation to *n*-butanol did not differ signficantly from his bilateral self adaptation, and both were essentially the same as P. D.'s contralateral self-adaptation.

In human beings, a complete septum separates the two nostrils, which isolates them from one another aerodynamically. Therefore, contralateral adaption represents an effect that does not result from saturation of binding sites in olfactory receptor neurons, but must instead correspond to a phenomenon of the brain (Stuiver, 1958). H. M. and P. D. exhibit self adaptation that occurs in the CNS rather than in the peripheral nervous system.

This result illustrates the difference between a learned response and a conscious one. H. M. learned to respond to *n*-butanol—had I not asked for his judgment after each presentation, he would not have evinced a consistent behavioral pattern. Had I not told him whether he was right or wrong after each time he responded, he would have answered essentially at random, as I discovered in a separate set of trials. The same is true of normosmic subjects, because the odorant is so highly diluted that reinforcement greatly increases their *d*′ values. However, H. M. lacks the ability to recognize odors. When asked to make conscious choices, he confuses odor quality with intensity. He retains the physiological capacity, because his pattern of adaptation cannot be distinguished from that of a normosmic. His peripheral and central nervous systems function well enough that the following tactics could ascertain from his behavior whether an unknown odorant is β-phenethyl alcohol or *n*-butanol: (1) measure d' for the unknown; (2) adapt to *n*-butanol and remeasure d' for the unknown; (3) adapt to β-phenethyl alcohol and remeasure d' for the unknown. This procedure, albeit cumbersome, will identify the unknown odorant as the one for which adaptation caused a more profound decrement of d'. Hence, H. M. does indeed meet the fourth of the guidelines for displaying olfaction.

Because H. M. cannot consciously identify odors or remember them, an experimenter would have difficulty asking him to assess the degree of similarity of two smells. But it is possible to pose that problem to experimental animals. One can train an air-breathing vertebrate to respond to a given odorant and then ask how frequently it gives the same response to a different one. This kind of generalization has been studied (Mason, et al., 1987b, for example). Investigations of the olfactory abilities of nonhuman species also permit a comparison of behavior with the electrical responses of the nervous system. The outcome of such investigations illustrates the limits of current understanding.

Olfaction in Nonmammalian Vertebrates: Behavior versus Electrophysiology

H. M. has convinced me that he uses olfaction to detect odors, without the necessity of probing the consequences of olfactory nerve transection, an obviously impossible experiment to perform on a human being. To some extent, my conviction arises from introspection: H. M. behaves as I would imagine myself doing, were I unable to recognize or remember odors. Similarly, the belief that many mammals use olfaction derives in part from outward behaviors, such as sniffing, which mimic our own. Anthropocentric interpretation of these behaviors might deceive, however. A rabbit's quivering nostrils could serve to bring nonvolatile stimuli to the VNO rather than volatiles to olfactory receptors.

Interpretation becomes even more complicated for nonmammalian vertebrates, such as amphibians and birds, which do not exhibit stimulus-seeking behaviors that resemble those of human beings. For such species olfactory nerve transection (ONX) provides a pivotal piece of evidence in determining whether an animal uses olfaction to guide its actions. Classical conditioning associates a stimulus with a reinforcement. ONX eliminates the responding, presumably by blocking perception of the stimulus. For instance, tiger salamanders avoid bright light. If presented with an airstream containing an odorant, a salamander will not usually evince any response. If a flash of bright light follows the delivery of odorant, the animal will learn to avoid the odorant. Over a period of about one week a salamander will learn to avoid eight to nine conditioned presentations out of 10 (Mason et al., 1980, 1987b). Animals do not avoid odors that are not followed by reinforcement, and their behavior extinguishes if reinforcement is discontinued. Sham surgery does not affect their responding, nor does ONX abolish their responding to bright light.

A more rapid conditioning paradigm measures galvanic skin response instead of avoidance (Dorries et al., 1997). Tiger salamanders have been conditioned to respond to *n*-butanol by both methods. Tiger salamanders cannot be conditioned to respond to camphor by either method, despite the fact that their peripheral neurons and olfactory bulbs exhibit pronounced electrical activity in response to this odorant. Although camphor might conceivably exert some peculiar anesthetic effect, the disjunction between behavior and electrophysiology implies that experimenters have not yet identified the nerve impulses that code for olfactory information.

ONX produces general anosmia. While no surgical procedure is known that produces selective hyposmia (specific odor blindness), chemical treatments can do so. If presentations of cyclohexanone are followed by reinforcement but presentations of *n*-butanol are not, the salamanders avoid the former but do not respond to the latter. After having been conditioned in this fashion a salamander will avoid cyclopentanone, demonstrating that the animal perceives cyclopentanone as being more like cyclohexanone than like *n*-butanol. Animals can be trained to avoid two different odorants, for examples, cyclohexanone and dimethyl disulfide. Lavage of the receptor mucosa (see figure 14.2.) with a solution of cyclohexanone temporarily diminishes the frequency with which the animals respond to that odorant, but does not affect their responding to dimethyl disulfide. Lavage of the receptor mucosa with dimethyl disulfide diminishes the frequency with which the animals respond to dimethyl disulfide, but does not affect their responding to cyclohexanone (Mason & Morton, 1982). These partial impairments represent selective hyposmias, and their duration depends on the concentration of the lavage solution. By contrast, lavage with *n*-butanol has no significant effect on responding to either cyclohexanone or to dimethyl disulfide. Finally, ONX drastically reduces responding to both odorants.

Lavage produces a temporary impairment in animals analogous to adaptation in humans. One can run experiments on animals, though, which cannot be done on human subjects. Lavage with dilute (1 mM) ethyl acetoacetate does not affect tiger salamanders' responding, nor does lavage with more concentrated (50 mM) sodium cyanoborohydride. Performed sequentially, however, those two lavages produce the same selective hyposmia as does lavage with cyclohexanone, and they also impair the animals' generalization to cyclopentanone. This result is congruent to the chemical specificity of that sequence of reagents (Mason et al., 1984a, 1985), even though salamanders do not appear to be able to smell ethyl acetoacetate. However, electrophysiology shows no differential alteration in the electrical responses of the receptor neurons or of single neurons in the olfactory bulb as a consequence of lavages. The electrical responses of the receptor mucosae of immobilized animals to airborne cyclohexanone (cf. Arzt et al., 1986) were found to be significantly reduced by lavage for 2 days after treatment, as were the electrical responses to dimethyl disulfide, but after 3 days both had recovered completely (Schaefer & Winegar, 1988). Because the selective behavioral impairment lasted nearly a week, no correlation could be established between behavior and electrical activity. Similarly, a total of 52 individual neurons in the olfactory bulb were examined over a period of 6 days. Before treatment, 60% of the cells responded to dimethyl disulfide only, none to cyclohexanone only, 20% to both odorants, and 20% to neither odorant. Six days after treatment, only 10–20% of the cells responded to dimethyl disulfide only, 20–30% responded to both odorants, and a few percent responded to cyclohexanone only.

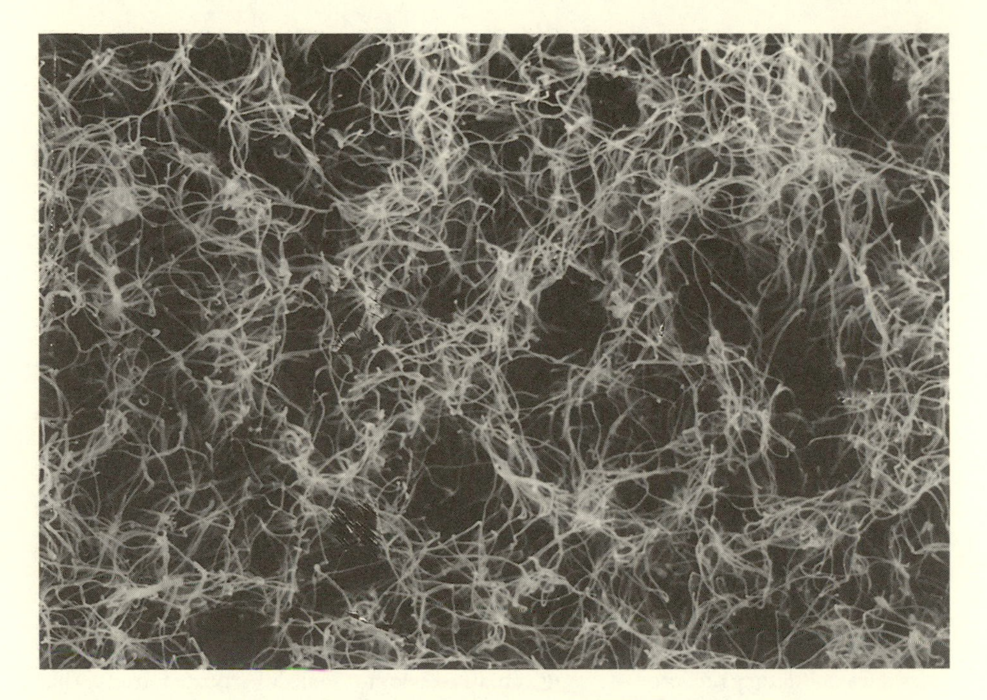

Figure 14.2 Scanning electron micrograph of the surface of olfactory receptor mucosa of a tiger salamander. Viewed from this aspect, the end of each individual neuron has an area of approximately 1 μ, but all that can be seen are the cilia (0.2 μm in diameter and approximately 10 μm long), of which each neuron has a few projecting from its dendritic knob. The dashed line at bottom indicates a 10-μm scale.

Monitoring the electrical activity of the nervous system does not appear to give results that accord with the behavior of conditioned animals. On the one hand, electrical responses are observed for volatile chemical stimuli that tiger salamanders apparently do not smell. On the other hand, no selective attenuation of their electrical responses takes place under conditions that reproducibly impair their behavioral responding to one odor but not to another.

Olfactory Coding: Still a Mystery

If I can ascertain what another organism detects via olfaction, then I can perform experiments upon it, which cannot be performed on human subjects. The objective of such experiments—to find out how odor is coded—has yet to be achieved. Suppose the olfactory code were unraveled. Reproducing an odor would become a matter of replicating the pattern of neural responses without having to duplicate the chemical stimulus (much as cinematography appears to reproduce color without necessarily matching the complete spectroscopic profile of the original scene) (Robertson, 1992).

Can a method for archiving an odor—so that it can be passed on to posterity with fidelity—be developed without such knowledge? Arguably, the answer is yes, though discovering the molecular mechanism of coding would simplify that task considerably.

If the odors of specific objects translate into unitary percepts, which constitute the basic entities in linguistic descriptions of olfaction, then the question follows as to whether these unitary percepts take shape at the level of the receptor neurons or in the olfactory bulb or elsewhere in the brain. That question remains unanswered, as of this writing. Because the sense of smell does not correlate perfectly with externally monitored patterns of electrical response from the receptor neurons or the olfactory bulb, the nature of olfactory coding remains unknown. Outside the laboratory unitary percepts rarely equate to pure compounds. Two vocabularies coexist, one of smells (which varies from individual to individual, and which refers to other inputs besides olfaction) and the other of chemical structures.

Theories abound, which relate those two vocabularies, but no one of them has emerged as predominant. Many of the theories suppose the existence of specific receptor sites on the surface of the receptor neurons. One hypothesis posits a set of odors of specific objects (e.g., camphor, sperm, urine, fish) that correspond to pure compounds and represent fundamental submodalities (Beets, 1982). Another (based on molecular biology) proposes dozens—perhaps hundreds—of different types of cell surface receptor proteins, each of which is tuned to a specific odorant compound or class of compounds (Buck, 1996; Zhao et al., 1998).

Although the molecular mechanisms of olfaction remain unknown at present, two sets of assumptions seem likely to be true if olfactory receptor proteins do indeed exist. First, olfactory receptor sites ought to bind odorant molecules quite loosely; that is, with dissociation constants $K_d \geq 10^{-7}$M with reference to the odorant concentration in the surrounding mucus. If this be not true, then receptors should become easily saturated. Furthermore the on and off rates for binding must be fast, leading to what has been termed "fast and loose" binding (Mason & Morton, 1984). Secondly, high sensitivity does not require tight binding. As originally noted for rate-sensitive detectors (Paton, 1961), a given type of receptor site can exhibit useful responses for a wide domain of concentrations below its dissociation constant.

These speculations present a meager collection of precepts about what the nose should do. No less meager is the list of what the nose cannot do. Olfaction cannot rely entirely on emission or absorption of electromagnetic radiation by isolated odorant molecules, because optical isomers can have different odors (Friedman & Miller, 1971; Laska & Teubner, 1999). Nor can the characteristic stench of organosulfur compounds depend on the reaction depicted in in equation 1, because lavage of the olfactory mucosa of experimental animals with iodoacetamide or with methymercury hydroxide followed by iodoacetamide (which would irreversibly modify the hypothesized receptor site) does not affect their responding to dimethyl disulfide (Mason et al., 1987a).

$$\text{\textasciitilde S–S\textasciitilde} \text{ or } \text{\textasciitilde SH} + \text{RSH or RSSR} \xrightarrow{\text{binding event}} \text{\textasciitilde SSR} \quad (1)$$

hypothetical receptor sites *odorant molecules* *receptor-bound odorant*

The paucity of acknowledged impossibilities hampers understanding of how chemical structure translates into a sequence of nerve impulses. Not until experiment falsifies many more plausible suggestions can a coherent theory of olfactory coding take shape.

Reconstructing Odors

In perfumery a mixture that smells different from its components is an example of an "accord" (Calkin & Jellinek, 1994). In this respect, olfaction exhibits "synthetic processing," just as color vision cannot distinguish a mixture of red and green light from yellow light (Robertson, 1992). In another respect, olfaction exhibits "analytic processing": human subjects can name individual unitary percepts in a mixture of object odors, even though each object odor is a complex mixture of pure compounds (Livermore & Laing, 1998). Not surprisingly, chemists have chosen to develop instrumentation whose output lends itself exclusively to analytic processing.

A spectrometer and an organ of perception both function in the same general way: a detector converts an input into an electrical signal, which is then processed. Despite this similarity between contemporary scientific apparatus and the nose, one cannot easily draw further analogies between them. Mass spectrometry (MS) illustrates a technique whose sensitivity rivals that of human olfaction. When combined with gas chromatography (GC), GC/MS is currently the method of choice for analyzing unknown mixtures of volatile compounds. Curiously, no set of first principles exists for predicting the fragmentation pattern that a new compound will give in its MS. In that sense, GC/MS resembles the sense of smell: the instrument conveys a large amount of information, which chemical understanding cannot yet fully comprehend. However, GC retention times and MS fragmentation patterns can be archived on a computer and reproduced at any time in the future. This cannot yet be accomplished for odors.

For archiving odors no instrumental or electronic method presently matches the ability of a trained perfumer. An accord depends on the proportions of its components and is highly sensitive to the presence of minor components. GC/MS analyzes common flavors and fragrances, each of which contains hundreds of chemically distinct constituents, but does not necessarily record sufficient information for an odor to be reconstructed from pure compounds. A component that appears by GC/MS to represent a negligible fraction of a mixture might, nevertheless, play an important role in its odor.

The perfumer's art passes from generation to generation via transmission from master to apprentice. But how is one to know whether a scent (particularly one from a natural source) has changed over the span of years? One point of view holds that archiving an odor simply requires more information, such as GC/MS might obtain, with ever greater resolution in separating the constituents and ever greater sensitivity in characterizing the trace components. This approach is consistent with the notion that each of the 10^7–10^8 receptor neurons transmits its own pattern of response to a chemical stimulus, a "holistic code" (Lettvin & Gesteland, 1965), in which each receptor cell responds to most (if not all) chemicals. That means that the nose sends a huge volume of information to the brain, a large fraction of which may prove redundant. Because there is no way of recognizing a priori the redundant part, even the most thorough GC/MS analysis might not contain any more information than the nose actually transmits to the olfactory bulb. In other words, from this point of view solving the problem of archiving odors resides in pursuing chemical analysis to an increasing degree of refinement. The two methods of profiling an odor—holistic coding postulated for the nervous system versus separation and analysis of the constituents by a chemist—acquire very different kinds of information in very different ways, but approach the same limit.

An alternative viewpoint holds that the key to archiving an odor involves simulating the way the olfactory bulb handles inputs, rather than the brute force method outlined above. Developing a biomimetic "artificial nose" (Dickinson, et al., 1998), therefore, hinges upon finding a suitable processing algorithm. The artificial detectors themselves can have "imperfect selectivity," just as do the receptor neurons in the holistic coding model (Hirschfeld, 1986). After connecting an array of such detectors in parallel to an appropriate computer, the system "learns" to recognize mixtures. The Hirschfeld model does not require any knowledge of chemical structure or even any understanding of how chemicals interact with the detectors. Current investigations of "artificial noses" based on this model explore the time course of the array's response as an additional source of information, but it remains to be seen if an array with ≤10,000 detectors (the number of glomeruli in the olfactory bulb, the junction boxes that receive the inputs from receptor neurons) can discriminate odor intensity as well as odor quality without confusing the two. If so, the outputs of a collection of Hirschfeld detectors could be stored in a computer. Although such a database would not tell how to reconstruct an odor, it would provide a comparison standard for evaluating the similarity of one smell to another.

How Do Odors Travel?

What does the cartoon in figure 14.3 depict? The cartoon in figure 14.4?

I drew the cartoon in figure 14.4 on my computer and asked my teenaged daughter what it looked like. She immediately replied, "stinky cheese." Then I asked her what figure 14.3 represented. "a shining cheese." When I showed figure 14.3 to a group of students, none of them thought of an odor emanating from the central wedge, and most of them thought of something luminous—one student called it a picture of "golden cake." Then I asked them what figure 14.4 depicted, and none thought of luminosity. About half gave answers that related to smell: for example, "aroma wafting from a piece of pie."

Straight lines radiate from the center of figure 14.3 uniformly in all directions—isotropically, as a physicist might say. This universally symbolizes rays of light. By contrast, many viewers interpret the wavy lines in figure 14.4 as symbolizing odor. The picture of an odor shows it drifting in a variegated, nonuniform pattern—aeolotropically, to borrow a term from nineteenth-century physics. The difference between figures 14.3 and 14.4 represents the distinction between light, which radiates from its source, and vapors, which are transported via convection. If molecular motions alone conveyed odor it would diffuse isotropically, but so slowly that would take an hour to travel a few centimeters. Instead, odors are carried as plumes in currents of air. How could it happen that, despite the invisibility of these plumes, they are recognizable as symbolic of odors? Could it be that olfaction senses the heterogeneity of the stimulus, as it is ordinarily presented? And that, as a consequence, we comprehend odors as traveling by convection—like smoke or waves of heat—rather than by passive diffusion?

Animals that are celebrated for their keen sense of smell—dogs and rats, for example—have whiskers that exhibit astonishing sensitivity to slight displacements.

Figure 14.3 What does this depict?

To these species the structure of air currents is not invisible. A considerable portion of their brains has evolved to process information from their whiskers (or vibrissae), and one may plausibly suggest that their olfactory sensitivity derives, at least in part, from the ability to monitor the structure of air and to situate a scent within the currents that eddy about their snouts (Cain, et al., 1985). If, even in the absence of vibrissae, human olfaction can, nevertheless, sense the heterogeneity of odors, this adds another complexity to our experience of smell. As argued above, diffusion through mucus limits the temporal resolution of olfaction to about 0.1 second. If humans have the capacity

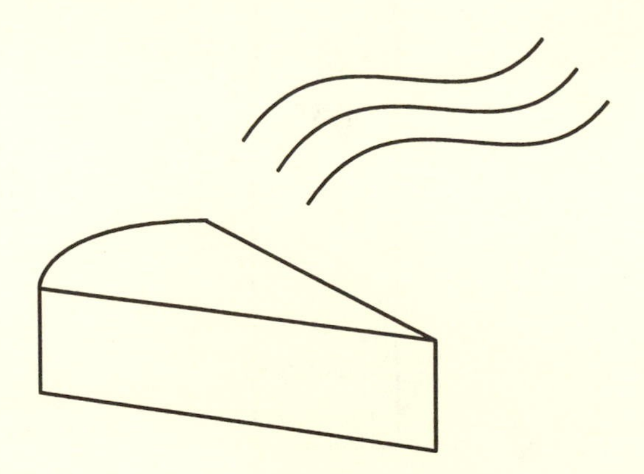

Figure 14.4 What does this depict?

to sense whether a scent is well mixed with air or arrives in a plume, that ability would have to depend on fluctuations that take place on such a timescale.

Catullus's Nose

The opening paragraphs of this chapter imagined that the gods of love might itemize the ingredients of the aroma that Catullus praised. Would it prove adequate for them to give a head space analysis (that is, the chemical composition of the vapors in equilibrium with the odor source, as drawn schematically in figure 14.5 for a liquid in a bottle)? Could Catullus's nose apprehend any difference between a sample of air that replicated the head space and the fragrance that so pleased him?

If human olfaction can discern the filamentous character of scent plumes in air, then a headspace analysis, no matter how complete, might not suffice to reconstruct the fragrance. The distribution of molecules would play a role, as would the rate with which they are replenished after they have been depleted by sniffing. If, on the other hand, olfaction (independent of other sensory inputs) cannot differentiate a heterogeneous stimulus from one that has been well mixed with air, then a complete chemical analysis could serve to archive odors.

Let me relate a piece of anecdotal evidence that suggests the dependence of olfaction upon the way that odors arrive at the nose. One of my students had been a subject in numerous signal detection experiments using weak solutions of β-phenethyl alcohol in water as a stimulus, in which she was inhaling its vapors in a continuous

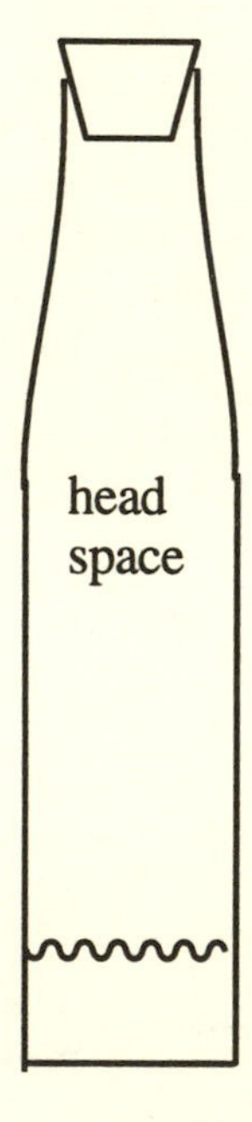

Figure 14.5 "Head space" refers to the vapors in equilibrium with the other contents of the vessel.

sniff of about 1–2 seconds duration. She consistently averred that the solution did not possess any floral odor whatsoever. Then one afternoon she happened to walk by an open bottle containing a somewhat more concentrated solution, and unexpectedly caught a whiff of it. Suddenly she perceived the smell of roses. Because she was about a meter away from the open bottle it is hard to believe that more molecules entered her nose during that breath than she had been inhaling during any of her previous exposures to β-phenethyl alcohol. Was her revelation due to a difference in how her nose transmitted information to her brain? Could odor quality depend on whether you take a sniff or catch a whiff?

The sense of smell often elicits a sense of epiphany that other senses do only rarely (Laird, 1935; Chu & Downes, 2000). Does this depend on whether you smell a scent plume or an odorant well mixed with air? Experiment can address that issue. Were we to probe the olfactory abilities of rats or dogs temporarily deprived of their vibrissae (difficult experiments, which have not, to the author's knowledge, been reported), considerable insight would be gained. An experimenter can design detection tasks under more rigorously controlled conditions for animals than for humans, so as to rule out alternative sensory inputs. However, people give a much more detailed report of their perceptions. Continued interplay between studies of humans and of animals stands at the center of piecing together a molecular picture of chemoreception.

Chemical understanding interprets phenomena in terms of the positions and movements of atoms and the forces acting upon them. The Oxford English Dictionary defines phenomenon as "that of which the senses or mind directly takes note; an immediate object of perception." This chapter has argued that behavioral performance permits an assessment of what another creature can detect by means of olfaction alone, thus extending the notion of "senses or mind" beyond the realm of the human. The belief that all terrestrial vertebrates share similar olfactory mechanisms implies that these will ultimately prove susceptible to chemical understanding. The confirmation of such understanding will be the ability to archive odors.

Acknowledgment

The author is grateful to the National Science Foundation for continued funding for his research. This work was supported by NSF grant CHE-9983610.

References

Arzt, A. H., Silver, W. L., Mason, J. R. & Clark, L. 1986. "Olfactory Responses of Aquatic and Terrestrial Tiger Salamanders to Airborne and Waterborne Stimuli." *Journal of Comparative Physiology A*, 158: 479–487.

Baek, I., Linforth, R. S. T., Blake, A., & Taylor, A. J. 1999. "Sensory Perception Is Related to the Rate of Change of Volatile Concentration in-Nose During Eating of Model Gels." *Chemical Senses*, 24: 155–160.

Beets, M. G. J. 1982. "Odor and Stimulant Structure." In E. T. Theimer, ed. *Fragrance Chemistry, the Science of the Sense of Smell* (pp. 27–122). New York: Academic Press.

Betcher, S. A., & Doty, R. L. 1998. "Bilateral Detection Thresholds in Dextrals and Sinistrals Reflect the More Sensitive Side of the Nose, Which Is Not Lateralized." *Chemical Senses*, 23: 453–457, 761.

Brower, K., & Schafer, R. 1975. "The Recognition of Chemical Types of Odor: The Effect of Steric Hindrance at the Functional Group." *Journal of Chemical Education*, 52: 538–540.

Buck, L. B. 1996. "Information Coding in the Mammalian Olfactory System." *Proceedings of the Cold Spring Harbor Symposia on Quantitative Biology*, 56: 147–155.

Cain, W. S., Mason, J. R., & Morton, T. H. 1985. "Use of Animals for Detection of Land Mines and Other Explosives: A Review and Critique of Prospects." Final report for U.S. Army Contract DAAK70-84-K-008.

Calkin, R. R., & Jellinek, J. S. 1994. *Perfumery, Practice and Principles*. New York: Wiley-Interscience.

Chastrette, M., Thomas-Danguin, T., & Rallet, E. 1998. "Modelling the Human Olfactory Stimulus-Response Function." *Chemical Senses*, 23: 181–196.

Chu, S., & Downes, J. J. 2000. "Odour-Evoked Autobiographical Memories: Psychological Investigations of Proustian Phenomena." *Chemical Senses*, 25: 111–116.

Corbit, T., & Engen, T. 1971. "Facilitation of Olfactory Detection." *Perception and Psychophysics*, 10: 433–436.

Czerny, M., Mayer, F., & Grosch. W. 1999. "Sensory Study on the Character Impact Odorants of Roasted Arabica Coffee." *Journal of Agricultural and Food Chemistry*, 47: 695–699.

Dickinson, T. A., White, J., Kauer, J. S., & Walt, D. R. 1998. "'Artificial Nose' Technology." *Trends in Biotechnology*, 16: 250–258.

Dorries, K. M., Adkins-Regan, E., & Halpern, B. P. 1997. "Sensitivity and Behavioral Responses to the Pheromone Androstenone Are Not Mediated by the Vomeronasal Organ in Domestic Pigs." *Brain, Behavior and Evolution*, 49: 53–62.

Dorries, K. M., White, J., & Kauer, J. S. 1997. "Rapid Classical Conditioning of Odor Response in a Physiological Model for Olfactory Research, the Tiger Salamander." *Chemical Senses*, 22: 277–286.

Eichenbaum, H., Morton, T. H., Potter, H., & Corkin, S. 1983. "Selective Olfactory Deficits in Case H.M." *Brain*, 106: 439–442.

Engen, T. 1978. "The Origin of Preferences in Taste and Smell." In J. H. A. Kroeze, ed. *Preference, Behavior, and Chemoreception* (pp. 263–273). London: IRL Press.

Engen, T. 1988. "The Acquisition of Odour Hedonics." In S. Van Toller & G. H. Dodd, eds. *Perfumery, the Psychology and Biology of Fragrance* (pp. 79–90). London: Chapman and Hall,.

Engen, T. 1991. *Odor Sensation and Memory*. New York: Praeger.

Friedman, L., & Miller, J. G. 1971. "Odor Differences Between Enantiomeric Isomers." *Science*, 172: 1043–1046.

Herrick, C. J. 1948. *The Brain of the Tiger Salamander* Chicago: University of Chicago Press.

Hirschfeld, T. 1986. "Remote and *In Situ* Analysis." *Fresenius Zeitschrift für Analytische Chemie*, 324: 618–624.

Hornung, D. E., Chin, C., Kurtz, D. B., Kent, P. F., & Mozell, M. M. 1997. "Effect of Nasal Dilators on Perceived Odor Intensity." *Chemical Senses*, 22: 177–180.

Hummel, T., Sekinger, B., Wolf, S. R., Pauli, E., & Kobal, G., 1997. "'Sniffin' Sticks': Olfactory Performance Assessed by the Combined Testing of Odor Identification, Odor Discrimination and Olfactory Threshold." *Chemical Senses*, 22: 39–52.

Kaissling, K.-E. 1998. "Flux Detectors Versus Concentration Detectors: Two Types of Chemoreceptors." *Chemical Senses*, 23: 99–111.

Koshland, D. E., Jr. 1977. "A Response Regulator Model in a Simple Sensory System." *Science*, 196: 1055–1063.

Koshland, D. E., Jr., Goldbeter, A., & Stock, J. B., 1982. "Amplification and Adaptation in Regulatory and Sensory Systems." *Science*, 217: 220–225.

Laird, D. A. 1935. "What Can You Do With Your Nose?" *Scientific Monthly*, 41: 126–130.

Laska, M., Distel, H., Hudson, R. 1997. "Trigeminal Perception of Odorant Quality in Congenitally Anosmic Subjects." *Chemical Senses*, 22: 447–456.

Laska, M., & Teubner, P. 1999. "Olfactory Discrimination Ability of Human Subjects for Ten Pairs of Enantiomers." *Chemical Senses*, 24: 161–170.

Lettvin, J. Y., & Gesteland, R. C. 1965. "Speculations on Smell." *Proceedings of the Cold Spring Harbor Symposia on Quantitative Biology*, 30: 217–225.

Livermore, A., & Laing, D. G. 1998. "The Influence of Chemical Complexity on the Perception of Multicomponent Odor Mixtures." *Perception and Pschophysics*, 60: 650–661.

Mason, J. R., Clark, L., & Morton. T. H. 1984. "Selective Deficits in the Sense of Smell Caused by Chemical Modification of the Olfactory Epithelium." *Science*, 226: 1092–1094.

Mason, J. R., Clark, L., & Morton. T. H. 1987a. "Covalent Modification of Schiff Base-Forming Proteins. *In Vitro* Evidence for Site Specificity and Behavioral Evidence for Production of Selective Hyposmia *In Vivo*." In S. Roper ed. *Olfaction and Taste IX* (vol. 510, pp. 468–471). New York: *Proceedings of the New York Academy of Sciences*.

Mason, J. R., Johri, K. K. & Morton, T. H. 1987b. "Generalization in Olfactory Detection of Chemical Cues Containing Carbonyl Functions by Tiger Salamanders." *Journal of Chemical Ecology*, 13: 1–18.

Mason, J. R., Leong, F.-C., Plaxco, K., & Morton, T. H. 1985. "Two-Step Covalent Modification of Proteins. Selective Labelling of Schiff-Base-Forming Sites and Selective Blockade of the Sense of Smell." *Journal of the American Chemical Society*, 97: 6075–6084.

Mason, J. R., Meredith, M., & Stevens, D. A. 1981. "Odorant Discrimination by Tiger Salamanders After Combined Olfactory and Vomeronasal Nerve Cuts." *Physiology and Behavior*, 27: 125–132.

Mason, J. R., & Morton, T. H. 1982. "Temporary and Selective Anosmia in Tiger Salamanders Caused by Chemical Treatment of the Olfactory Epithelium." *Physiology and Behavior*, 29: 709–714.

Mason, J. R., & Morton, T. H. 1984. "Fast and Loose Covalent Binding of Ketones as a Molecular Mechanism in Vertebrate Olfactory Receptors. Chemical Production of Selective Anosmia." *Tetrahedron*, 40: 483–492.

Mason, J. R., & Stevens, D. A., 1981. "Discrimination and Generalization Among Reagent Grade Odorants by Tiger Salamanders (*Ambystoma tigrinum*)." *Physiology and Behavior*, 26: 647–653.

Mason, J. R., Stevens, D. A., & Rabin, M. D. 1980. "Instrumentally conditioned avoidance by tiger salamanders to chemically pure odorants." *Chemical Senses*, 5: 99–105.

Mennella, J. A., & Beauchamp, G. K. 1998. "Infants' Exploration of Scented Toys: Effects of prior Experiences." *Chemical Senses*, 23: 11–17.

Meredith, M. 1998. "Vomeronasal Function." *Chemical Senses*, 23: 463–466.

Nachbar, R. B., & Morton, T. H. 1981. "A Gas Chromatographic (GLPC) Model for the Sense of Smell. Variation of Olfactory Sensitivity with Conditions of Stimulation." *Journal of Theoretical Biology*, 89: 387–407.

Paton, W. D. M. 1961. "A Theory of Drug Action Based on the Rate of Drug-Receptor Combination." *Proceedings of the Royal Society (London) Series B*, 154: 21–69.

Rehn, T. 1978. "Perceived Odor Intensity as a Function of Airflow Through the Nose." *Sensory Processes*, 2: 198–205.

Robertson, A. R. 1992. "Color Perception." *Physics Today* 45 (December): 24–29.

Schafer, R. R., & Winegar, B. D. 1988. "Electrophysiological Correlates of Two-Step Covalent Modification." Final report, NSF Grant CHE 8509557.

Sobel, N., Prabhakaran, V., Hartley, C. A., Desmond, J. E., Glover, G. H., Sullivan, E. V., & Gabrieli, J. D. E. 1999. "Blind Smell: Brain Activation Induced by an Undetected Air-Borne Chemical." *Brain*, 122: 209–217.

M. Stuiver, M. 1958. "*Biophysics of the Sense of Smell.*" Ph.D thesis, University of Groningen.

Takahashi, N., Kawamura, M., Shinotou, H., Hirayama, K., Kaga, K., & Shindo, M. 1992. "Pure Word Deafness Due to Left Hemisphere Damage. *Cortex*, 28: 295–303.

Yee, K. K., & Costanzo, R. M. 1998. "Changes in Odor Quality Discrimination Following Recovery from Olfactory Nerve Transection." *Chemical Senses*, 23: 513–519.

Zhao, H. Q., Ivic, L., Otaki, J. M., Hashimoto, M., Mikoshiba, K., & Firestein, S. 1998. "Functional Expression of a Mammalian Olfactory Receptor." *Science*, 279: 237–242.

15

The Slighting of Smell

(with a Brief Word on the Slighting of Chemistry)

WILLIAM G. LYCAN

> Man really has so little appreciation of olfaction that the English language itself in that connection is atrophied. It is impossible to make the simple declarative statement, "He smells," without risk of double entendre . . .
>
> Ralph Bienfang, *The Subtle Sense* (1946, p. 3)

Philosophy articles on propositional attitudes, attitude content, and related matters sometimes begin by reminding us that there are many propositional attitudes besides *belief*: there are desire, intention, remembering, guessing, speculating, wondering, hoping, wishing, resenting, regretting, being embarrassed, and whatnot. Those same philosophy articles then immediately serve notice that, for simplicity, they will discuss only belief. Desire, intention, and memory very occasionally receive attention from philosophers of mind and philosophers of psychology; the other attitudes almost never. It is an interesting question why belief holds this overwhelming social preeminence over the other attitudes.

Similarly, philosophy articles on perception sometimes begin by reminding us that there are other senses besides *vision*; but those articles then immediately serve notice that, for simplicity, they will discuss only vision. The other senses—especially the chemical senses, smell, and taste—almost never receive the slightest attention from philosophers (though noteworthy exceptions are Perkins, 1983, and Clark, 1993). Psychologists and neuroscientists have, of course, done work on smell, but there, too, not nearly as much as they have on vision and usually in pursuit of research on something else, such as memory.[1] My question here is how the philosophy of perception would be different if *smell* had been taken as a paradigm rather than vision. I argue that vision is not a *typical* sense modality at all, but a very unusual, unrepresentative example of one. If I am right about that, then we may expect to find that the philosophy of perception has been warped and skewed by its persistent focus on vision. And—now that the Decade of Neurophilosophy is over—I contemplate ushering in the era of Rhinophilosophy (or more grandly, Osphresiophilosophy). But the claims I shall actually make here will be only rudimentary.

The Preeminence of Vision

There is a difference between the social preeminence of belief and that of vision. For it is arguable that, in fact, belief is conceptually prior to the other attitudes. The more rarified propositional attitudes may well be defined partly in doxastic terms; for example, a correct analysis of regret or wish or even memory would very probably make mention of belief, but not vice versa. More crucially, the *perceptual* propositional attitudes seem obviously to be analyzed in terms of belief, whether or not we agree with the more radical thesis of Armstrong (1968) that the analysis is exclusive—that is, that perceiving itself *is* simply a matter of forming beliefs in a certain way. Finally, it is not implausible (though it may be false) to hold that even to desire that P is simply to be disposed to act in a way that would bring it about that P assuming all one's beliefs are true (cf. Stalnaker, 1984). Besides, belief is at bottom propositional information storage, and it is only to be expected that all the attitudes would presuppose a background of propositional information storage.

But vision cannot be defended in this way against the other senses. Hearing, smell, taste, touch, kinaesthesis, other forms of proprioception, and the rest do not even hint of conceptually presupposing vision in any way at all. What, then, lends vision its social preeminence? I hazard that vision's pride of place is owed to its vast superiority in informational richness for normal human beings. In whatever way one might venture to quantify "information" about the external world, more of that stuff comes our way through vision than through any other mode of sensing.[2] (Touch may run a close second. And it may also be that touch would get you through times of no vision better than vision would get you through times of no touch, if total numbness somehow befell you; but that is because of the constant urgency of your need to fit and orient securely within your contiguous physical surround, not because touch supplies *more information* than does vision. As Aristotle emphasized, we need to distinguish a sense modality's *cognitive* value from its *survival* value, and here the two conspicuously diverge. Any animal must have a sense of touch, while many animals lack sight and many more, including ourselves, can get along pretty well without using it. Thus, although vision supplies vastly more information and surpasses touch in cognitive value, it lags far behind touch in survival value; and in that sense, vision is a luxury.[3])

Thus, my line on vision will be very like Churchland and Churchland's (1983) line on language: from the biological perspective and considered from the evolutionary point of view, language is a very special trait that for obscure historical reasons evolved in just one species and got wildly out of hand—it is a luxurious excrescence on the part of *Homo sapiens*, in wretched excess. Naturally, human beings themselves have come to think it the greatest, most central, yea essential feature of themselves, the definer of what it is to be human, etc.; but that is an outrageous distortion, however understandable. Likewise, vision is a very special trait that evolved in a comparatively small number of species and produced an unexpected and unexpectable wretched excess of information; but because of its extraordinary and extraordinarily atypical informational richness, vision is thought of by human philosophers as perception itself, other sense modalities being conceived as vastly inferior imitations of perception.

Directness

There is perhaps another reason for the social preeminence of vision, probably a spin-off from the first: common sense awards a kind of *directness* to visual perception that it does not grant to smell, and possibly not even to hearing. (The directness I have in mind is folk-epistemological, *not* the sort that figures in technical debates over "direct" versus "indirect" sensing in the philosophy of perception or in Gibsonian psychology. (These I will discuss later.) If I come home and smell in the living room traces of a perfume that is my wife's perfume, I will instantaneously believe my wife is home.[4] But I will believe that only in virtue of having an olfactory experience and making something like an instantaneous inference. Even if I hear my wife's voice singing, down the hall, common sense rules that I know or believe she is home *just in virtue* of having that aural experience, whether or not what is properly called "inference" is involved. But if I walk into the room and simply see my wife face to face, common sense counts me as simply having seen that she is home, here with me, and there is neither any "in virtue of" nor indirectness of any sort about it. (I haven't checked, but I am pretty sure there is a corresponding legal distinction, in the law of evidence: if I had seen my wife in the house, I would "know of my own knowledge" that she was home, while hearing or smell would count as "inference" on my part.[5])

Common sense is probably naively mistaken on this point about "directness," simply having failed to notice that visual contact with a real external object involves mediating visual experience just as crucially. The reason for this failure to notice the mediation is the enormous gap in informational richness, mentioned before; vision supplies so highly articulate a representation of what is before us that in all but the most unusual situations we "see right through" the representation to the array of objects correctly represented.[6] Direct realism about spatial arrangement and even about color is overwhelmingly plausible and attractive to the untutored, until we come under the influence of philosophers and are forced to consider the Argument from Illusion. ("Direct realism" is roughly the view that perception either simply acquaints us with real qualities of external objects or, when faulty, manages to mislead us without interpolating special objects or qualities of sensing that mediate our misperception of the objects—admittedly quite a trick. I shall refine this terminology shortly. The Argument from Illusion is roughly the argument that when we are perceptually deceived, we have experiences *as of* whatever falsely seems to be going on, and so *there are* such reality-neutral experiences that even in the veridical case mediate between us as subjects and the external world. These experiences are at least inner representings, and may further be argued to involve "sensa" or sense-data of Russell's phenomenalist sort.)

Some version of Direct realism may be true of vision, despite the Argument from Illusion. But on the other hand, other senses are not so tractable: either indirect realism or out-and-out subjectivism may hold for the informationally stingier sense modalities. It might be said that Direct realism is halfway plausible for vision but unattractive for hearing and hopeless for smell. Fair enough; there is no reason to assume that one metaphysical account must be common to all the various sensory systems there are. But notice a key dialectical point: if Direct realism fails drastically enough for *even one* mode of sensing, and we are forced to posit inner sensory representations or special sensory qualia or even phenomenal entities such as sense-data, proper to that modality, then our main reason for resisting such posits *for the case of vision* is undercut. Or vice versa. (We shall see that a key distinction must be made for both cases later in this chapter.)

The main reason I have just mentioned is the general *excrescencehood* of sensa: inner sense-impressions—especially genuine phenomenal individuals—are, at best, cumbersome theoretical baggage whose causal role is difficult or impossible to specify and, at worst, dubiously intelligible and contrascientific rubbish, probably in violation of at least one law of physics. We have every reason to hope that science and philosophy can get along without anything that bad, and I would add that if science can, then philosophy can; but if either science or philosophy cannot entirely dispense with inner sense-impressions, for the case of so much as one mode of sensing, then there is little reason to deny such sense-impressions to other modes, including vision, as long as there remains any positive reason at all for accepting them. *Ab esse ad*: not only *posse*, but no principled reason why not elsewhere.

Before proceeding, however, let us recognize the distinction I mentioned a short while ago. Following all too many philosophers of perception, I have been using terms like "experiences," "sensory representations," "inner sense-impressions," "sensa," and "sense-data," doing hardly a thing to specify the ontological category I have in mind. On one understanding, the most notorious, such items can be understood as being phenomenal individuals, sense-data of the sort dear to Russell, Carnap, and Ayer, having their qualia as bonafide monadic first-order properties. But more cautiously, they can be conceived as being merely internal representations, presenting aspects of the things they represent but *not* themselves bearing any qualia-like properties—as far as they have qualitative "characters," those characters are merely classificatory, as in "a visual representation of the blue(-indicating) sort."[7]

The distinction is ontologically important, especially for the question of metaphysical naturalism or materialism and also for epistemology. For actual sense-data are *merely* phenomenal, immaterial objects; moreover, they are objects of consciousness, of acquaintance, and (on some accounts) of sensing itself. By contrast, mere perceptual representations can be physical, neurophysiological items, and they need not themselves be objects of consciousness, acquaintance, or (certainly) sensing.

Thus, a difference between two senses of "Direct" in "Direct realism": "Direct" means *unmediated*, but unmediated by what?[8] A *weak* form of Direct realism merely repudiates the sense-datum theory, and holds only that we perceive external objects without having to perceive or be acquainted with excrescent phenomenal ones. A *stronger* form holds that we perceive external objects without even hosting internal representations or being in states characterized according to phenomenal features such as color. Hereafter, I shall call these just "Weak" and "Strong" Direct realism, though other forms of directness are philosophically interesting as well.[9]

Distinguishing the Senses

Of course, there are many possible philosophical and psychological departures from Strong Direct realism, though fewer from Weak Direct realism. To explore the possibilities for just one mode of sensing would take a long time, and that is what I propose to start to do for smell, in comparison to vision. I believe the best way to begin is to list some *general respects of similarity and difference* between senses, and then ask where smell falls among the other senses regarding *its* pattern of similarity and difference. But I shall be able actually to discuss only a few of these respects of comparison.

1. *Informational richness*, as before. Vision is king. Smell, in humans, is informationally very poor. (Imagine having neither sight nor hearing, but *only* smell in addition to touch. Even more vividly, think of the perception of *language*: we read written language with our eyes—a bountiful source of information about faraway things and events. Less efficiently, we can hear people speak, with our ears. Still less efficiently, some can understand Braille by touch. But try to imagine a Braille of smell. It *could* be done, but just barely.)

2. *Degree of objectification*, that is, the degree to which the (ostensibly) external object and its qualities are psychologically treated as separate from the subject and from the subject's mental condition. In vision—even in visual hallucination—we construct perceptual objects conceived as separate from ourselves, and psychologically we locate the colors of those objects in the objects themselves, as if they existed unperceived. In smell, we do far less of that sort of objectification; considered only phenomenologically, a smell seems a modification of our own consciousness rather than a stable property of a perceptual object that would exist unperceived, even though we can, to some extent, put the smell back into the external object once we stop bracketing the object and feel the smell as emanating from the object.

Objectification is a function of informational richness, but also of a kind of robustness internal to a single sense modality. Vision is a paradigm in the second way as well as the first. It offers a multitude of different perspectives that are, to some extent, under the subject's control. Looking at an object from a single point of view, one forms some expectations of how the object would appear from a different angle, and one can then move to a different vantage point and test those expectations. And whether or not they are confirmed, the move yields different (still purely visual) representations of the object, which may, in their turn, be checked against the original appearances; more perspectival views of what is ostensibly the same object may be fit into a coherent file whose label purports to name that object. Visual checking and filing may continue indefinitely, and (barring anomaly) will continue to reinforce the idea of a perduring external object which is as it is independently of the subject's circumstantial view of it at the time. (It may or may not be a significantly further step to the idea of the object's existing entirely unperceived.)

Smell lacks informational richness, as I've said. I think it also lacks intramodal robustness. We cannot easily see how one could *entirely by smell* check and recheck an ostensible external object's identity or character by gaining successively different olfactory perspectives on that object. The idea of an "olfactory perspective" makes little sense unless founded on an extraneous mode of sensing. (The same object could exhibit different smells in different parts and from different directions, but we could establish this only by distinguishing the parts using some sense other than smell; thus, the perspective in question would not be purely olfactory.)

Perhaps this is too hasty, for we have not considered the idea of a somewhat different environment: David Sanford (1983) reminds us that the external world could have lacked ordinary physical objects but divided itself into discrete, physical object-sized odor regions. We may suppose that for some reason the boundaries of the odor regions would stay fairly stable, so that the inhabitants of Sanford's world could move about and orient themselves in the environment according to the odor regions they had passed through. This stable complexity might afford a notion of olfactory perspective; one could approach the same odor region from different olfactory points of view, the points of view being individuated according to the smells of their neighbors. This notion of perspective could further be enriched if, in addition to the odor region one was occupying at the time, one could smell various nearby regions, at a distance, in proportionate strengths. But the latter question demands a well-motivated view of odor individuation in cases of odor mingling, and to date, such a view eludes psychological consensus. (What we ordinarily think of as a single smell, say, of a perfume, is in fact a complex accretion, in the case of a perfume a very deliberate, sophisticatedly engineered amalgam, of more fundamental odors functioning as various "notes." But the style of accretion exhibited by smell is disputed as between the "analytic" model of complexity that applies to visual color and the "synthetic" or "chord" model for complexity in hearing tones.)

I think we should concede that a Sanford world might make sense of the notion of (purely) olfactory perspective. But Sanford worlds are fantasy, not even science fiction, and very likely nomologically impossible.

It has been pointed out to me[10] that the real world offers one candidate for a purely olfactory perspective: in wine tasting. One can identify a wine by first smelling it as it is poured, then smelling it in the glass, then smelling it from the tip of one's tongue, then smelling it from the back of one's throat through the postnasal passages, and perhaps further vantage points. But this is a very rare case, because we do not often engage an object both with the nose and with different parts of the mouth. For human beings, in all but a few circumstances, there is no purely olfactory perspective.

Further Distinguishing Features

3. Phenomenological location—that is, *where* the relevant sense qualities feel as though they are located. Vision intrinsically presents us with an apparently three-dimensional external world; even if we ruthlessly bracket the apparent external objects and focus just on our visual field as we introspect it, the field has a vivid geometry—its elements are spatially located with respect to each other, and color patches still *seem* to be out in the external environment, on view at a distance. Hearing, too, comes to us spatially organized, even phenomenally speaking; normally, to hear a sound at all is to hear it as coming from this direction or that and from such-and-such a distance. By contrast, smell is aspatial. Phenomenologically, an odor is just with us, happening right in the center of our minds. If we bracket the external source of the smell and waive our contingent knowledge of that source, smell has no directional or other spatial aspect.

It is easy to miss that fact, for (a) we can and often do determine, by sniffing, that a smell is coming from a particular direction, and (b) we can also locate physical objects *by* smell. But from neither of these truths does it follow that smell is phenomenologically spatial, in the strict sense of "phenomenology" I have in mind.

Regarding (a), I submit that such directional determinings require active *and diachronic* search. In particular, *you have to move your head*. The search works by intensity gradients; as your nose moves, it finds the direction in which the target odor grows stronger, and you follow it. Your rudimentary spatial odor map is generated by patterns of intensity that can be discovered only through motion. My phenomenological claim is rather that if one merely concentrates on one's olfactory sensations *passively and at a time*, they present no geometry or other spatial character. [11–15]

Regarding (b): to locate a physical object by its smell, we normally have to determine the odor's direction. But as we have seen, to make that determination we, in turn, have to move our heads and track intensity gradients; it cannot be done phenomenologically, in my passive sense of "phenomenology."[16]

There is one other way of locating a physical object by its smell, which is to *identify* the object by its smell when we can figure out independently where the object would be, given that it is present at all (if my wife is approaching, as I can tell by her perfume she is, she must be coming in the room's only door). But inference from collateral information about the physical layout of one's environment is not phenomenology either.

To see the overall point, consider vision again, and hearing. To have a visual experience (even one known to be nonveridical) *is* to be appeared to as by something in space. There is simply no such thing as a visual experience without geometry or at least spatial properties. The same is *nearly* true of hearing; phenomenal sounds are almost always intrinsically heard as coming from a particular direction and often as being at such-and-such a distance. (There are exceptions, e.g., if one is using a particularly good Walkman, a sound *can* be produced that one will feel aspatially and right in the "center" of one's mind, without any even apparent directional character.) But a phenomenal smell at a time just suffuses consciousness.

4. *Degree of presumed resemblance or at least correspondence between a sensory field and the reality it presents*: this is very strong for vision, so strong that the "seeing right through" phenomenon is the normal case, and the resemblance or correspondence is (falsely) felt as *identity* of field elements with external objects. No such thing obtains for smell, because an olfactory field at a time is information-poor and unstructured. Things could have been a little better, for in a Sanford world, there would be (given our present olfactory systems) at least a *diachronic* correspondence between smells and the structure of the external world. But whether a synchronic correspondence would be possible in the Sanford world would depend on whether one could smell various nearby regions, at a distance, in addition to the region one was occupying at the time, and the latter question, in turn, demands a well-motivated view of odor individuation through mingling; again, we have no such view as yet.

5. *Proper vs. common sensibles*.[17] Does the sense modality in question present common sensibles in addition to proper sensibles? For vision, yes; for smell it seems not.

6. *Relation between object and messenger*. Visual information reaches us through a very general and object-neutral medium, viz., light waves governed by the laws of optics. Whatever the object seen, its image is produced in us by the same mode of transmission. But with smell, one cannot separate the object from the messenger, for the messenger is simply a fragmented small bit of the object itself; the object must actually penetrate our body and interact chemically with it. Nalini Bhushan—who suggested this respect of comparison to me—also calls attention to the possibility that the actual

molecules carry more deeply embedded messages, something beyond mere indication of their commonsensical sources such as roses or rotten eggs; in lower organisms, molecules often carry very specific and simple messages.

7. *Characteristic features of the quality space*: for example, how many dimensions has the space in question? How many primaries underlie the profusion of distinguishable qualities? Color vision is the paradigm here, as in the color order system devised early in this century by Munsell, and those questions and others like them have been answered fairly satisfactorily for color. Sound–space is almost as well understood. Somewhat less definitive but useful work has been done on tastes. But smell lags far behind in this particular race. Researchers agree neither on a fixed number of dimensions nor on a fixed number of primaries. (Guesses as to primaries range from four through twelve or thirteen. But at least the researchers provide vivid taxonomic names. In 1756 Linnaeus called his seven odor categories "aromatic," "fragrant," "ambrosial," "alliacious," "hircine," "foul," and "nauseating." Kandel and Schwartz also canonify Amoore's seven categories of 1952, although these are not the same: "camphoraceous," "musk," "floral," "peppermint," "ethereal," "pungent," and "putrid." But it may be that even if categories such as these do correspond to specific receptors, none of them should count as primaries, because on one prominent usage a genuine primary must be a component of every complex quality in the modal space.)

Pictures have been drawn of the olfactory space; for example, the University of California at Davis has produced a fine "Aroma Wheel," that organizes over a hundred different smells around a circle, dividing them into species and superspecies.[18] But the empirical bases and the theoretical foundations of such pictures are very much in question (even when oenological ulterior motive is less obvious). I suspect that that fundamental unclarity is due to the many unresolved questions about the basics of the olfactory sense: we do not know whether quality is independent of changes of intensity; and again, it is unclear how mixtures of substances are perceived—the nature of the amalgamation, whether there is a cancellation phenomenon, and so forth.

Investigation of sensory quality spaces has benefitted greatly in recent years from the technique of multidimensional scaling based on overlapping judgments of indistinguishability, though the MDS technique is not always easily applied, and has not been done with enough subjects or enough different odors.[19] MDS has helped the mapping of smell, in particular, to catch up a bit.

Moreover, MDS has been used on neural responses to the same stimuli (Erickson & Schiffman, 1975), and shows a nice match between the emerging olfactory quality space and the pattern of neural responses. Thus, smell may catch up to color and hearing in the orderliness of its quality space and in the accompanying explanatoriness of the underlying neurophysiology. But that is not *overwhelmingly* likely.

8. *Variety and kinds of besetting illusions*. Optical illusions are rife and familiar to all; auditory illusions are well known. But it is hard to think of olfactory analogues (however, see Bhushan, 1995).

9. *Credentials of sense impressions as representations*. Visual experience represents; it represents external objects, truly or falsely, as having geometrical properties, colors, and various relations to other objects. It is more difficult to say whether olfactory experience represents; I believe most philosophers would initially deny that smellings

represent any more than pains or itches do. Nor is it entirely clear what is being asked. Some useful test questions are as follow: Does a smell actually have semantical properties (reference, a truth or satisfaction condition)? Can a smell be treated formally, á la Hintikka, as a function from possible worlds to sets? And can it be given a linguistic-functional "dot" characterization, á la Sellars? Can a smell be incorrect, a *mis*representation? Positive answers to questions like these would argue that smells are indeed representations.

It does not initially seem that those questions do have positive answers. Phenomenally speaking, a smell is just a modification of our consciousness, a qualitative condition or event in us. Even if we infer the presence of natural gas from the characteristic foul and pungent smell, the smell does not *itself present* natural gas as its own intentional object. We infer gas only because we already know by induction that that smell is typically produced by ambient natural gas. (We also know that the smell is not always produced by gas—indeed, strictly it is *never* produced by pure natural gas itself, because it is actually given off by an adulterant manufactured by a chemical company.) A smell is just a *quale*, whether we take the quale strictly to be a first-order property of a phenomenal individual or an "adverbial" modification of an event or state of sensing. Conventional wisdom has it that qualia merely linger on uselessly in the mind, and do not represent.

Nonetheless, I have argued at length (Lycan, 1996, chap. 7) that smells do represent; what they represent first and foremost are odors, which I understand to be clouds of molecules diffusing in the air. Only by a kind of deferred referring do smells represent the external things by reference to which they are usually classified. But I shall say no more on this topic here.

10. Credentials as presenting a phenomenological individual. The sense-datum theory is the view that in perception we are most directly acquainted with phenomenal individuals having monadic and relational qualitative properties. Philosophy since Logical Empiricism has conspired to make out that sense-datum theory is entirely unmotivated and was accepted by empiricists only on the basis of certain arguments that are embarrassingly fallacious. Those arguments were primarily epistemological, and they certainly were fallacious; moreover sense-data themselves are felt to be both metaphysically sordid, as nonphysical intruders in a naturalistic universe, and epistemologically troublesome, in that they generate a notoriously aggravated "problem of the external world." Accordingly, most contemporary philosophers either shun sense-data or simply ignore them, and seem to assume that phenomenal individuals have obligingly faded away along with the luminiferous ether and the Easter Bunny.

In fact, there are other, far better arguments in defense of phenomenal individuals. A powerful one is Frank Jackson's (1977) semantical argument, which points out that the language most naturally used to describe visual experience treats that experience as a color mosaic, breaking it up into individual patches having phenomenal shape and color. The visual language exhibits just the structure of reference, quantification, and number that we find in any discourse having as its subject matter a domain of individuals and their properties and relations. Proponents of "adverbial" accounts of sensing have responded by trying to explain away the semantical appearance of that referential and quantification structure.

I think such accounts fail utterly, for reasons I have elaborated twice (Lycan, 1987, chap. 8; 1996, chap. 4) and cannot even summarize here. I also think that, fortunately, the admission of the referential and quantificational structure itself fails to demonstrate the actual existence of sense-data (which is an even longer story), and that there aren't really any phenomenal individuals; but my claim for now is that the visual field at least is *experienced as* a mosaic of patches or regions each having both shape and color, and that this creates a powerful prima facie case for sense-data. (The region boundaries can, of course, be vague, like the boundaries of a topography of the earth's surface.)

I take that claim to be indisputable for the case of vision, and nearly so for that of hearing (we hear phenomenal sounds that individually have properties such as pitch and volume). The claim is far less obvious for the case of smell, for smell is experienced just as a state of consciousness, having neither geometry nor articulate individuation. It is unclear whether an olfactory expert placed in a smell-rich environment could distinguish and attend selectively to different smells (without diachronic head movement), again because the business of odor mingling is ill-understood. *If* an expert could do that, s/he might say that each of the different smells has its distinctive quale; but even so, that attending is more readily described adverbially, in terms of its smelling thisly or thusly when one attends this way or that way.[20]

Denouement

Thus, a moral: alone among the senses, vision builds a wall. It is an all-enclosing wall, an articulated color mosaic to be sure, but without chink or leakage, and it is a *trompe l'oeil* as well. Prephilosophically, we incline toward Strong Direct realism because we take the wall for an open window, and project the color mosaic onto our physical environment without even noticing that we have visual *experience* at all. Then along comes the philosopher and belts us with the Argument from Illusion. We see the wall for the first time, but now because of its all-encompassing chinklessness, we see each tile of the mosaic as an actual individual thing conspiring to separate us from the external world. Of course, such objects are not themselves physical objects, and so Russell gets us to believe in sense-data, not as philosophical theory but as simple fact, and we give up Weak Direct realism as well as Strong.

I think that if the philosophy of perception had begun with smell rather than with the wildly atypical sense vision, there would have been little temptation toward sense-datum theory, at least not the powerful temptation induced by concentrating on vision. Thus, Weak Direct realism would have been taken as fairly obvious, not as the contentious and rather naive doctrine that "Direct realism" is often taken to be. But at the same time, if the philosophy of perception had begun with smell instead of with vision, Strong Direct realism would fairly readily have been seen to be false, because smell involves representation, and our detection of external states of affairs by olfactory means requires internal representation; no one (even prephilosophically) would "look [or sniff] right through" a smell sensation and project it onto an external object without even noticing that an experience was what one was having. Thus, the whole tempest over "Direct realism" would have been confined to its proper philosophical teapot.

The Slighting of Chemistry

There is an amusing if fortuitous analogy between philosophers' neglect of smell and their neglect of chemistry in philosophy of science.

Having grown up in a chemist's household[21] and been at least temporarily a chemistry major in college, I soon noticed the slighting of chemistry now documented by the editors in their introduction to this volume. Consider, in particular, the wording of job advertisements in the American Philosophical Association publication, *Jobs for Philosophers*. "Philosophy of science" listed as an "AOS" (area of specialization) means, neologistically, philosophy of physics. Philosophers of biology and philosophers of economics might apply, but would rarely be considered. If a philosopher of biology or a philosopher of economics is wanted, the ad will so specify, instead of listing "philosophy of science." Philosophy of chemistry is simply never mentioned.

But the analogy I mentioned has specifically to do with questions of realism. This century has seen three main positions regarding the ontological status of scientific entities (for obvious reasons of space, I present them in cartoon form): First, the instrumentalism of the Logical Positivists, driven by the verificationist theory of meaning. Second, the historicist relativism associated with the works of Paul Feyerabend and Thomas Kuhn. Third, a commonsensical sort of explanationist realism according to which posited unobservables really exist, and are much as they are literally characterized—for example, electrons really are tiny subvisible charged particles that flow through copper wire and jump through Wilson cloud chambers—and our belief in them is justified by the fact that it best explains great masses of observable phenomena.

Explanationist realism has been a minority position well back in the nineteenth-century. Positivism dominated until the early 1960s, when in a rather startling over-reaction to the Positivists' mania for rigor, precision, and epistemological sanctity, historicist relativism came to the fore, distracting attention from the realist criticisms of Positivism made by Quine, Sellars, Putnam, and Smart. Common sense realist opposition to both instrumentalism and historicism has been heard since, and is, I believe, on the increase, but it remains highly controversial.

I have said that had smell rather than vision been taken as a paradigm in the philosophy of perception, Weak Direct realism would have been taken as fairly obvious. So, too, I believe that if chemistry rather than physics had been taken as a paradigm in the philosophy of science, commonsense scientific realism would never have been seriously in dispute (though, of course, and rightly, it would have been disputed by some).

Physics is highly theoretical. Most of it is very abstract, and as Cartwright (1983) has emphasized, it is heavily and irremediably idealized. Further, much of physics deals with the unimaginably large or with the unimaginably small and ill-behaved. The mathematics of general relativity and of quantum theory are strange and do not encourage realist interpretation (for one thing, they are mutually incompatible). Despite the heroic efforts of many theorists with IQs double or quadruple mine, there is no plausible ontological interpretation of quantum field theory, even if we all are persuaded by authority that the theory is basically correct. Very probably, philosophers should not be realists about all of physics, perhaps not about much of it at all.

Philosophy of biology is more realist at bottom because no one denies or paraphrases away the existence of organisms, populations of organisms, or biological mechanisms operating within and around those. But evolutionary biology is contaminated by politics. It is not just that claims and counterclaims have been made about biological "superiority" and about what is "natural"/"unnatural"; even the units-of-selection controversy is affected by theorists' conservative or collectivist political views. Moreover, the idea of natural teleology encourages ideological interest.

Contrast chemistry. (I mean classical chemistry, not work conducted in chemistry departments that is really physics.) It is very concrete, having to do with what various familiar substances are made of. As far as I know, it is untouched by politics; at least I have not heard of Marxist chemistry or feminist chemistry. The classical chemical equations are the best-confirmed and most robust theoretical generalizations known to science, and they dramatically explain myriads of striking macrophenomena. No working chemist would think of denying that familiar substances are composed of molecules, which, in turn, are very uniform and regular constellations of atoms, or that atoms are tiny subvisible particles.[22] It would be hard to take an instrumentalist attitude toward hydrolysis.

Moreover, chemists *use* items characterized in chemical terms reliably to produce various macroscopic effects; synthesis is not only a huge area of academic chemistry, but the job of any of a number of whole commercial and other worldly industries. Chemists take various molecules and cut and paste them together into useful substances never found in nature, most notably pharmaceuticals and plastics. It would never occur to any such worker to doubt that her/his raw materials were real or that they had most of the features ascribed to them by classical chemical theory. Historicist claims, and postmodernist vaporings, seem irrelevant and silly here.

Finally, as this volume's editors observe, instrumentation technology makes ever greater advances in micro-*observation*, as with the scanning tunnelling microscope. Chemical objects such as molecules, atoms, and even some subatomic particles no longer count so obviously as "unobservables" as they did in Positivist times. The technological relativity of the "theory"/"observation" distinction was always a Positivist weakness (consider the prevalent use of eyeglasses); here it threatens to leave Positivist instrumentalism very short of subject matter.[23]

If smell had been taken as a paradigm in the philosophy of perception, and chemistry as one in the philosophy of science, then in philosophy generally a representationalist and explanationist realism would be doing more nearly as well as it deserves.

Appendix: A Few Quotations

If the whole body were an eye, where were the hearing? If the whole were hearing, where were the smelling? *I Corinthians* 12:17

Great Turns are not always given by strong Hands, but by lucky Adaption, and at proper Seasons; and it is of no import, where the Fire was kindled, if the Vapor has once got up into the Brain. For the *upper Region* of Man, is furnished like the *middle Region* of the Air; The Materials are formed from Causes of the widest Difference, yet produce at last the same Substance and Effect. Mists arise from the Earth, Steams from Dunghils, Exhalations from the Sea, and Smoak from Fire; yet all Clouds are the same in Composition, as well as Consequences: and the Fumes issuing from a Jakes, will furnish as comely and useful a Vapor, as Incense from an Altar. Thus far, I suppose, will easily be granted me; and then it will follow, that as the Face of Nature never produces Rain, but when it is overcast and disturbed, so Human Understanding, seated in the Brain, must be troubled and overspread by Vapours, ascending from the lower Faculties, to water the Invention, and render it fruitful. (Jonathan Swift, *A Tale of a Tub* 2nd ed. [A.C. Guthkelch and D. Nichol Smith, eds., Oxford: Clarendon Press, 1958], pp. 162–163)

Have you an ambition to found a new science? Why not measure a smell? Can you measure a smell? Can you measure the difference between one kind of smell and another? Until you can measure their likenesses and their differences you can have no science of odor. Find out what odor is—weight, or a vibration, and therefore capable of being reflected. Odors are becoming more and more important in the world of scientific experiments and in medicine—and the need of more knowledge will bring more knowledge, as surely as the sun shines. (Alexander Graham Bell, in *Scientific Monthly*, vol. 25, No. 6 [December, 1927], p. 481)

Consider . . . how many olfactory items are subject to taxation: cigarettes, cosmetics, perfumes, cigars, and pipe tobacco. It appears that our devotion to olfactory pleasures is the basis for a large part of our taxation structure. (Bienfang, pp. 12–13)

In eighteenth-century France . . . there reigned in the cities a stench barely conceivable to us modern men and women. The streets stank of manure, the courtyards of urine, the stairwells stank of mouldering wood and rat droppings, the kitchens of spoiled cabbage and mutton fat; the unaired parlours stank of stale dust, the bedrooms of greasy sheets, damp featherbeds, and the pungently sweet aroma of chamber-pots. The stench of sulphur rose from the chimneys, the stench of caustic lyes from the tanneries, and from the slaughterhouses came the stench of congealed blood. People stank of sweat and unwashed clothes; from their mouths came the stench of rotting teeth, from their bellies that of onions, and from their bodies, if they were no longer very young, came the stench of rancid cheese and sour milk and tumorous disease. The rivers stank, the marketplaces stank, the churches stank, it stank beneath the bridges and in the palaces. The peasant stank as did the priest, the apprentice as did his master's wife, the whole of the aristocracy stank, even the king himself stank, stank like a rank lion, and the queen like an old goat, summer and winter. For in the eighteenth-century there was nothing to hinder bacteria busy at decomposition, and so there was no human activity, either constructive or destructive, no manifestation of germinating or decaying life, that was not accompanied by stench.

And of course the stench was foulest in Paris, for Paris was the largest city of France. (Patrick Süskind, *Perfume: The Story of a Murderer*, tr. John E. Woods [London: Hamish Hamilton, 1985])

They call it the smell of the sea, Morris thought, *but really it's not. It's the smell of the land.* It came from the tidal marshes—all the things that lived and died and rotted at the water's edge, all the smells that fermented in the marginal wetlands and when released blew out to sea. Sailors considered it a friendly odor because it meant that land, port, home, family were near. Otherwise it was something to be neutralized with Lysol. (Tom Clancy, *Red Storm Rising* [New York: Berkley Books, 1987] p. 431)

The history of civilization is the story of man's emancipation from a lot that was harsh, brutish, and short. Every step of that upward climb to a sophisticated way of life has been paralleled by a corresponding advance in the art of perfumery. (Eric Maple, quoted in Tom Robbins' *Jitterbug Perfume* [New York: Bantam Books, 1984], p. iv)

After five days of experimenting, he hit upon what seemed the ideal mixture: one part beet to twenty parts jasmine to two parts citron, a ratio that inspired him to name the scent K23. . . .

Like a lobster with a pearl in its claw, the beet held the jasmine firmly without crushing or obscuring it. Beet lifted jasmine, the way a bullnecked partner lifts a ballerina, and the pair came on stage on citron's fluty cue. As if jasmine were a collection of beautiful paintings, beet hung it in the galleries of the nose, insured it against fire or theft, threw a party to celebrate it. Citron mailed the invitations. (Robbins, p. 212)

Notes

1. In the neurophysiologists' bible of the past decade, Kandel and Schwartz (1989), "The Chemical Senses" take up one chapter out of sixty-two, and smell in particular occupies nine pages (417–425) out of 981. The situation is not improving: in the encyclopedic *The Cognitive Neurosciences* (Gazzaniga, 1995), smell takes up seven pages (110–116) out of 1447.

For that matter, in my search through *Bartlett's* and other quotation encyclopedias for a suitable epigraph to this Chapter 5, looking under smell, odor, olfaction, perfume, and the like, I found very few quotations at all, and those I did find were really about something else (e.g., "A rose by any other name would smell as sweet"). Somewhat spitefully, I have made do with Bienfang. For the reader's delectation, I have appended a brief selection from my own more extensive catalogue of specifically olfactory quotations.

A recent lapse from the great smell-ignoring conspiracy was *Odeurs . . . une odyssée*, the gigantic olfaction Expo or fair held in Paris in June 1997. I thank Richard A. (Red) Watson for sending me its imposing and fragrant catalogue—and for his most helpful conversation and correspondence on this topic since 1991. I am also grateful to Peter Machamer for some excellent source materials provided on very short notice, and to the editors for their helpful comments on an earlier draft.

2. The editors have raised several good questions about this impressionistic claim, and at least three qualifications are required. First, it may be that vision furnishes not more total information but only more immediately usable information. Second, I am speaking of normal (20–20) human vision and normal human olfaction, unaided by instruments. (Suitably fantasized instruments, of course, could change any estimate of amount of information.) Third, smell has the power as sight does not to detect and possibly identify visually hidden sources.

3. I owe that way of putting the point, and much else in this chapter, to discussions with Aaron Ben-Ze'ev.

There is an interesting issue over the relative cognitive authority of sight and touch (though I can only footnote it because this is an article about smell). Aristotle argued that sight is cognitively superior to touch, presumably because of its greater information payload. Berkeley disagreed, on the grounds that touch corrects sight but sight does not correct touch. Common sense, I believe, sides with Berkeley. Macbeth tests his vision of the dagger by clutching; if he hath the dagger not, but sees it still, the seeing is hallucinatory. Suppose, on the other hand, Macbeth could feel and heft a dagger manually but could not see it: most people would take that as prima facie evidence of an invisible dagger. Curiously, psychological research appears to show otherwise (Rock & Harris, 1967).

4. This is a fictional example; my actual wife, Mary Lycan, does not need to use perfume.

5. However, compare *metaphorical* expressions such as "I smell a rat" and (from a James Bond novel) "You both got guns, I can smell 'em." They carry at least two connotations: what is smelled is sinister or bad, and the sense is feral instinct rather than "cerebral." But they do not particularly connote that the perception is indirect, except perhaps to suggest *suspecting* rather than perceiving at all, almost nonperceptual. (I thank Dr. Jehangir Chubb for this observation.)

6. What exactly might "see right through" mean here? There are any number of possibilities.

7. I gather there is scholarly dispute over which of these characterizations fits the classical British Empiricists: Were, for example, Berkeley's ideas *blue*, or were they merely "of blue"?

8. The airline industry imposes a conventional usage here: a flight can be "direct" without being "nonstop."

9. For example, epistemological directness, *one* form of which is noninferentialness. Among those philosophers who accept Weak but *not* Strong Direct realism in my terminology, some hold that our perceptual beliefs about external objects are justified only by being inferred from prior perceptual representations (such as its visually seeming to one that a blue thing is present, or one's recognition that one is appeared to bluely), while others grant that justified perceptual beliefs may be based on representations in a way that falls short of being inferential.

10. I seem to remember that it was Alan Fuchs. He thinks not, but at least two of his colleagues at the College of William and Mary side with me in this.

11. No audience to whom I have ever presented this article has bought this claim. (I am not sure why not; read on.) But note that at least the claim is entirely independent of everything else in the chapter. If you are pressed for time, you can skip down to item 4. The hell with it.

12. There is an obvious neurophysiological explanation of the fact claimed here: although we have two nostrils and two olfactory bulbs, this is a superfluity; the morphological separation has no analogue of the stereoptic or stereophonic functions found in the visual and aural systems, respectively, and the olfactory system performs no mapping coordinate transformation such as are accomplished there (see Shepherd, 1995). But do not take this remark as an argument; because my claim is phenomenological, I would like to convince you without argument.

13. If you are not yet convinced, try a blindfold experiment on yourself. Blindfolded, seat yourself in a fixed position, immobilize your head, and while holding your nose, have someone place a strongly odored object somewhere in the room. Then open your nostrils and try to say where the object is located with respect to you. Repeat ad lib, for different odors and different locations. (Eva Feder Kittay tried this, with kitchen laboratory assistance from Jeff Kittay and me, after a presentation of this article at SUNY Stony Brook which—of course—had failed to convince her. She scored no better than chance. Hah!)

14. Yes, I know that if we strictly immobilize a subject's eyeballs by injecting the eye musculature with curare, the subject will soon lose articulate visual experience. So *in principle*, someone might protest, there is no spatial phenomenological difference between vision and smell. But (i) we have to use an invasive and toxic drug, thus disrupting the *normal* phenomenology of vision, and (ii) it is unclear that what "experience" remains is properly called *vision* at all.

15. One genuine qualification is needed (I owe this point to Holly Smith): arguably certain odors may present themselves as being located *in the nose*; indeed, if they are made too intense, they may be felt as physical sting in the nose. This is most obviously the case with odors having either *pungent* or *putrid* components, and I believe the reason is that those two apparent olfactory primaries in particular have partially electrical neurophysiological bases (Kandel & Schwartz, 1989, pp 418–419).

16. Edward Casey has reminded me that there are numerous schools of phenomenology and many senses of "phenomenological." It may be that at most one of those senses supports my claim here. I thank Casey for his generous and forbearing comments on my would-be phenomenological material.

17. I originally entitled this item "Primary/secondary distinction." But Durland (1996) has convinced me that "the" primary/secondary distinction is highly vexed, and also not to be identified with that between common and proper sensibles. (I take a proper sensible to be a feature that can be detected by only one sense modality, and a common sensible to be one that can be detected by more than one sense modality.)

18. In their introduction to this volume, the editors mention "Henning's 'smell prism.' " It is a triangular solid whose six points represent six alleged primaries: "fragrant," "putrid," "ethereal," "burned," "resinous," and "spicy" (see Henning, 1916).

19. Austen Clark (1993) expounds this at length, and makes an excellent case for the usefulness of MDS's results to the philosophy of mind.

20. Several more vectors for distinguishing sense modalities are worth extensive discussion, although I shall say almost nothing about them.

(11) The nature of the modal field qua *field.*

(12) Dependence on real-world concepts. In one way, this means dependence of public description on real-world concepts, which for most senses is total. In another way, it means that to have the sense at all requires some Kantian real-world association, which is not at all true of smell.

(13) Phenomenal distinctness from other senses. Smell gets massively mistaken for taste.

(14) Cognitive penetrability.

(15) Relations to memory of various sorts. The triggering phenomenon is most pronounced with smell. Perhaps this might be explained in terms of the ratio of information in the trigger to information recalled.

(16) Effect on affect. There is a triggering phenomenon here, too. That is because smells go directly to your limbic system, often making you say something in Limbic.

21. My father (William H. Lycan, 1903–1994) was a chemist, trained during the 1920s at the University of Illinois by the great organic chemists, Roger Adams and "Speed" Marvel. Although he began his career briefly as an academic, he spent most of it as an inventor of pharmaceuticals with the Johnson & Johnson company. He was aware, in a dim, abstract, and distant way, that there are people who do not work as chemists but nonetheless manage to tolerate their lives—much as Ronald Reagan must have been aware that there are poor people in the United States—yet he could neither understand that fact nor make it real to himself.

22. In my undergraduate physics courses, circa 1962–1964, I was taught various Positivist instrumentalist doctrines not only *as fact* but as constitutive of basic physical thinking. I do not recall any suggestion of instrumentalism in any of my chemistry courses, although I would be interested to learn how chemistry was taught in Europe and in the United States during the 1940s and 1950s.

23. The editors note that the aggravation of technology-relativity is not all. The new instrumentation threatens to erode the distinction between an instrument, such as a magnifying glass, that is clearly distinct from the observed object, and one that itself mingles with the object.

References

Armstrong, D. M. 1968. *A Materialist Theory of the Mind.* London: Routledge and Kegan Paul.

Bhushan, Nalini. 1995. "In Search of a Genuine Smell Illusion: The Challenge Posed by Smell to the Representational Theory of Perception." Smith College, unpublished manuscript.

Bienfang, Ralph. 1946. *The Subtle Sense.* Norman, OK: University of Oklahoma Press.

Cartwright, Nancy. 1983. *How the Laws of Physics Lie.* Oxford: Clarendon Press.

Churchland, Paul M., & Smith Churchland, Patricia 1983. "Stalking the Wild Epistemic Engine." *Noûs* 17(March), 5–18.

Clark, Austen. 1993. *Sensory Qualities.* Oxford: Clarendon Press.

Durland, Karánn. 1996. "Primary and Secondary Qualities: Common Sense, Science, and Berkeley." Doctoral dissertation, Department of Philosophy, University of North Carolina.

Erikson, Robert P., & Schiffman, Susan S. 1975. "The Chemical Senses: A Systematic Approach." In M. S. Gazzaniga & C Blakemore, eds. *Handbook of Psychobiology* (pp. 393–426). New York: Academic Press.

Gazzaniga, Michael S., ed. 1995. *The Cognitive Neurosciences*. Cambridge, MA: Bradford Books/MIT Press.

Henning, Hans. 1916. *Der Geruch*. Leipzig: Barth.

Jackson, Frank. 1977. *Perception*. Cambridge: Cambridge University Press.

Kandel, E. R., & Schwartz, J. H. 1989. *Principles of Neural Science* (2nd ed.). New York: Elsevier.

Lycan, William G. 1987. *Consciousness*. Cambridge, MA: Bradford Books/MIT Press.

Lycan, William G. 1996. *Consciousness and Experience*. Cambridge, MA: Bradford Books/MIT Press.

Perkins, Moreland. 1983. *Sensing the World*. Indianapolis: Hackett.

Rock, I., & Harris, C. S. 1967. "Vision and Touch." *Scientific American* 216 (May). 94–104.

Sanford, David H. 1983. "The Perception of Shape." In C. Ginet & S. Shoemaker, eds. *Knowledge and Mind* (pp. 130–158). Oxford: Oxford University Press.

Shepherd, Gordon M. 1995. "Toward a Molecular Basis for Sensory Perception." In Gazzaniga, M. S., ed. *The Cognitive Neurosciences*. Cambridge, MS: Bradford Books/MIT Press.

Stalnaker, Robert. 1984. *Inquiry*. Cambridge, MA: Bradford Books/MIT Press.

Index